Sea, River, Lake Ice Properties and Their Applications in Practices

Sea, River, Lake Ice Properties and Their Applications in Practices

Editors

Zhijun Li
Tomasz Kolerski
Li Zhou
Xiaohong Shi
Zhengyong Zhang
Fang Li

MDPI • Basel • Beijing • Wuhan • Barcelona • Belgrade • Manchester • Tokyo • Cluj • Tianjin

Editors

Zhijun Li
State Key Laboratory of Coastal and Offshore Engineering
Dalian University of Technology
Dalian
China

Tomasz Kolerski
Faculty of Civil and Environmental Engineering
Gdańsk University of Technology
Gdańsk
Poland

Li Zhou
School of Naval Architecture, Ocean and Civil Engineering
Shanghai Jiao Tong University
Shanghai
China

Xiaohong Shi
Water Conservancy and Civil Engineering College
Inner Mongolia Agricultural University
Hohhot
China

Zhengyong Zhang
College of Sciences
Shihezi University
Shihezi
China

Fang Li
School of Naval Architecture, Ocean and Civil Engineering
Shanghai Jiao Tong University
Shanghai
China

Editorial Office
MDPI
St. Alban-Anlage 66
4052 Basel, Switzerland

This is a reprint of articles from the Special Issue published online in the open access journal *Water* (ISSN 2073-4441) (available at: www.mdpi.com/journal/water/special_issues/Sea_River_Ice).

For citation purposes, cite each article independently as indicated on the article page online and as indicated below:

LastName, A.A.; LastName, B.B.; LastName, C.C. Article Title. *Journal Name* **Year**, *Volume Number*, Page Range.

Contents

Preface to "Sea, River, Lake Ice Properties and Their Applications in Practices"

This reprint is a collection of research articles focusing on the geographical scale of glacier, lake, river and sea ice variations, as well as the engineering scale of ice properties and their practical applications via laboratory tests and numerical modeling from the Special Issue. Additionally, it includes research on ecosystems under lake ice. The main goal is successfully achieved through the joint efforts of authors, anonymous reviewers, and editorial managers. In total, 1 review article and 15 research articles are included in this reprint. These articles cover a wide range of topics, including water resources from Chinese mountain glacier variation; lake ice phenology at different latitudes and altitudes around the world; ice properties from laboratory experiments and numerical modeling; ice engineering with different purposes in China and the Arctic; and ecosystem under lake ice at different temporal and spatial scales. This reprint received contributions from researchers from different parts of China and from Chinese international cooperation partners because of its focus on "higher temperature ice"under global warming. All papers presented are innovative and of high quality, which can promote research on ice properties and their applications in practices ranging from mountains to sea, especially in popular water ecosystem environments under ice during seasonal ice period. This reprint could be taken as textbook or training material for postgraduate students, researchers and advanced technicians.

Zhijun Li, Tomasz Kolerski, Li Zhou, Xiaohong Shi, Zhengyong Zhang, and Fang Li
Editors

water

MDPI

Editorial

Sea, River, Lake Ice Properties and Their Applications in Practices

Zhijun Li [1,*], Tomasz Kolerski [2], Li Zhou [3], Xiaohong Shi [4], Zhengyong Zhang [5] and Fang Li [3]

1 State Key Laboratory of Coastal and Offshore Engineering, Dalian University of Technology, Dalian 116024, China
2 Faculty of Civil and Environmental Engineering, Gdańsk University of Technology, 80-233 Gdańsk, Poland
3 School of Naval Architecture, Ocean and Civil Engineering, Shanghai Jiao Tong University, Shanghai 200240, China
4 Water Conservancy and Civil Engineering College, Inner Mongolia Agricultural University, Hohhot 010018, China
5 College of Sciences, Xinjiang Production and Construction Corps Key Laboratory of Oasis Town and Mountain-Basin System Ecology, Shihezi University, Shihezi 832003, China
* Correspondence: lizhijun@dlut.edu.cn

Abstract: This Special Issue aims to highlight research articles focusing on the geographical scale of glacier and lake ice variations, as well as the engineering scale of ice properties and their practical applications via laboratory tests and numerical modeling. Additionally, it includes research on ecosystems under lake ice. The main goal is successfully achieved through the joint efforts of authors, anonymous reviewers, and editorial managers. In total, 1 review article and 15 research articles are included in this Special Issue. These articles cover a wide range of topics, including water resources from Chinese mountain glacier variation; lake ice phenology at different latitudes and altitudes around the world; ice properties from laboratory experiments and numerical modeling; ice engineering with different purposes in China and the Arctic; and ecosystem under lake ice at different temporal and spatial scales. This Special Issue received contributions from researchers from different parts of China and from Chinese international cooperation partners because of its focus on "higher temperature ice" under global warming. All papers presented are innovative and of high quality. This Special Issue can promote research on ice properties and their applications in practices ranging from mountains to sea, especially in popular water ecosystem environments under ice during seasonal ice period.

Keywords: water resources from melted glaciers; lake ice phenology; ice properties; ice engineering; ecosystem under lake ice

Citation: Li, Z.; Kolerski, T.; Zhou, L.; Shi, X.; Zhang, Z.; Li, F. Sea, River, Lake Ice Properties and Their Applications in Practices. *Water* **2023**, *15*, 899. https://doi.org/10.3390/w15050899

Received: 9 February 2023
Revised: 1 January 1970
Accepted: 21 February 2023
Published: 26 February 2023

1. Introduction

Ice and snow exist as the solid states of water at a low temperature and are highly related to atmosphere and hydrosphere. They are parts of human living environments in cold regions as well. In the context of global warming, ice and snow exhibit significant changes due to their sensitivity to temperature. The influences of these variations on human activity may seem to stand independently but are in fact integrated within the cryosphere if the scope is enlarged to a geographical scale. The IPCC WGI AR4 (2007) states that the main components of the cryosphere are snow, river and lake ice, sea ice, glaciers and ice caps, ice shelves, ice sheets, and frozen ground [1]. Figure 1 illustrates their connections at the spatial and temporal scales, showing both the independency and integration of these components. Being located at a medium latitude, China has no ice caps, but there are mountain glaciers in regions of high altitudes. The melting of glaciers provides water sources to rivers and lakes. Global warming has promoted the melting of glaciers and increased the water level of lakes in Northwest of China [2,3]. The water in these rivers and lakes then enter the sea via runoffs either above- or under- ground. Ice occurs in the Bohai Sea and the northern part of the Yellow Sea every winter, but most of the ice on Earth exists at polar regions of high latitudes.

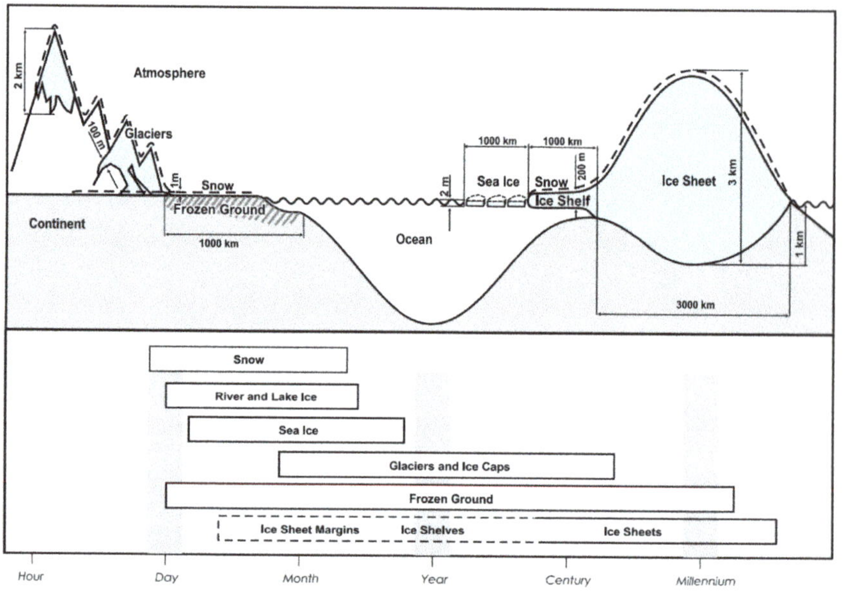

Figure 1. Components of the cryosphere and their time scales (modified from Ref. [1]).

Research on ice engineering is driven by human activities and is essentially focused on the properties of ice, either at a small or a large scale. With sea ice as an example, Ding (1999) overviewed engineering properties of sea ice [4] and introduced the tetrahedron proposed by a Japanese scientist summarizing the relationships between ice physical properties, climate, and human activities. This tetrahedron can be opened from the top to form a big triangle composed of four small triangles [4], as shown in Figure 2. Ice thermology, ice mechanics, and ice dynamics are the three basic apexes. Although this sketch does not cover all the aspects, it concisely conveys the interconnections between many ice-related research fields and reflects the fact that the physical property of sea ice is the fundament of other research topics. In recent years, increased navigation through the northern sea route in the Arctic regions has incurred a large number of research studies focusing on the interaction between ice and its structures [5,6].

Due to global warming, lake ice, river ice, and sea ice are all reducing. These cause delayed formation of ice, formation of thinner ice, and shorter duration of seasonal ice covers. Therefore, the geographical ice margin moves further away from the equator, multiyear ice becomes thinner and decreases in extent, and, as a result, the fraction of first-year ice becomes higher [7]. There are also other phenomena, including an increase in ice temperature, a decrease in ice salinity, a reduction in ice density, and a widened varying range of ice density [8]. It has become a new topic around the world as to whether these would induce an increase in under-ice radiation flux and water temperature [9,10], thereby affecting under-ice environments during freezing seasons. In China, many lakes are utilized for aquaculture, which is tightly related to these issues. It is important to investigate the changes in underwater ecosystems and form a holistic understanding of the changes at different latitudes and altitudes around the world. On the basis of this, it is of significance to connect the prediction of climate change to under-ice ecosystems in lakes. The Special Issue, entitled "Sea, River, Lake Ice Properties and Their Applications in Practices", is proposed under such context, which invites researchers from different fields to contribute their investigation on scientific and technological problems at different temporal and spatial scales.

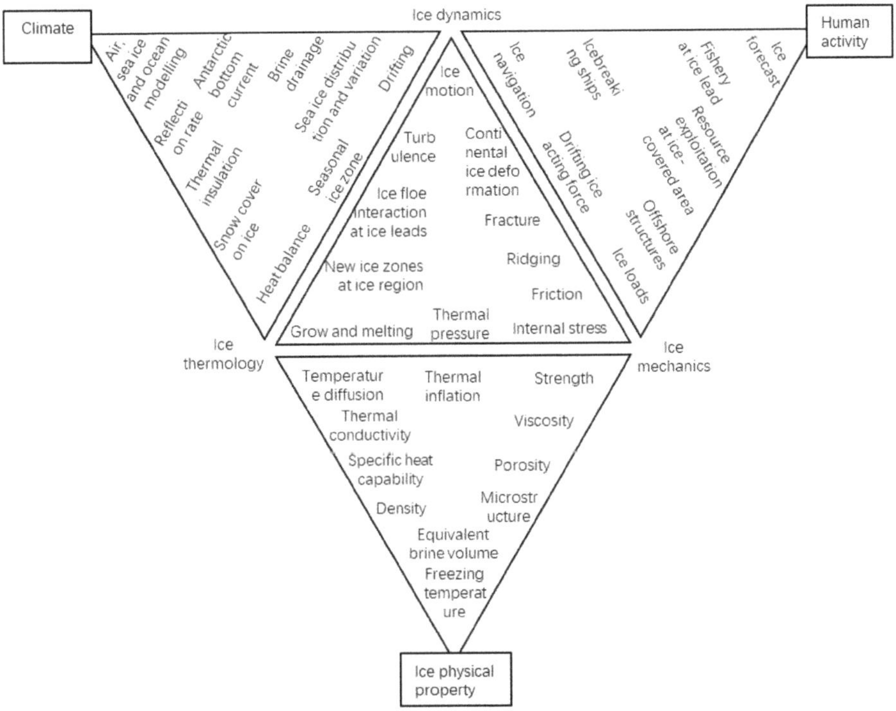

Figure 2. Relationships between the research fields related to ice (modified from Ref. [4]).

In this Special Issue, we focus on "higher temperature" ice properties and their applications. This Special Issue can guide future ice science and engineering in polar regions under climate change. Therefore, it covers the physical, thermal, mechanical, and electrical properties of any kind of crystal ice. The scope of this Special Issue also includes research on remote sensing, engineering, ecosystem, and entertainment through the use of multi-approaches of testing, theoretical analysis, numerical modeling, and physical modeling.

2. Summary of This Special Issue

This Special Issue received 24 manuscripts, of which 15 papers are published. The contents of these published papers are water resources from mountain glaciers, lake ice phenology, ice properties and their application in practices, sea ice engineering, and newly developed ecosystem environments under lake ice.

Two papers are published on water resources from mountain glaciers and mountain glaciers' variation in the past 50 years in Xinjiang Area. Liu et al. [11] proposed that the dynamic change in the glacier material balance in Tianshan Mountain is of significant scientific value for assessing the impact of climate on water resources. These water resources are the ecological barrier of the oasis in the northwest arid region in Xinjiang. The authors used a spatially distributed degree-day model using multi-source remote sensing data, such as MOD11C3 and TRMM3B43, to simulate the mass balance in the Tianshan Mountains' south and north slope basins, and they compared and attributed the spatiotemporal heterogeneity of the mass balance by using a geographical detector. They obtained the changes in glacier size, distribution, material balance, as well as their formation causes, in typical basins on the northern and southern slopes of the Tianshan Mountains. The results show the characteristics of water resource changes and water resource regulation in the north and south slope basins of the Tianshan Mountains. The other contribution is from Zhang et al. (2022) [12]. This paper combines GIS spatial analytical techniques to explore the characteristics of glacier changes in Xinjiang and its response to topography and climate change in the past 50 years. The results show that the glaciers in Xinjiang during these 50 years are mainly receding and splitting. In particular, small glaciers are more sensitive to climate change. Phenomena such as glacier extinction and splitting result in an increasing number of glaciers but a decreasing total area of glaciers in the southern mountains. Topographic relief is the most important factor controlling the spatial differentiation of glacier changes among the topographic factors. The lack of a corresponding increase in precipitation is one of the reasons for the apparent shrinkage of glacial resources in the study area. The changes in glacier resources and their meltwater runoff contributions in Xinjiang are of great significance to hydrological processes and water resource utilization.

On lake ice phenology, there are two papers published in this Special Issue. One paper covers the Eurasian Lakes, and the other paper focuses on the case of Wuliangsu Lake, about 200 km^2 in area located near the Yellow River in Inner Mongolia, China. The Eurasian freezing lakes has a vast scope, covering nearly one quarter of Earth, which ice phenology varies widely. The degree-day method is generally used to investigate variations in ice phenology, but it only pays attention to temperature without considering another critical factor—radiation. Thus, a mathematically complete and physically meaningful analytical model is introduced in the study by Leppäranta and Wen (2022) [13], which provides a first-order tool for evaluating climate sensitivity of ice seasons. Their results show that local temperature, radiation, and their gradients are the key factors influencing ice phenology, especially in the Tibetan Plateau with strong radiation. Freezing depends on lake depth, and breakup depends on accumulated ice thickness The freezing date and breakup date both change by around five days per one-degree shift in air temperature away from the climatological ice margin. This work helps re-recognize freezing lakes' ice phenology in the two dimensions of altitudes and spatial distribution and provides a supplement to our understanding of land–air interaction. Huoet al. (2022) [14] studied the ice phenology of Wuliangsu Lake using optical satellite data. Considering the possible effect of clouds above the lake [15], the authors developed a new algorithm based on remote sensing modeling to tackle this challenge. Finally, they effectively removed the influence of aquatic plants and clouds on lake ice identification and obtained satisfactory results with daily monitoring.

Ice properties and their applications in different practices, except for ice engineering, constitute a large part of the research in this Special Issue. There are four papers that rely on full-scale measurement data concerning Chinese river ice and the Antarctic ice sheet. In river management, to solve the problems of existing time-consuming and inaccurate ice velocity monitoring methods, Wang et al. (2022) [16] proposed a new ice velocity extraction method based on a UAV-based remote sensing technology. High-resolution orthoimages were obtained with a UAV during the ice flood period, and the feature points in the drift ice images were then extracted using a scale-invariant feature transform (SIFT) algorithm. With the use of the random sample consensus (RANSAC) algorithm, the authors tracked the motion trajectories of these feature points, and an ice displacement rate field was finally established. Meanwhile, research on river ice thickness also calls for improvements of existing measuring techniques. Ground-penetrating radar (GPR) has proven to be a very effective method for examining ice thickness. Han et al. (2022) [17] carried out a series of GPR surveys using dual-channel host 200 MHz shielded antennas at the Toudaoguai Hydrological Station on the Yellow River, China, and collected samples to analyze ice impurities and physical properties. A combination of GPR data, as well as ice impurities, porosity, density, and temperature, provides insights into the variability of radar transmission speed and the equivalent dielectric permittivity in river ice. The results show that the radar transmission speed falls between 0.141 m/ns and 0.164 m/ns, and the equivalent dielectric permittivity of river ice increases in concert with ice temperature. Lu et al. (2022) [18] analyzed the ice fabrics and microstructures of 23 one-meter-long ice cores drilled from blue ice areas at the shear margin of Dalk Glacier, Antarctica. Their results show the type and evolution of ice fabrics under shear stress of ice flow and bedrock. The characteristics of grain and bubble properties are revealed, and the mechanism of the formation of ice layers in these study sites are discussed. The stable water isotopes of these ice cores are also analyzed to study the correlations with ice fabrics and microstructures.

Under laboratory conditions and with the use of non-natural ice, two papers present newly researched results. To understand disc ice floes' surface melting, bottom melting, and lateral melting during the melting season, tests under controlled conditions were performed in a cold laboratory [19]. The results show that the bottom melting rate is larger than the lateral melting rate, followed by the surface melting rate, and the smaller the size of an ice floe, the higher the lateral melting rate. According to the experimental data, a new indicator was created using the ratio of the diameter to the thickness of an ice sample. Therefore, the relationships between the surface/bottom melting rates and temperature gradient were formulated; the relationship among the lateral melting rate, the temperature difference, and the ratio of the diameter to the thickness was also quantified. These two equations can be applied to the melting parametrization scheme of ice for a range of diameters up to 100 m, which covers simulations of the energy and mass balance values of the Arctic sea ice and coastal fresh-water ice during the summer melting season [19]. Zhang et al. (2022) [20] studied the uniaxial compressive strength of distilled-water ice grown at different temperatures in a cryogenic laboratory, which differs from previous research that has mainly focused on the effect of test temperature. The results show that the uniaxial compressive strength of ice is a power function of the strain rate and has a linear relationship with the -1/2 power of grain size. Additionally, it is found that the peak compressive strength has a logarithmic relationship with the growth temperature. In addition, it is deduced that the growth temperature affects the strength of ice by controlling the grain size.

Liu et al. (2002) [21] is the only paper that uses computer-assisted tools to apply ice properties to practice. Liu et al. (2022)'s paper combines various ice blasting theories and engineering practice results; specifically, the authors applied the blasting theory of Euler's theorem and the JWL (Jones–Wilkins–Lee) state equation, and they used the ANSYS/LSDYNA software to simulate ice blasting effects under different ice conditions. Through an analysis of ice simulation data and a verification of the measured data, this paper summarizes and deduces the blasting vector mode which is suitable for the main stream of Heilongjiang Province, China.

Three papers examining the navigation problems of polar ships are included in this Special Issue. In the paper by Zhou et al. (2022a) [22], the authors calculated ice resistance using the circumferential crack method and the Monte Carlo methods in a probabilistic way for the first time, considering the effect of both ice thickness and bending strength based on the measured data from the Chinese Xuelong Icebreaker. The simulation results show that the Burr distribution is the most suitable model to describe the distribution of the calculated ice resistance and the ice-force amplitude concentrated at a lower level. The numerical tool can provide a significant reference for ice-going ships sailing in level ice in most scenarios with regard to ice resistance and operational safety with high accuracy and computational efficiency. Meanwhile, Zhou et al. (2022b) [23] proposed an overlapping grid method based on CFD to analyze propeller performance during a typical ice blockage condition. The ice blockage model is simulated by using a water-insoluble rectangular solid block, and ice blockage is measured by the distance between the solid block and the propeller. Compared to the model test data, it shows the numerical difference is less than 5%, which is quite accurate. The present method is reasonable and vital to propeller design and optimization. Other than ice forces on the ship body and propellers, icing is possible to occur in polar navigation vessels. In the paper by Chen et al. [24], a study on the generation and distribution of sea spray is carried out to address the problem of ice accumulation forecasting for polar navigation vessels, and new research ideas and directions are proposed. In order to solve the problem that sea spray cannot be quantified using the traditional finite element grid calculation method, this research mainly adopts the calculation method of smoothed-particle hydrodynamics and uses the Lagrangian particle tracking idea to accurately capture the motion state of each particle and to calculate the relevant physical quantity in real time, which innovatively achieves an accurate measurement of sea spray, breaking through the traditional sea spray droplet on sailing ship coverage and the proportion of the difficult problem that cannot be quantified. The research results can serve as a guidance for other fields, such as marine structure design, polar ship icing forecast, and others.

The mooring system of a floating offshore structure in the Arctic region is another kind of engineering application in the Arctic. For instance, Zhang et al. (2022) [25] investigated an optimal mooring system for the Kulluk platform operating in the Beaufort Sea under ice-free and ice-covered conditions during a whole year. In order to complete the layout design of the mooring system to satisfy the year-round operation, both the effects of wave loads and ice loads should be considered. This research establishes a coupled numerical production system composed of a platform and a mooring system. The nonlinear finite element method, the discrete element method, and empirical formula are used to analyze the ice loads. The simulated results are found to compare reasonably with the field measurements. A quantitative time-domain analysis is carried out to study the mooring line configurations. This research work will provide a reference for the optimal design of mooring systems of the platforms operating in the Arctic Sea.

In addition to traditional ice property application in ice engineering, water ecosystem environment under lake ice has become a new topic in Cryosphere Science [26]. Although it is difficult to solve the collaborative carrying capacity of water resources, water environment, and water ecology in Wuliangsuhai Lake due to the long ice period, strong evapotranspiration, and intensified interference of human activities, Ren et al. [27] innovatively construct an evaluation index system and method to evaluate the collaborative carrying capacity of Wuliangsuhai Lake water resources–water environment–water ecology, including the calculation module of ice period, and realizes the dynamic simulation and regulation of the collaborative carrying capacity under multiple scenarios. Their results have certain extensibility and portability. The primary productivity beneath the ice is an important variable for understanding of how temperate lake ecosystems are changing due to global warming.

Author Contributions: Writing—original draft preparation, Z.L.; writing—review and editing, T.K., L.Z., X.S., Z.Z. and F.L. All authors have read and agreed to the published version of the manuscript.

Funding: This Special Issue originated from the ideas of ecosystems under ice-covered waters through ice properties driven by meteorological and hydrological conditions. It was funded by the National Key Research and Development Program of China (2019YFE0197600) and by the National Natural Science Foundation of China (52171259, 52260028).

Conflicts of Interest: The authors declare no conflict of interest.

References

1. IPCC. *Summary for Policymakers (SPM), Climate Change 2007: The Physical Science Basis: Contribution of Working Group I to the Fourth Assessment Report of the Intergovernmental Panel on Climate Change*; Solomon, S., Ed.; Cambridge Univ. Press: New York, NY, USA, 2007.
2. Li, X.; Zhang, T.; Yang, D.; Wang, G.; He, Z.; Li, L. Research on lake water level and its response to watershed climate change in Qinghai Lake from 1961 to 2019. *Front. Environ. Sci.* **2023**, *11*, 1130443. [CrossRef]
3. Zhang, G.; Yao, T.; Xie, H.; Yang, K.; Zhu, L.; Shum, C.K.; Bolch, T.; Yi, S.; Allen, S.; Jiang, L.; et al. Response of Tibetan Plateau lakes to climate change: Trends, patterns, and mechanisms. *Earth-Sci. Rev.* **2020**, *208*, 103269. [CrossRef]
4. Ding, D. *Introduction to Sea Ice Engineering*; China Ocean Press: Beijing, China, 1999. (In Chinese)
5. Sun, Q.; Zhang, M.; Zhou, L.; Garme, K.; Burman, M. A machine learning-based method for prediction of ship performance in ice: Part I. ice resistance. *Mar. Struct.* **2022**, *83*, 103181. [CrossRef]
6. Xie, C.; Zhou, L.; Ding, S.; Liu, R.; Zheng, S. Experimental and numerical investigation on self-propulsion performance of polar merchant ship in brash ice channel. *Ocean Eng.* **2023**, *269*, 113424. [CrossRef]
7. Alfredsen, K.; Hendrikse, H.; Høyland, K.V.; Kolerski, T.; Leppäranta, M.; Lu, P.; Guo, X. How does climate change affect ice formation and presence in rivers, lakes and oceans, as well as its impact on infrastructure. *Hydrolink* **2022**, *3*, 77–79.

8. Wang, Q.; Lu, P.; Leppäranta, M.; Cheng, B.; Zhang, G.; Li, Z. Physical properties of summer sea ice in the Pacific Sector of the Arctic during 2008-2018. *J. Geophys. Res. Oceans* **2020**, *125*, e2020JC016371. [CrossRef]
9. Huang, W.; Zhao, W.; Zhang, C.; Leppäranta, M.; Li, Z.; Li, R.; Lin, Z. Sunlight penetration dominates the thermal regime and energetics of a shallow ice-covered lake in arid climate. *Cryosphere* **2022**, *16*, 1793–1806. [CrossRef]
10. Wang, M.; Wen, L.; Li, Z.; Leppäranta, M.; Stepanenko, V.; Zhao, Y.; Niu, R.; Yang, L.; Kirillin, G. Mechanisms and effects of under-ice warming water in Ngoring Lake of Qinghai–Tibet Plateau. *Cryosphere* **2022**, *16*, 3635–3648. [CrossRef]
11. Liu, L.; Tian, H.; Zhang, X.; Chen, H.; Zhang, Z.; Zhao, G.; Kang, Z.; Wang, T.; Gao, Y.; Yu, F.; et al. Analysis of spatiotemporal heterogeneity of glacier mass balance on the Northern and Southern Slopes of the Central Tianshan Mountains, China. *Water* **2022**, *14*, 1601. [CrossRef]
12. Zhang, X.; Liu, L.; Zhang, Z.; Kang, Z.; Tian, H.; Wang, T.; Chen, H. Spatial and temporal variation characteristics of glacier resources in Xinjiang over the past 50 years. *Water* **2022**, *14*, 1057. [CrossRef]
13. Leppäranta, M.; Wen, L. Ice phenology in Eurasian Lakes over spatial location and altitude. *Water* **2022**, *14*, 1037. [CrossRef]
14. Huo, P.; Lu, P.; Cheng, B.; Zhang, L.; Wang, Q.; Li, Z. Monitoring ice phenology in lake wetlands based on optical satellite data: A case study of Wuliangsu Lake. *Water* **2022**, *14*, 3307. [CrossRef]
15. Li, Z.; Wang, Q.; Tang, M.; Lu, P.; Li, G.; Leppäranta, M.; Huotari, J.; Arvola, L.; Shi, L. Diurnal cycle model of lake ice surface albedo: A case study of Wuliangsuhai Lake. *Remote Sens.* **2021**, *13*, 3334. [CrossRef]
16. Wang, E.; Hu, S.; Han, H.; Li, Y.; Ren, Z.; Du, S. Ice velocity in upstream of Heilongjiang based on UAV low-altitude remote sensing and the SIFT algorithm. *Water* **2022**, *14*, 1957. [CrossRef]
17. Han, H.; Li, Y.; Li, W.; Liu, X.; Wang, E.; Jiang, H. The influence of the internal properties of river ice on ground penetrating radar propagation. *Water* **2023**, *15*, 889. [CrossRef]
18. Lu, S.; Zhang, N.; Wang, D.; Shi, G.; Ma, T.; Ma, H.; An, C.; Li, Y. Spatial variations of fabric and microstructure of blue ice cores at the Shear Margin of Dalk Glacier, Antarctica. *Water* **2023**, *15*, 728. [CrossRef]
19. Li, Z.; Wang, Q.; Li, G.; Lu, P.; Wang, Z.; Xie, F. Laboratory studies on the parametrization scheme of the melting rate of ice–air and ice–water interfaces. *Water* **2022**, *14*, 1775. [CrossRef]
20. Zhang, Y.; Qian, Z.; Lv, S.; Huang, W.; Ren, J.; Fang, Z.; Chen, X. Experimental investigation of uniaxial compressive strength of distilled water ice at different growth temperatures. *Water* **2022**, *14*, 4079. [CrossRef]
21. Liu, D.; Dai, C.; Yu, C.; Ma, Y. Numerical simulation and experimental study on single point blasting of ice jam of Heilongjiang River based on ANSYS/LSDYNA. *Water* **2022**, *14*, 2808. [CrossRef]
22. Zhou, L.; Gu, Y.; Ding, S.; Liu, R. Prediction of ice-resistance distribution for R/V Xuelong using measured sea-ice parameters. *Water* **2022**, *14*, 517. [CrossRef]
23. Zhou, L.; Zheng, S.; Diao, F.; Ding, S.; Gao, J. Experimental and numerical study on ice blockage performance of propeller in cavitation flow. *Water* **2022**, *14*, 1060. [CrossRef]
24. Chen, J.; Bai, X.; Wang, J.; Chen, G.; Zhang, T. Research on sea spray distribution of marine vessels based on SPH-FEM coupling numerical simulation method. *Water* **2022**, *14*, 3834. [CrossRef]
25. Zhang, A.; Chuang, Z.; Liu, S.; Chang, X.; Hou, L.; He, Z.; Liu, S. Research on mooring system design for Kulluk Platform in Arctic Region. *Water* **2022**, *14*, 1762. [CrossRef]
26. Qin, D.; Ding, Y.; Xiao, C.; Kang, S.; Ren, J.; Yang, J.; Zhang, S. Cryospheric Science: Rresearch framework and disciplinary system *Natl. Sci. Rev.* **2018**, *5*, 255–268. [CrossRef]
27. Ren, B.; Sun, B.; Shi, X.; Zhao, S.; Wang, X. Analysis of the cooperative carrying capacity of Ulan Suhai Lake based on the coupled water resources–water environment–water ecology system. *Water* **2022**, *14*, 3102. [CrossRef]

water

MDPI

Article

Analysis of Spatiotemporal Heterogeneity of Glacier Mass Balance on the Northern and Southern Slopes of the Central Tianshan Mountains, China

Lin Liu [1,2,†], Hao Tian [1,†], Xueying Zhang [1], Hongjin Chen [1], Zhengyong Zhang [1,2,*], Guining Zhao [1,*], Ziwei Kang [1], Tongxia Wang [1], Yu Gao [1], Fengchen Yu [1], Mingyu Zhang [1], Xin Yi [1] and Yu Cao [1]

1 College of Sciences, Shihezi University, Shihezi 832000, China; liulin779@163.com (L.L.); haotian1996_china@163.com (H.T.); zxy970716@163.com (X.Z.); zlyzxlx@163.com (H.C.); kangziwei0808@163.com (Z.K.); wtx0428@163.com (T.W.); gyiks1225@163.com (Y.G.); yufengchen678@163.com (F.Y.); zmy_zz1015@163.com (M.Z.); jiuwuxin1331@126.com (X.Y.); cyfqq1230@163.com (Y.C.)

2 Key Laboratory of Oasis Town and Mountain-Basin System Ecology, Xinjiang Production and Construction Corps, Shihezi 832000, China

* Correspondence: zyz0815@163.com (Z.Z.); 13345492733@163.com (G.Z.); Tel.: +86-18040830081 (Z.Z.); +86-13345492733 (G.Z.)

† These authors contributed equally to this work.

Abstract: Glacier mass balance can visually indicate the degree of glacier response to climate change. The mountain glaciers are an essential source of recharge for rivers in arid regions and play a vital role in maintaining regional ecological stability and production life. This paper drives a spatially distributed degree-day model using multi-source remote sensing data such as MOD11C3 and TRMM3B43 to simulate the mass balance in the Tianshan Mountains' south and north slope basins. The spatiotemporal heterogeneity of the mass balance was compared and attributed using a Geographical detector. The results show that: (1) The glaciers in the north and south basins are mainly distributed at an altitude of 3900–4300 m, and the total glacier area accounts for 85.71%. The number of less than 1 km^2 glaciers is the most in the whole region. (2) During the study period, the glaciers in the north and south basins were negative (−465.95 mm w.e.) an entire interannual change rate was −28.36 mm w.e./a. The overall trend of ablation can be divided into two stages: from 2000 to 2010a, the persistence increased, and from 2010 to 2016a, the volatility decreased. (3) In the attribution of mass balance, the factors affecting glacier mass balance can be divided into two parts: climate and topography. The cumulative contribution rate of climate factors in Kaidu is nearly 20% higher than that of topographic factors, but the contribution rate of climate factors in Manas is only 7.3% higher. Therefore, the change of glacier mass balance in the Kaidu river basin is more driven by climate factors, while the glacier mass balance in the Manas river basin is more affected by the combination of climate and topographic factors. (4) The climate accumulation is the dominant factor in the Manas river basin (69.55%); for the ablation area, the Kaidu river basin is dominated by climate (70.85%), and the Manas river basin is dominated by topographic factors (54.11%). Due to the driving force of climate and topographic factors and the different coupling modes, glacier mass balance's spatiotemporal heterogeneity in the north and south slope basins is caused. This study contributes to analyzing the mechanism of regional changes in the glacier mass balance. It provides a scientific basis for investigating the characteristics of water resource changes and water resource regulation in the north and south slope basins of the Tianshan Mountains.

Keywords: Manas river basin; Kaidu river basin; glacier mass balance; Geographical detector; attribution analysis

Citation: Liu, L.; Tian, H.; Zhang, X.; Chen, H.; Zhang, Z.; Zhao, G.; Kang, Z.; Wang, T.; Gao, Y.; Yu, F.; et al. Analysis of Spatiotemporal Heterogeneity of Glacier Mass Balance on the Northern and Southern Slopes of the Central Tianshan Mountains, China. *Water* **2022**, *14*, 1601. https://doi.org/10.3390/w14101601

Academic Editor: Achim A. Beylich

Received: 5 April 2022
Accepted: 13 May 2022
Published: 17 May 2022

1. Introduction

With global warming, as significant freshwater resources are known as "solid reservoirs", most glaciers are in a state of retreat and ablation. Some studies have also shown that mountain glaciers are more sensitive to climate warming [1,2]. At the same time, a series of disaster phenomena caused by the change and movement of mountain glaciers have also been widely valued by various research institutions [3]. As an indicator of climate change, glaciers mass balance is a critical junction connecting glaciers and climate, and it also comprehensively reflects the hydrothermal conditions of glacier development [4,5]. Glacier meltwater generated by the change of glacier mass balance has a stabilizing effect on regional water resources. Given the importance of glacier mass balance, it has been an important research content in glacier change detection worldwide. The World Glacier Monitoring Service (WGMS) aims to monitor global glacier changes continuously, systematically compile global glacier monitoring data, make long-term observations of 10 key reference glaciers, and publish their mass balance data every two years [6]. Urumqi glacier No.1, located in the Tianshan Mountains in Central Asia, is one of the reference glaciers included in the ten essential monitoring items of WGMS, which can reflect the basic characteristics of the changes in central Asian glaciers to a certain extent [7,8]. However, affected by topography and regional climate conditions, the monitoring data of a single glacier cannot comprehensively reflect the overall change characteristics of regional glaciers. Selecting suitable regions to reflect the spatial heterogeneity of glacier mass balance changes can further explain the internal factors affecting the changes in the mass balance of mountain glaciers, which will be more conducive to understanding the mechanism of their recharge to rivers.

The research methods about glacier mass balance mainly include field observation, remote sensing monitoring, model simulation, etc. [9–11], glacier surface observation can provide detailed parameters. Still, it is difficult to reflect the spatial heterogeneity of the region due to the significant human interference in the observation process, which is usually confined to a single glacier in some areas. In addition, remote sensing monitoring has the advantages of a wide range and multiple time series. For example, the target decomposition technology using polarimetric SAR data has broad application prospects in ice research. Polarimetric decompositions have become an effective method for identifying glacier areas, especially over debris-free glaciers [12–14]. The above techniques can realize the accurate measurement and detection of single or multiple glaciers. Still, it is difficult to reflect the spatial heterogeneity of glacier change in a large area [7]. Compared with glacier observation and remote sensing monitoring, model simulation can better reflect the physical process mechanism of mass balance [15,16]. It can be divided into the empirical model based on the degree-day factor and the physical model based on energy balance. Compared with glacier observation and remote sensing monitoring, model simulation can better reflect the physical process mechanism of mass balance [17] and can be divided into empirical models based on degree-day factors and physical models based on energy balance. The energy balance model is based on the energy balance principle, considers the energy input and output items, and focuses more on the mechanism of the mass balance process. However, the input parameters and variables are complex and challenging to obtain. There are high requirements for data accuracy or resolution, which limits the popularization and application of the model. The degree-day model is simple and easy to obtain, its results are accurate and reliable in large-scale glacier studies, and it is most widely used. By comparing and analyzing numerous models and methods, the degree-day model regards the degree-day factor as a statistical variable of albedo, latent sensible heat flux, relative humidity, and other factors. The degree-day model adopts a constant representation according to the difference in time and space, which effectively reduces the model's complexity. Therefore, if a large-scale glacier mass balance study is to be carried out, taking into account factors such as topography and climate, and taking into account the advantages of the simplicity and universality of the degree-day model and balance of the ice melt/snowmelt module, a degree-day model based on the water balance principle is more suitable for the model. [18].

As an ecological barrier of an oasis in the arid area of northwest China, the dynamic change of glacier mass balance in Tianshan Mountain is of great scientific value in assessing the impact of climate on water resources [9]. The Tianshan Mountains in China are 250–350 km wide from north to south. The most extensive independent zonal mountain system globally, the farthest mountain system from the ocean globally, and the most major mountain system in the arid region of the world. At present, the study of glacier mass balance in The Tianshan Mountains of China is mainly limited to the analysis of measured data of a single glacier, the simulation of glacier mass balance in different regions or the whole basin, and the impact of climate change on regional mass balance [19–22]. There are few studies on the difference and comparison of glacier mass balance in the same mountain system under different site conditions. The regional differences in ecological and geographical characteristics on the northern and southern slopes of the Tianshan Mountains affect the development and evolution of glaciers in different regions, leading to varying factors of glacier mass balance changes in the different areas. Comparison and attribution analysis of the mass balance of glaciers on the north and south slopes will make a comparison to explain further the mechanism of the change of mountain glaciers' mass balance. In this paper, glaciers in the Manas river basin on the northern slope of the central Tianshan Mountains and the Kaidu river basin on the southern slope as research objects. Using MOD11C3 and TRMM3B43 as driving data, a spatially distributed degree-day model was constructed to analyze and compare the mass balance of glaciers in typical watersheds on the north and south slopes of the Geographical detector to make an attribution analysis of their spatiotemporal heterogeneity characteristics. This study is helpful to analyze the mechanism of regional changes in mountain glacier mass balance and provides the scientific basis for exploring the features of water resource change and water resources regulation in the northern and southern slopes of the Tianshan Mountains.

2. Study Area

Tianshan Mountain, located in the hinterland of Eurasia, is the most extensive independent latitudinal mountain system in the world, which stretches for more than 1700 m in China. It crosses the central part of Xinjiang Uygur Autonomous Region and is the boundary between northern and southern Xinjiang [23]. It consists of parallel mountain ranges divided into north, middle, and south zones. The study area is located in the central Tianshan Mountains, including the Yilianhabierga Mountains, Saarming Mountains, Bortoula Mountains, etc. Yilianhabierga Mountains are about 320 km long in the east and west, 110 km wide in the north and south, and the mountains are generally above 4000 m above sea level [24]. The Manas River on the northern slope and the Kaidu River on the southern slope selected in this study are derived from Mount Ylianhabirga (Figure 1). The south slope of Tianshan Mountain has less precipitation, is deep inland, and has an arid climate. The north slope of Tianshan Mountain has more precipitation than the southern slope because of the humid airflow from the Arctic Ocean and the Atlantic Ocean. The climate is relatively moist. The southern slope has sufficient sunlight, but the northern slope is less, and the temperature of the south slope is higher than that of the north slope. Manas river basin, located on the north slope of the central Tianshans Mountain, is the basin with the most significant number and scale of glaciers in the inland area of Junggar and has a moderate temperate continental arid climate [25,26]. The Kaidu River, located on the southern slope of the central Tianshan Mountains, is one of the four primary sources of the Tarim River Basin, the longest inland river in China, and belongs to the warm temperate continental arid climate [25,26].

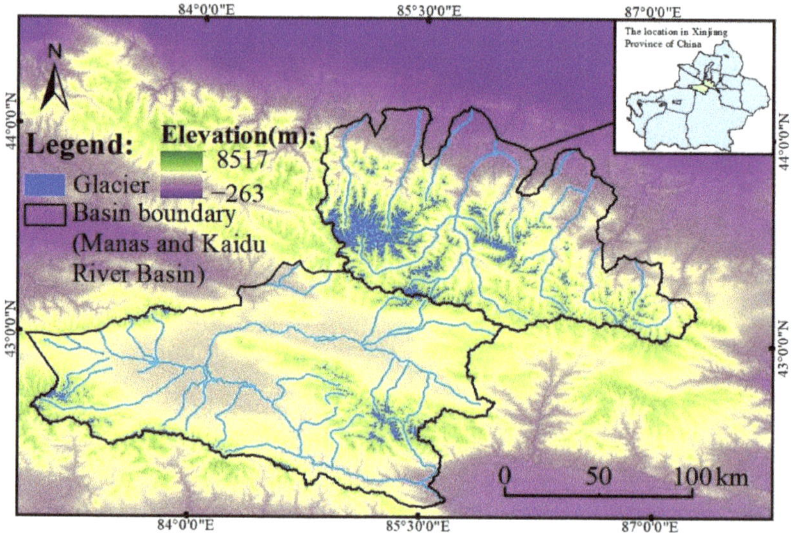

Figure 1. Location of the study area.

3. Data and Methods

The data of this paper are divided into multi-source remote sensing data, glacier area, and boundary vector data of the Manas river basin and the Kaidu river basin. Remote sensing raster data include DEM, MOD11C3, TRMM3B43, ≥0 °C accumulated temperature, precipitation, air temperature, solar radiation, etc. (Table 1). The specific applications of the data in this paper are as follows: DEM can reflect the topographic characteristics of mountainous areas and can be used to obtain topographic factors in mountainous regions, which is the basis for the spatial discretization of various factors of glacier mass balance to construct degree-day model, and it is also a factor that affects the change of glacier mass balance. MOD11C3 and TRMM3B43 are monthly data from 2001 to 2016a, the primary meteorological data driving the distributed degree-day model. The climatic background data such as ≥0 °C accumulated temperature, precipitation, air temperature, and solar radiation are put into the Geographical detector and the terrain factors as climate factors to explore the reasons for the changes in the glacier mass balance. Among them, air temperature and precipitation are also important indicators to construct the ice/snow degree-day factor. The glacier area vector data come from the Second China Glacier Catalog Dataset.

Table 1. Data sources.

Type of Data	Resolution	Data Sources (Accessed on 17 November 2021)
DEM	30 m × 30 m	RESDC (https://www.resdc.cn/data.aspx?DATAID=217)
MOD11C3	0.05° × 0.05°	NASA (https://lpdaac.usgs.gov/products/mod11c3v006/)
TRMM3B43	0.25° × 0.25°	NASA (https://disc.gsfc.nasa.gov/datasets/TRMM_3B43_7/)
Solar radiation	10 km × 10 km	TPDC (https://data.tpdc.ac.cn/)
≥0 °C accumulated temperature Air temperature Precipitation	0.5 km × 0.5 km	RESDC (https://www.resdc.cn/DOI/doi.aspx?DOIid=39)
The Second China Glacier Catalog Dataset	—	NCDC (http://www.ncdc.ac.cn/)

3.1. Simulated Glacier Mass Balance

MODIS and TRMM have an effective spatiotemporal resolution. They have a better objective and scientific description of the mountainous areas with few measuring stations. MODIS and TRMM are involved in the climate data inversion of the Himalayas and Alps, such as using MOD10A1 and MYD10A1 to study the snow line change in the Alps [27]; MODIS land surface temperature data are also used to retrieve the temperature of the Qinghai Tibet Plateau and the central Himalayas [28,29]. In the relevant studies using TRMM, most of them focus on the spatial distribution characteristics of precipitation and the accuracy of precipitation data [30,31]. The above research should prove the feasibility of MODIS and TRMM in describing temperature and precipitation in mountainous areas and has a high reference for the study.

This paper used the distributed degree-day model to simulate the mass balance of glaciers in typical basins on the north and south slope of the central Tianshan Mountains, to analyze the temporal and spatial differences in glacier mass balance between the Kaidu river basin and Manas river basin and drive the geographic detector with climate and topographic factors to explore the leading factors affecting mass balance change in both basins and make a comparative analysis. The methodology flowchart is as follows (Figure 2):

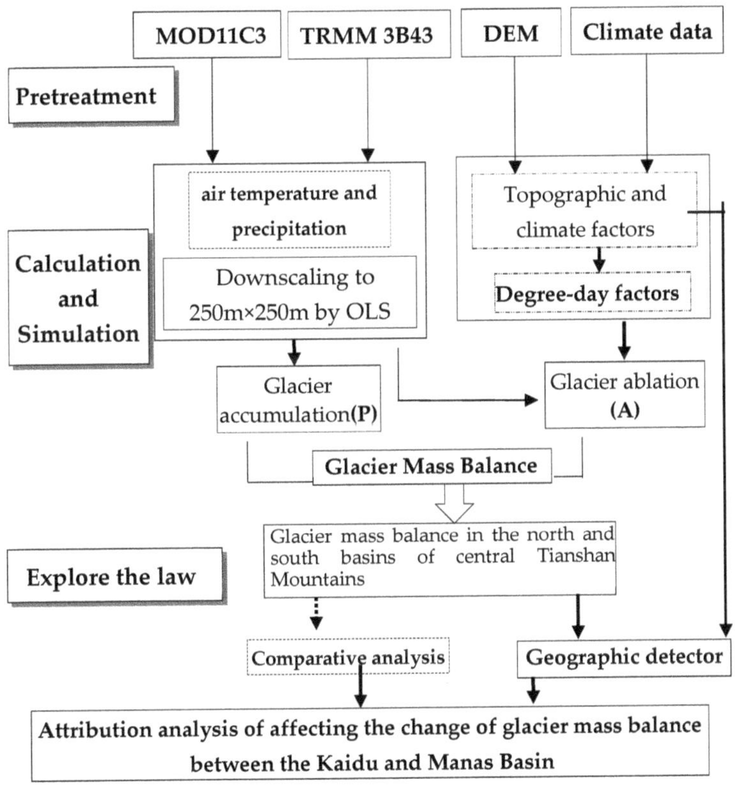

Figure 2. Methodology flowchart.

The degree-day model simplifies the process mechanism of glacier ablation and accumulation. Still, the traditional degree-day model is only a conceptual model. In contrast, the distributed model improves the ability to express the spatial characteristics of each element in the degree-day model, enhances its physical basis, and divides the glacier area into grids. The spatial variation of the overall glacial mass balance is illustrated by describing the specific variation of each grid cell in terms of a set of factors that affect the variation of the glacial mass balance. This way shows the spatial change of the entire glacier mass balance [32]. Due to the different resolutions of the multi-source remote sensing data driving the distributed degree-day model, it is difficult to describe the ice surface climate with too low resolution. Therefore, in this study, the OLS model was constructed in the process of modeling the distributed degree-day model to downscale MOD11C3 and TRMM3B43 to 250 m × 250 m to meet the model accuracy requirements. In contrast, the distributed degree-day model reflects the spatial variation characteristics of the mass balance through the grid.

For the ablation of glacier ice and snow, the amount of ablation in a certain period is expressed as follows [15,16,33,34]:

$$A = D \cdot PDD \tag{1}$$

where A is the ablation of water equivalent of glacier ice and snow in a certain period (mm w.e.). D is the degree-day factor of glacier ice/snow (mm d^{-1} °C^{-1}). PDD is the positive accumulated temperature in a certain period. It can be obtained by the following formula [35]:

$$PDD = \int_{N_1-1}^{N_2} \frac{1}{\delta\sqrt{2\pi}} \int_0^{+\infty} T_m e \frac{-(T_m - T_a)^2}{2\delta^2} dTdt \tag{2}$$

where it is assumed that the monthly average temperature (°C) within the year is usually distributed, Ta is the annual average temperature (°C), δ is the standard deviation of the temperature distribution, N_1, N_2 are the start and the end dates of the calculation, and the period is $N = N_2 - N_1 + 1$.

The net mass balance at a point (or height) is [24]:

$$Bi = P + A + f \tag{3}$$

where Bi is the glacier mass balance in a certain period (mm w.e.). P is the accumulation amount of the glacier surface in a certain period, that is, the solid precipitation (mm). A is the amount of ablation. f is the amount of meltwater infiltration and freezing or internal supply (mm w.e.), usually calculated as 10% of the ablation volume.

The solid precipitation can be calculated according to the limiting temperature method [36]:

$$P_s = \begin{cases} P & T \le T_S \\ \frac{T_L - T}{T_L - T_S} P & T_S < T < T_L \\ 0 & T \ge T_L \end{cases} \tag{4}$$

$$P_L = P - P_S \tag{5}$$

where P_s and P_L are the solid and liquid precipitation (mm), P is the total monthly precipitation (mm). T is the monthly average temperature (°C), T_S and T_L are the limiting temperatures of solid and liquid precipitation (°C). Usually, it can be determined according to the observation results of Urumqi glacier No.1 in the Tianshan Mountains and related literature, and take −0.5 °C and 2 °C, respectively.

3.2. Attribution of Glacier Mass Balance Change

The Geographical detector can effectively explain the spatial coupling between geographic phenomena and multiple factors [2,37,38]. In this paper, the factor detection module of the Geographical detector is used to quantitatively detect whether the factors affect the spatial heterogeneity of the changes in the glacier mass balance in the study area and determine the degree of influence of each factor. The formula is as follows:

$$q = 1 - \frac{\sum_{h=1}^{L} N_h {\sigma_h}^2}{N\sigma^2} = 1 - \frac{SSW}{SST} \tag{6}$$

where $h = 1, 2, \ldots, L$ is the stratification of the variable Y or factor X, the classification or partition. N_h and N are the number of units in the h stratum and the whole area, respectively. ${\sigma_h}^2$ and σ^2 are the variances of the stratum h and the whole area Y. SSW and SST are the sums of the conflict within the layer and the total friction of the whole area, respectively.

The factors affecting the change of glacier mass balance are diverse, and the internal mechanisms are complex, but they can be roughly classified into two major factors: topography and climate. Topographic factors mainly include altitude, slope, slope aspect, etc., and climatic factors mainly comprise air temperature, precipitation, accumulated temperature, solar radiation, etc. [2,39–41]. Each factor is coupled and influenced by each other, which affects the changes in the glacier mass balance. In particular, the hydrothermal conditions and terrain complexity of the north and south slopes of the Tianshan Mountains will further cause spatial differences in the glacier mass balance on the north and south slopes [42,43]. Through the factor detection of the Geographical detector, the q-value of the above factors in each geographical area is calculated and sorted. The higher the q-value is, the stronger the explanatory power of the factor for the change of glacier mass balance is. According to the contribution rate, the driving factors of glacier mass balance change in northern and southern slopes can be identified and attributed.

4. Results

4.1. Analysis of Glacier Area in the Central Tianshan Mountains

Glacier area data are derived from the Second China Glacier Catalog Dataset, which is used to determine the distribution of glaciers in each basin during the mass balance calculation, and also as the basis for calculating the glaciers mass balance of individual glaciers. In order to know the distribution of glacier's area in the vertical direction, all the glaciers in both basins were arranged according to their average elevation. Most of the large glaciers are located at high elevations and are concentrated. Still, the number is generally small, while the small glaciers are found at low elevations but are widely distributed and numerous (Figure 3). At the same time, glacier area is the most intuitive manifestation of glacier size. The glacier size of each elevation zone can also be known by counting the glacier area in each elevation zone, which influences the change of mass balance. When the glacier size is large, it can effectively restrain the glacier melting. While the glacier size is small [44], it is easier to melt. The lowest elevation range of glaciers' distribution in both basins is 3500–3700 m, the highest is 4500–4800 m, and most of the large-scale glaciers are in the range of 3900–4300 m. The total area of glaciers at 4100–4300 m is the largest, about 560 km^2, the 3900–4100 m is the second elevation range, and the area at 3500–3700 m is the least, which is less than 20 km^2. The glaciers in the Kaidu river basin are mainly concentrated in the southwest and northeast of the basin. In contrast, the glaciers in the Manas river basin are focused on the southern part of the basin. The number of glaciers in the Manas river basin is 2.7 times that of the Kaidu river basin. The maximum area of a single glacier in the Manas river basin is 36 km^2, while that in the Kaidu river basin is 16 km^2. Overall, the total volume of glaciers in the Manas river basin is more extensive than those in the Kaidu river basin.

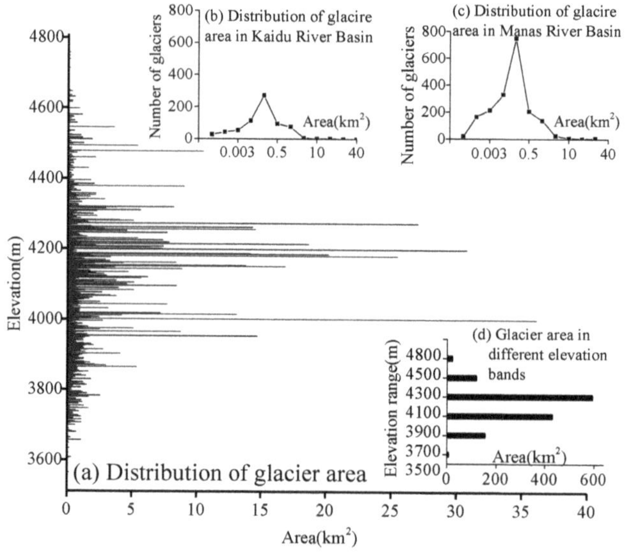

Figure 3. Elevation distribution characteristics of a single glacier area in both basins of central Tianshan Mountains. (**a**) Distribution of glacier area; (**b**) Distribution of glacier area in Kaidu River Basin; (**c**) Distribution of glacier area in Manas River Basin; (**d**) Glacier area in different elevation bands.

Generally speaking, small glaciers are the most sensitive to climate change, and their change of mass balance is the most pronounced. Therefore, the glaciers in both basins are divided into 11 classes according to the area size, with 0.001, 0.003, 0.005, 0.1, 0.5, 1, 5, 10, 15, 20, and 40 km^2. Although there are some differences in the number of glacier bars between both basins, glacier size distribution is very similar. There are a lot of glaciers in both basins, which are smaller than 1 km^2, with peak numbers occurring between 0.01 and 0.03 km^2. At the same time, it can be found that the larger area of glaciers in both basins, the fewer number with an area of more than 10 km^2. There is no glacier with an area of more than 20 km^2 in the Kaidu river basin. In addition, the small glaciers are more likely to melt because of their isolation and distribution in low altitudes. So the key to maintaining the mass balance of glaciers in both basins is the existence of a small number of medium and large area glaciers. The cooling effect on the surrounding environment due to the large size of the glacier can have a mitigating effect on glacier melt.

4.2. Comparison of Glacier Mass Balance between Both Basins in the Central Tianshan Mountains

Based on China's Second Glacier Cataloguing data, the mass balance of a single glacier in the Manas river basin and the Kaidu river basin was compared and analyzed (Figure 4). Overall, the mass balance of glaciers in the central Tianshan Mountains is mainly distributed between 0–1000 mm w.e., and the number of glaciers in −1000–1500 mm w.e. is significantly lower than former, and the number of glaciers between −2000 mm w.e. and 0 mm w.e. is the least. It can be seen that the mass balance of the glaciers in the central Tianshan Mountains is dominated by negative balance, and the distribution characteristics are similar to the normal distribution. The glaciers in a positive balance in the Kaidu river basin are significantly less than those in the Manas river basin. Still, the mass balance of most glaciers in the Kaidu river basin is between 0 and −500 mm w.e., while the mass balance of most glaciers in the Manas river basin is between 0 and −1000 mm w.e. There are more glaciers in the Manas river basin with a mass balance of over −1500 mm w.e. than in the Kaidu river basin. The maximum value of the mass balance of glaciers in the Kaidu river basin is comparable to that of the Manas river basin. Although the glaciers' mass balance spans in both basins are similar, the Manas river basin has more glaciers and a higher negative balance.

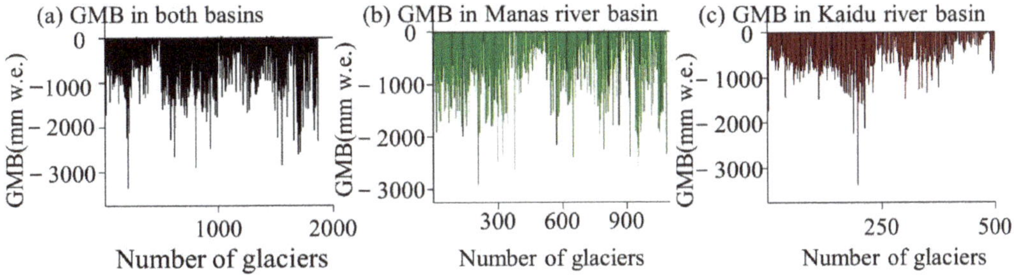

Figure 4. The mass balance of a single glacier in both basins of the central Tianshan Mountains. (**a**) GMB in both basins; (**b**) GMB in Manas river basin; (**c**) GMB in Kaidu river basin.

The interannual changes in the mass balance of glaciers can show the overall changes in the watershed and more clearly reflect the regional changes in mass balance (Figure 4). During the study period, the glaciers in both basins were in a negative balance state and continued to melt (−465.95 mm w.e.), with an interannual variation rate of −28.36 mm w.e./a. The most obvious glacier ablation was in 2010, followed by 2014; The melting trend can be divided into two stages: 2001–2010a is a continuous increase, and 2010–2016a is weakened volatility. Comparing both basins, from 2001 to 2016, the glacier ablation in the Kaidu river basin was 3.79% higher than that in the Manas river basin, but the change rate of the glacial mass balance in the Manas river basin (−28.85 mm w.e./a) slightly higher than the Kaidu tiver basin (−26.34 mm w.e./a). Glacier ablation in both basins reached the highest value in 2010, and the ablation in the Kaidu river basin (878.89 mm w.e.) was higher than that in the Manas river basin (811.26 mm w.e.). The trajectories of mass balance in both basins are similar with time. Still, the volatility of the trajectory in the Manas river basin is higher than that of the Kaidu river basin, with only a 1.07 mm w.e. difference. According to its spatial distribution map (Figure 5), the high value of glacier increased mass balance in both basins appears at the glacier aggregation region. The low mass balance values are mainly distributed at the end of glaciers or in clusters of small glaciers. The melting risk of dispersed glaciers is higher than that of glaciers with a high aggregation degree. It is not difficult to find that the glaciers on the southern slope are smaller in size. Still, the average ablation volume is higher than that on the northern slope, and the glaciers on the north slope are larger in scale but have higher ablation rates than those on the southern slope. This regional difference is closely related to the hydrothermal conditions and topographic conditions of the northern and southern slopes of the central Tianshan Mountains.

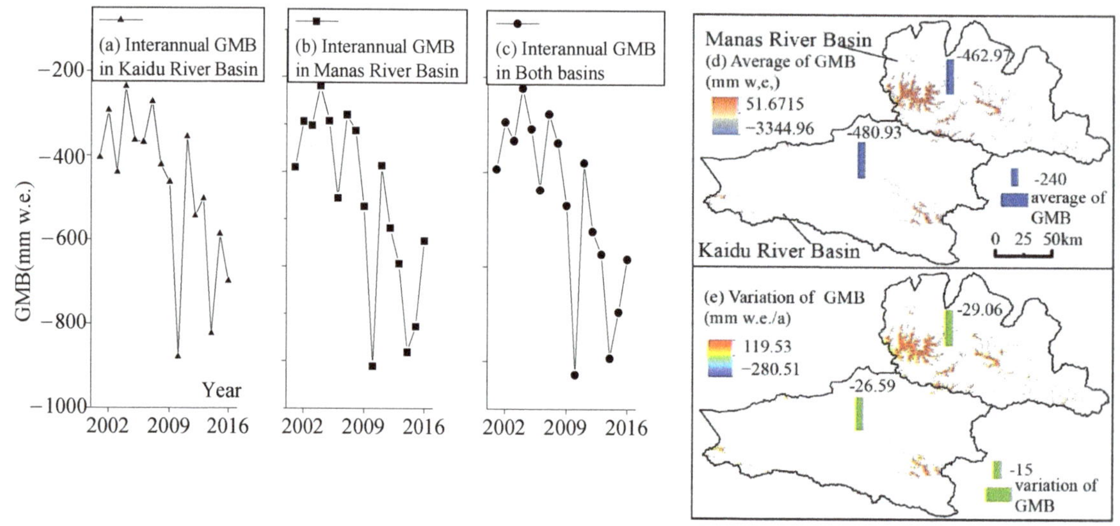

Figure 5. Glacier mass balance in both basins in central Tianshan (2001–2016). (**a**) Interannual GMB in Kaidu River Basin; (**b**) Interannual GMB in Manas River Basin; (**c**) Interannual GMB in Both Basin; (**d**) Average of GMB; (**e**) Variation of GMB.

In mainly glacier research, the Tianshan glaciers, as the representative of continental glaciers, play a pivotal role. Many scholars have carried out many types of research on changes in the mass balance of some glaciers in various basins in the Tianshan Mountains. Among them, Wucheng, X. et al. estimated the mass balance of the Tianshan glaciers in China based on the Landsat and ICESat data from 2003 to 2009a. They concluded that the average mass balance of the Tianshan Mountains during the period was −120 mm w.e. [45]. However, their estimated results were small compared with this study. Xiaohui, H. et al. used the geodetic method to estimate that the average mass balance of glaciers in the Tu-Haalik area from 1972 to 2016a was 180 mm w.e. Still, during the period from 1999 to 2016a, it reached 360 mm w.e. [46], which is far greater than the previous value of glacier mass balance, demonstrating the accelerated loss of glaciers since 2000. Ninglian, W. et al. collected the latest results of recent research on the mass balance of glaciers in the Tianshan Mountains. They found that the average mass balance of the Junggar basin was 600 mm w.e. [47], which was slightly larger than that of the present study. In addition, Puyu, W. et al. estimated that from 1972 to 2011a, glacier No. 6 in Yushugou, Miaoergou, and No. 1 glacier in Urumqi Heyuan in the East Tianshan Mountains were −510 and −510 mm w.e. [48]. Yanjun, C. et al. obtained the average mass balance of the Qingbingtan No. 72 glacier in the South Tianshan Mountains from 2008 to 2014a to be −510, −510, and −380 mm w.e., respectively [49], which are similar to the results of this study.

4.3. The Attribution of Glacier Mass Balance Change in the North–South Slope Basin in the Central Tianshan Mountains

By using factor detection and interaction detection based on a Geographical detector, the dominant factor affecting the mass balance can be determined by ranking the factor *q*-value (Table 2). The detector found that the dominant factors affecting the change of mass balance in the basin can be classified into topography and climate. Altitude is an important factor affecting the mass balance in the study area, followed by >0 °C accumulated temperature and annual average temperature. The contribution of annual average precipitation to solar radiation is equivalent. Topographic relief, slope, and aspect have limited influence on mass balance change. The contribution rate of topographic factors to both basins reached 44.87%, slightly lower than that of climatic factors (51.27%).

Therefore, the climate factor is still the dominant factor affecting the change of glacier mass balance in both basins. Although independent climatic factors have less influence than altitude, the interaction and coupling of each factor will have a stronger effect on glacier mass balance change.

Table 2. Contribution rate of mass balance factor affecting glacier (%).

Basins	Elevation	Topographic Relief	Slope	Aspect	Solar Radiation	Air Temperature	Precipitation	>0 °C Accumulated Temperature
Both Basins	42.75	1.06	1.07	3.85	11.08	14.41	10.14	15.64
Kaidu	30.88	0.77	0.79	2.09	14.73	11.31	26.90	12.53
Manas	40.44	1.02	1.02	3.85	11.00	14.70	8.02	19.95

The most critical factor affecting mass balance in the Kaidu river basin is altitude, followed by precipitation and its contribution rate is equivalent to elevation, reaching 26.9%. It shows that the precipitation factors greatly influence the glacier mass balance in the Kaidu river basin. The contribution of solar radiation, air temperature, and accumulated temperature is not much different. The contribution rate of slope direction is slightly higher than that of topographic relief and slope, but the contribution rate of the three is the low value of the whole factors. The contribution rate of topographic relief to the slope is less than 1% in the Kaidu. The contribution rate of factors affecting glacier mass balance in the Manas is higher than 1%. The first factor is still altitude, with the contribution rate of 40.45%, followed by the accumulated temperature of >0 °C. The contribution rate of solar radiation to temperature is equivalent, and the precipitation's contribution rate is slightly lower than the first two factors. Similar to the Kaidu river basin, the contribution rate of topographic relief, slope, and aspect are also low. Still, the contribution rate of the three in the Manas river basin is slightly higher than that in the Kaidu river basin, and the contribution rate of the slope is 3.85%. In the overall role of topographic factors and climatic factors, the Kaidu and Manas river basins are the same, still dominated by climatic factors. However, in the Kaidu river basin, the contribution rate of the climate factor is nearly 20% higher than that of the topographic factors. The Manas river basin's climate factors contribution rate is only 7.3% higher than the topographic factors, which is closely related to the more complex topographic environment in the Manas river basin. As the Manas river basin is located on the northern slope of the Tianshan Mountains, the precipitation is relatively abundant (annual average precipitation is 257.5 mm, Figure 6), reducing the warming effect of solar radiation on glaciers. The average elevation in the basin is 4094 m. The large glaciers are primarily distributed at high altitudes, which is more conducive to their accumulation, alleviating the Manas river basin's glacier ablation. There are few large glaciers in the Kaidu river basin, mainly distributed as small glaciers, and the average elevation of a single glacier in the basin is 3984 m (Figure 6). Glacier positive equilibrium is limited. Although the precipitation is less than in the Manas river basin, precipitation is liquid mainly at low altitudes, so it is difficult to stay on small glaciers. Higher heat in water droplets can accelerate the melting of small glaciers and increase their negative balance. According to the calculation, the ice climate data of both basins are obtained, and the average annual temperature of the glacier in the Manas river basin is −13.1 °C. In comparison, the Kaidu river basin is −9.7 °C. The main reason is that the Kaidu river basin is located on the southern slope of the Tianshan Mountains. Its solar radiation is higher than that of the Manas river basin, with a value of 179.29 W/m^2. The accumulated temperature above 0 °C is 337.6 °C, which is significantly higher than the Manas river basin (182.5 °C) on the northern slope of the Tianshan Mountains (Figure 6). Under the combined effect of topography and climate factors, the negative balance intensity of the Kaidu river basin is higher than that of the Manas river basin.

Under the dual influence of climatic factors and topographic factors, the glaciers in both basins of the central Tianshan Mountains are melting. Still, there are some differences in the dominant factors affecting the melting of glaciers on the northern and southern slopes. The terrain on the south slope of the Tianshan Mountains is relatively flat and has a less natural place for glacier development. The limited precipitation cannot effectively accumulate glaciers at high altitudes. Temperature and solar radiation are higher than those on the northern slope, which is more likely to promote glacier ablation. Glacier ablation is more than on the north slope. Due to the complex terrain and more large glaciers on the north slope, such conditions effectively alleviate the glacier's negative balance, and thermal energy mainly affects ablation factors. The precipitation affects large glaciers or upstream glaciers in the basin, but the influence ability is limited.

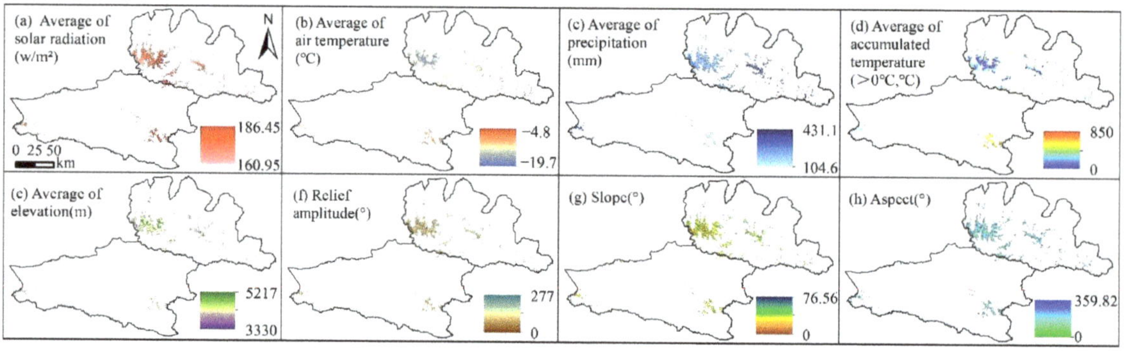

Figure 6. Driving factors affecting glacier mass balance. (**a**) Average of solar radiation (w/m^2); (**b**) Average of air temperature (°C); (**c**) Average of precipitation (mm); (**d**) Average of accumulated temperature (>0 °C, °C); (**e**) Average of elevation (m); (**f**) Relief amplitude (°); (**g**) Slope (°); (**h**) Aspect (°).

5. Discussion

The changes in the mass balance of glaciers are affected by various influencing factors. These factors will also produce corresponding eigenvalues in mass balance changes, and different degrees of mass balance will be adapted to other climatic and topographical conditions. If the attribution comparison is based only on the mean value of the mass balance of glaciers in the region, the attribution analysis of glacier accumulation and ablation (positive and negative mass balance) in the basin is very likely to be ignored. This paper takes the positive and negative balance as the breakthrough point, combining it with the factor detection of the Geographical detector and comparing the differences between both basins. The mechanisms for glacial mass balance changes on the north and south slopes of the central Tianshan Mountains can be further refined. The glaciers were divided into positive and negative balance areas based on the balance value by distributed degree-day model. Using the Geographical detector to detect the contribution rate of each factor. The driving force of the positive and negative balance areas in both basins will be further discussed.

5.1. Contribution Analysis of Influencing Factors Based on Positive and Negative Mass Balance Area

The part in positive balance is considered the accumulation area, and the part in negative balance is regarded as the ablation area based on the value of glacier mass balance. From the results of the Geographical detector (Figure 7), the factors affecting the accumulation of glaciers in the two basins are dominated by climatic factors. The cumulative contribution rate of climate factors is 69.37%, of which the precipitation contribution rate is the highest, followed by the contribution rate of temperature. The contribution rate of solar radiation is only 3.09%, and the effect of solar radiation on the accumulation of glaciers is minimal. Among topographic factors, altitude still has a high contribution rate (26.53%), while the contributions of the other three topographic factors are all low (Table 3). For glacier accumulation areas, hydrothermal conditions are critical factors. Mountains provide a basic place for glaciers to accumulate/develop. It is possible to have a certain accumulation effect on glaciers. In the context of climate warming, the continuous melting of glaciers is a significant trend; the accumulation area of glaciers is mainly concentrated on the main body of glaciers with higher altitudes and larger scale, and they have the characteristics of seasonal accumulation [50–52]. The climate factor is still dominant (52.2%), but its cumulative contribution rate is equal to the topographic factor (47.8%). Among the climatic factors, the precipitation factor has the highest contribution rate (17.73%), followed by the solar radiation factor (14.85%). Among the topographic factors, the altitude contribution rate is the highest, which is 38.32%. The remaining three topographic factors have a low contribution rate to the ablation area. Different from the accumulation area, the elevation factor is the factor with the highest contribution to the negative balance of glaciers in the whole basin.

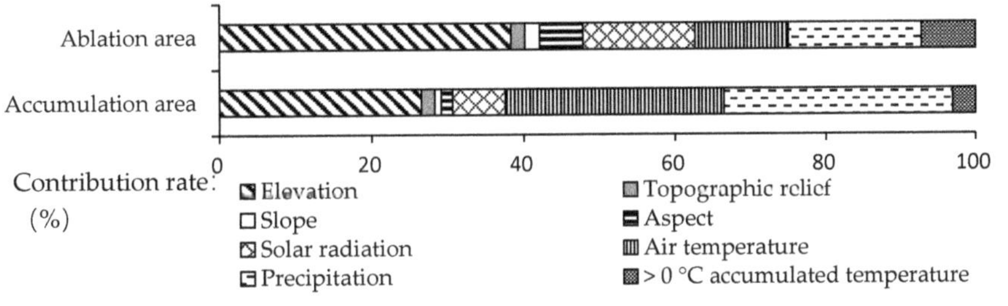

Figure 7. The contribution rate of each factor in the accumulation/ablation area.

Table 3. Each factor's statistical value in the accumulation/ablation area in the central Tianshan Mountains.

Area	GMB (mm/w.e.)	Elevation (m)	Topographic Relief (°)	Slope (°)	Aspect (°)	Solar Radiation (W/m^2)	Air Temperature (°C)	Precipitation (mm)	>0 °C Accumulated Temperature (°C)
Accumulation	30.21	4502.18	60.19	37.67	128.92	157.41	−14.16	241.94	128.84
Ablation	−718.97	3980.93	37.64	25.34	187.71	175.67	−11.45	254.88	260.91

Under the comprehensive action of multiple factors, there are significant differences in the natural environment in the positive and negative balance areas of the north and south slopes of the central Tianshan Mountains. According to statistics, the accumulation area is 33.3 km^2, the ablation area is 860.13 km^2, and the mass balance of the ablation area is dominant (Table 3). The topographic complexity of the accumulation area is higher than that of the ablation area, and the high altitude directly weakens the temperature of the environment in the accumulation area. The complex topographic relief is very likely to form the mountain's shadow to block the sunlight, thus eliminating the direct effect of solar radiation on the glacier surface and reducing the potential risk of glacier endothermic melting [53,54]. As shown in Table 3, the temperature and precipitation in the accumulation area are significantly lower than those in the ablation area. Most accumulation areas face the southeast, while the ablation areas are mostly facing the south. Therefore, the solar radiation received by the accumulation area is lower than that of the ablation area. The ablation area is mainly located in the lower altitude area, and the end of the glacier is straightforward to be ablated. The terrain here is relatively flat and more susceptible to solar radiation. Although the average temperature for many years is −11.45 °C, the accumulated temperature >0 °C is significantly higher than that in the accumulation area. It also has a higher risk of melting and is characterized by significant seasonal ablation.

5.2. Attribution Comparison of Both Basins Based on Positive and Negative Balance Area

The geographical environment of the Manas river basin on the northern slope of the Tianshan Mountains is quite different from that of the Kaidu river basin on the southern slope. The formation of positive and negative balance areas in the basins are also different, which can be further revealed through comparative analysis of the internal mechanism of the spatial heterogeneity of the mass balance of Tianshan glaciers. Topographic factors dominate the accumulation area and ablation area in the Kaidu river basin, and the ablation area is dominated by climatic factors, with cumulative contributions of 55.2% and 70.84%, respectively. The accumulation area in the Manas river basin is dominated by climatic factors, with a cumulative contribution rate of 69.38%, and the ablation area is dominated by topographical factors, with a cumulative contribution rate of 54.1% (Figure 8). The natural conditions of both basins control the patterns of glacier mass balance in the basins so that the dominant factors are distinctively regional.

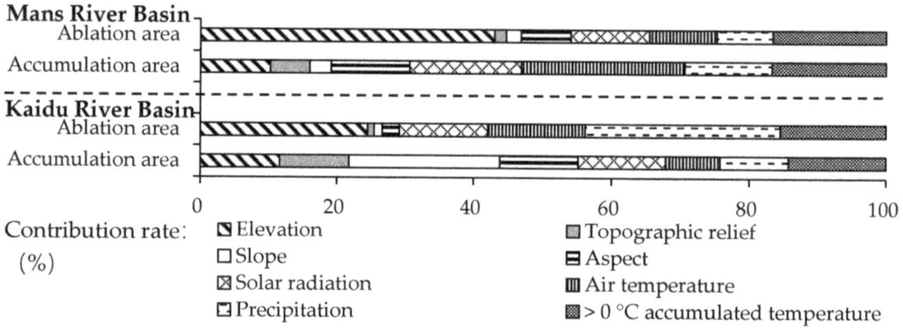

Figure 8. The contribution rate of each factor in the accumulation/ablation area in both basins.

In the accumulation area, the Manas river basin's positive mass balance and topographic complexity are higher than the Kaidu river basin (Table 4). It is worth noting that the slope factor has a very high explanatory power (22.06%) in the accumulation area of the Kaidu river basin, which may be related to the easy retention of solid snowfall on gentle slopes [55,56]. The accumulation area in the Manas river basin has the highest contribution rate of air temperature (23.73%). Taking into account factors such as more glaciers and larger scales in the Manas river basin, and then comparing climate data such as temperature and precipitation, it can be inferred that the stability of glaciers in the Manas river basin in the accumulation zone is higher and the surrounding environmental conditions are more conducive to the formation of positive mass balance of glaciers than those in Kaidu river basin. In the ablation area, the difference between both basins in the statistics value of the factors is slight. Still, the negative mass balance value of the Manas river basin is higher. The highest contribution rate of precipitation factor in the Kaidu river basin (28.33%) was slightly higher than that of elevation. The highest contribution rate of elevation factor in the Manas river basin (43.05 %). Glacier ablation is partly due to seasonal liquid precipitation at low altitudes [57–59]. The mean elevation of the ablation area in the Manas river basin is similar to that of the Kaidu river basin. The altitude range of about 3900–4000 m can be seen as a strong ablation zone in both basins. Although the dominant factors of the ablation/accumulation areas between both basins are different, in the final analysis, the interaction and coupling between topographic and climatic factors cause the differences in the mass balance of the glaciers in the basins.

In this study, comparing the differences in mass balance in the basins failed to go to a more detailed time scale. It focused on discussing natural environmental factors on the mass balance of glaciers between both basins without considering the direct or indirect effects of human activities on glaciers.

Table 4. Statistical value in the accumulation/ablation area of both basins.

Basins	GMB (mm/w.e.)	Elevation (m)	Topographic Relief (°)	Slope (°)	Aspect (°)	Solar Radiation (W/m^2)	Temperature (°C)	Precipitation (mm)	>0 °C Accumulated Temperature (°C)
					Accumulation Area				
Kaidu	20.37	4346.73	52.94	33.90	135.01	170.08	−9.16	138.38	400.76
Manas	31.18	4517.40	60.90	38.04	128.33	167.11	−14.65	252.08	102.21
					Ablation Area				
Kaidu	−591.73	3944.43	36.80	24.79	184.22	178.11	−9.78	219.20	333.47
Manas	−769.77	4006.34	38.06	25.62	189.70	174.71	−12.12	268.97	231.87

6. Conclusions

This paper uses distributed degree-day model driven by multi-source remote sensing data. It simulates the glacier mass balance of the south and north basins of the Tianshan Mountains from 2001 to 2016. Based on the analysis and comparison of the temporal and spatial changes, using eight factors, including terrain factors and climate factors, were introduced by a Geographical detector to conduct an attribution comparison from the basin scale and the positive and negative mass balance area scale in the basin. The conclusions are as follows:

(1) In terms of glacier distribution, the glaciers in both basins are mainly distributed in the altitude range of 3500–4800 m. Most glaciers are distributed in 3900–4300 m, and the total area in this altitude range accounts for 85.71%. Due to the altitude and mountain trend, the glaciers in Kaidu are mainly concentrated in the southwest and northeast, which are mainly concentrated in the south of Manas. In terms of glacier scale, the number of less than 1 km^2 glaciers is the most, and the number of glaciers over 10 km^2 is rare; in terms of glacier quantity and scale, the Manas river basin is more than the Kaidu river basin.

(2) During the study period, the glaciers in both basins were continuously melting. They were in negative balance (−465.95 mm w.e.), and the interannual change rate was −28.36 mm w.e./a. among them, the glacier melting was the most obvious in 2010. The melting trend can be divided into two stages: the persistence increased from 2000 to 2010a, and the volatility decreased from 2010 to 2016a. Compared with both basins, the amount of glacier melting in Kaidu is 3.79% higher than that in Manas, but the change rate of glacier mass balance in Manas (−28.85 mm w.e./a) is slightly higher than that in Kaidu (−26.34 mm w.e./a).

(3) In the attribution of mass balance, the factors affecting glacier mass balance can be divided into climate and topography. The total contribution rate of topographic factors is 44.87%, which is slightly lower than that of climatic factors (51.27%). Therefore, climatic factors are still the dominant factors affecting the difference in glacier mass balance between both basins; however, in Kaidu, the contribution rate of climatic factors is nearly 20% higher than that of topographic factors. The contribution rate of climate factors in Manas is only 7.3% higher, so the change of glacier mass balance in Kaidu is more driven by climate factors, while the glacier mass balance in Manas is more affected by the combination of climate and topographic factors.

(4) The factors affecting the positive/negative mass balance of glaciers in both basins are dominated by climate factors (69.37%, 52.2%). Still, the contribution gap between climate and terrain factors in the ablation area is smaller. In terms of accumulation area, the positive mass balance in Kaidu is dominated by topographic factors (55.2%), while that in Manas is dominated by climatic factors (69.37%). For the ablation area, climate factors are dominated in Kaidu (70.85%), and topographic factors are dominated in Manas (54.11%). The driving forces of climate and terrain factors and the different coupling modes lead to the spatiotemporal heterogeneity of glacier mass balance in the north and south slope basins in the central Tianshan Mountain.

Author Contributions: Conceptualization, L.L. and H.T.; Investigation, X.Z. and H.C.; Methodology, Z.Z. and G.Z.; Resources, X.Y. and Y.C.; Software, M.Z.; Validation, Y.G. and F.Y.; Visualization, Z.K. and T.W.; Writing—original draft, H.T.; Writing—review and editing, L.L. and H.T. All authors have read and agreed to the published version of the manuscript.

Funding: This research was supported by The National Natural Science Foundation of China (Grant No. 41761108, 41641003), The third Comprehensive Scientific investigation project in Xinjiang (Grant No. 2021xjkk08).

Data Availability Statement: Not applicable.

Acknowledgments: We acknowledge the research environment provided by Xinjiang Production and Construction Corps Key Laboratory of Oasis Town and Mountain-basin System Ecology.

Conflicts of Interest: The authors declare no conflict of interest.

References

1. Garg, V.; Kudekar, A.R.; Thakur, P.K.; Nikam, B.R.; Aggarwal, S.P.; Chauhan, P. Glacier Change Studies under Changing Climate Using Geospatial Tools and Techniques. *J. Indian Soc. Remote Sens.* **2021**, *49*, 2387–2406. [CrossRef]
2. Xingran, C.; Zhongqin, L.; Hui, Z.; Chunhai, X. Vulnerability of glacier change in the Tianshan Mountains region of China. *J. Geogr. Sci.* **2021**, *31*, 1469–1489.
3. Shugar, D.H.; Jacquemart, M.; Shean, D.; Bhushan, S.; Westoby, M.J.J.S. A massive rock and ice avalanche caused the 2021 disaster at Chamoli, Indian Himalaya. *Science* **2021**, *373*, eabh4455. [CrossRef] [PubMed]
4. Xingran, C.; Chunhai, X.; Zhongqin, L. Glacier changes and its effect on water resources in the upper reaches of Aksu River, Tien Shan, China, from 1989 to 2016. *Arab. J. Geosci.* **2022**, *15*, 1–11.
5. Liu, Q.; Liu, S.J.C.D. Response of glacier mass balance to climate change in the Tianshan Mountains during the second half of the twentieth century. *Clim. Dyn.* **2016**, *46*, 303–316. [CrossRef]
6. Braithwaite, R.J.; Hughes, P.D. Regional Geography of Glacier Mass Balance Variability Over Seven Decades 1946–2015. *Front. Earth Sci.* **2020**, *8*, 302. [CrossRef]
7. Duan, K.; Yao, T.; Wang, N.; Liu, H. Numerical simulation of Urumqi Glacier No. 1 in the eastern Tianshan, central Asia from 2005 to 2070. *Chin. Sci. Bull.* **2012**, *57*, 4505–4509. [CrossRef]
8. Xu, C.; Li, Z.; Wang, P.; Anjum, M.N.; Li, H.; Wang, F. Detailed comparison of glaciological and geodetic mass balances for Urumqi Glacier No.1, eastern Tien Shan, China, from 1981 to 2015. *Cold Reg. Sci. Technol.* **2018**, *155*, 137–148. [CrossRef]
9. Guining, Z.; Zhengyong, Z.; Lin, L.; Zhongqin, L.; Puyu, W.; Liping, X. Simulation and construction of the glacier mass balance in the Manas River Basin, Tianshan, China from 2000 to 2016. *J. Geogr. Sci.* **2020**, *30*, 988–1004.
10. Kenzhebaev, R.; Barandun, M.; Kronenberg, M.; Chen, Y.; Usubaliev, R.; Hoelzle, M. Mass balance observations and reconstruction for Batysh Sook Glacier, Tien Shan, from 2004 to 2016. *Cold Reg. Sci. Technol.* **2017**, *135*, 76–89. [CrossRef]
11. Wang, P.; Li, Z.; Li, H.; Wang, W.; Yao, H. Comparison of glaciological and geodetic mass balance at Urumqi Glacier No. 1, Tian Shan, Central Asia. *Glob. Planet. Change* **2014**, *114*, 14–22. [CrossRef]
12. Bhattacharya, A.; Muhuri, A.; De, S.; Manickam, S.; Frery, A.C. Data on Earth Observations and Remote Sensing Reported by Researchers at Federal University (Modifying the Yamaguchi Four-Component Decomposition Scattering Powers Using a Stochastic Distance). *IEEE J. Sel. Top. Appl. Earth Obs. Remote Sens.* **2015**, *8*, 3497–3506.
13. Ridha, T. Target Scattering Decomposition in Terms of Roll-Invariant Target Parameters. *IEEE Trans. Geosci. Remote Sens.* **2007**, *45*, 73–84.
14. Muhuri, A.; Manickam, S.; Bhattacharya, A. Scattering Mechanism Based Snow Cover Mapping Using RADARSAT-2 C-Band Polarimetric SAR Data. *IEEE J. Sel. Top. Appl. Earth Obs. Remote Sens.* **2017**, *10*, 3213–3224. [CrossRef]
15. Reeh, N.J.P. Parameterization of melt rate and surface temperature on the Greenland ice sheet. *Polarforschung* **1991**, *5913*, 113–128.
16. Jóhannesson, T.; Sigurdsson, O.; Laumann, T.; Kennett, M. Degree-day glacier mass-balance modelling with applications to glaciers in Iceland, Norway and Greenland. *J. Glaciol.* **2017**, *41*, 345–358. [CrossRef]
17. Yuhuan, C.; Baisheng, Y.; Jie, W.; Youcun, L. Influence of Degree-Day Factor Variation on the Mass Balance of Glacier No. 1 at the Headwaters of ürümqi River, China. *J. Earth Sci.* **2013**, *24*, 1008–1022.
18. Lihua, W.; Huilin, L.; Lin, W. Application of a Degree-Day Model for Determination of Mass Balance of Urumqi Glacier No. 1, Eastern Tianshan, China. *J. Earth Sci.* **2011**, *22*, 470–481.
19. Baumann, S. Comparison of Glacier Mass Balance Data in the Tien Shan and Pamir, Central Asia. *Arct. Antarct. Alp. Res.* **2017**, *49*, 133–146. [CrossRef]
20. Wang, P.; Li, Z.; Li, H.; Yao, H.; Xu, C.; Zhou, P.; Jin, S.; Wang, W. Analyses of recent observations of Urumqi Glacier No. 1, Chinese Tianshan Mountains. *Environ. Earth Sci.* **2016**, *75*, 720. [CrossRef]
21. Xu, M.; Wu, H.; Kang, S. Impacts of climate change on the discharge and glacier mass balance of the different glacierized watersheds in the Tianshan Mountains, Central Asia. *Hydrol. Processes* **2018**, *32*, 126–145. [CrossRef]

22. Yanjun, C.; Mingjun, Z.; Zhongqin, L.; Yanqiang, W.; Zhuotong, N.; Huilin, L.; Shengjie, W.; Bo, S. Energy balance model of mass balance and its sensitivity to meteorological variability on Urumqi River Glacier No.1 in the Chinese Tien Shan. *Sci. Rep.* **2019**, *9*, 13958.
23. Zhang, Z.; Liu, L.; He, X.; Li, Z.; Wang, P. Evaluation on glaciers ecological services value in the Tianshan Mountains, Northwest China. *J. Geogr. Sci.* **2019**, *29*, 101–114. [CrossRef]
24. Tongxia, W.; Zhengyong, Z.; Lin, L.; Zhongqin, L.; Puyu, W.; Liping, X.; Guining, Z.; Hao, T.; Ziwei, K.; Hongjin, C.; et al. Simulation of the Potential Distribution of the Glacier Based on Maximum Entropy Model in the Tianshan Mountains, China. *Water* **2021**, *13*, 1541.
25. Ren, L.; Xue, L.-Q.; Liu, Y.-H.; Shi, J.; Han, Q.; Yi, P.-F. Study on Variations in Climatic Variables and Their Influence on Runoff in the Manas River Basin, China. *Water* **2017**, *9*, 258. [CrossRef]
26. Zhang, F.; Bai, L.; Li, L.; Wang, Q. Sensitivity of runoff to climatic variability in the northern and southern slopes of the Middle Tianshan Mountains, China. *J. Arid. Land* **2016**, *8*, 681–693. [CrossRef]
27. Davide, F.; Veronica, M.; Antonella, S.; Guglielmina, D.; Maurizio, M. Snow Cover Variability in the Greater Alpine Region in the MODIS Era (2000–2019). *Remote Sens.* **2021**, *13*, 2945.
28. Yaping, C.; Yongjian, D.; Qiudong, Z.; Shiqiang, Z. A Comprehensive Evaluation of 4-Parameter Diurnal Temperature Cycle Models with In Situ and MODIS LST over Alpine Meadows in the Tibetan Plateau. *Remote Sens.* **2019**, *12*, 103.
29. Zhao, W.; He, J.; Wu, Y.; Xiong, D.; Wen, F.; Li, A. An Analysis of Land Surface Temperature Trends in the Central Himalayan Region Based on MODIS Products. *Remote Sens.* **2019**, *11*, 900. [CrossRef]
30. Shukla, A.K.; Ojha, C.S.P.; Singh, R.P.; Pal, L.; Fu, D. Evaluation of TRMM Precipitation Dataset over Himalayan Catchment: The Upper Ganga Basin, India. *Water* **2019**, *11*, 613. [CrossRef]
31. Xiao, P.; Yunfei, F.; Sen, Y.; Ying, G.; Deqin, L. Diurnal Variations of Precipitation over the Steep Slopes of the Himalayas Observed by TRMM PR and VIRS. *Adv. Atmos. Sci.* **2021**, *38*, 641–660.
32. Luo, Y. The mass balances modeling and the relationship with meteorological parameters GlaciersinTianshan. Master Thesis, Northwest Normal University, Lanzhou, China, 2012.
33. Yong, Z.; Shiyin, L.; Yongjian, L.; Jing, L.; Donghui, S.G. Preliminary Study of Mass Balance on the Keqicar BaxiGlacier on the South Slopes of Tianshan Mountains. *J. Glaciol. Geocryol.* **2006**, *28*, 477–484.
34. Braithwaite, R.J.; Zhang, Y.; Raper, S.C.B. Temperature sensitivity of the mass balance of mountain glaciers and ice caps as a climatological characteristic. *Z. Für Gletsch. Und Glazialgeol.* **2002**, *38*, 35–61.
35. Braithwaite, R.J.; Glaciology, O.O.J.J.o. Seasonal variation of ice ablation at the margin of the Greenland ice sheet and its sensitivity to climate change, Qamanrssp sermia, West Greenland. *J. Glaciol.* **1993**, *39*, 267–274. [CrossRef]
36. Kang, E.; Cheng, G.; Lan, Y.; Jin, H.S. A model for simulating the response of runoff from the mountainous watersheds of inland river basins in the arid area of northwest China to climatic changes. *Sci. China Ser. D Earth Sci.* **1999**, *42*, 52–63. [CrossRef]
37. Wang, G.; Peng, W. Quantifying spatiotemporal dynamics of vegetation and its differentiation mechanism based on geographical detector. *Environ. Sci. Pollut. Res.* **2022**, *29*, 32016–32031. [CrossRef]
38. Jinfeng, W.; Chengdong, X. Geodetector: Principle and prospective. *Acta Geogr. Sin.* **2017**, *72*, 116–134.
39. Arindam, C.; Chand, S.M.; Sunil, K.D.; Manasi, D. Glacier changes in the Chhombo Chhu Watershed of the Tista basin between 1975 and 2018, the Sikkim Himalaya, India. *Earth Syst. Sci. Data.* **2021**, *13*, 2923–2944.
40. Jing, H.; Ninglian, W.; An'an, C.; Xuewen, Y.; Ting, H. Glacier Changes in the Qilian Mountains, Northwest China, between the 1960s and 2015. *Water* **2019**, *11*, 623.
41. Zhang, Q.; Chen, Y.; Li, Z.; Li, Y.; Xiang, Y.; Bian, W. Glacier changes from 1975 to 2016 in the Aksu River Basin, Central Tianshan Mountains. *J. Geogr. Sci.* **2019**, *29*, 984–1000. [CrossRef]
42. Yajuan, C.; Junping, Y. A comparison on characteristics of climate change in the north and south of Tianshan Mountain. *J. Arid. Land Resour. Environ.* **2011**, *25*, 92–96. [CrossRef]
43. Hongzheng, P.; Tianding, H.; Peng, C.; Xiangying, L.; Keqin, J. Characteristics of Air Temperature in the Southern and Northern Slopes of Tianshan Mountains. *Plateau Meteorol.* **2015**, *34*, 753–761.
44. Zhang, X.; Liu, L.; Zhang, Z.; Kang, Z.; Tian, H.; Wang, T.; Chen, H. Spatial and Temporal Variation Characteristics of Glacier Resources in Xinjiang over the Past 50 Years. *Water* **2022**, *14*, 1057. [CrossRef]
45. Wucheng, X.; Zhongqin, L.; Hui, Z.; Mingjun, Z.; Pengbin, L.; Jianxin, M. Spatial-temporal variation of glacier resources in Chinese Tianshan Mountains since 1959. *Acta Geogr. Sin.* **2017**, *72*, 1594–1605.
46. Xiaohui, Y.; Jingdong, Z.; Hui, H. Study on glacier mass balance in the Karlik Range, East Tianshan Mountains, 1972–2016. *J. Glaciol. Geocryol.* **2019**, *41*, 1–11.
47. Ninglian, W.; Tandong, Y.; Baiqing, X.; Anan, C.; Weicai, W. Spatiotemporal Pattern, Trend, and Influence of Glacier Change in Tibetan Plateau and Surroundings under Global Warming. *Bull. Chin. Acad. Sci.* **2019**, *34*, 1220–1232. [CrossRef]
48. Puyu, W.; Zhongqin, L.; Huilin, L.; Lihua, W.; Shuang, J.; Ping, Z. Changes of Ice-thickness and Volume for Representative Glaciers in 'Tianshan Mountains in the Past 50 Years. *Acta Geogr. Sin.* **2012**, *67*, 929–940.
49. Yanjun, C.; Mingjun, Z.; Zhongqin, L.; Shuang, J.; Wenbin, W.; Shengjie, W. Understanding the mass balance characteristics of Qingbingtan Glacier No.72 during the period of 2008–2014. *J. Glaciol. Geocryol.* **2020**, *42*, 318–331.
50. Chaturvedi, R.K.; Kulkarni, A.; Karyakarte, Y.; Joshi, J.; Bala, G. Glacial mass balance changes in the Karakoram and Himalaya based on CMIP5 multi-model climate projections. *Clim. Change* **2014**, *123*, 315–328. [CrossRef]

51. Colucci, R.R.; Filippo, G.; Csaba, T. Unprecedented heat wave in December 2015 and potential for winter glacier ablation in the eastern Alps. *Sci. Rep.* **2017**, *7*, 7090. [CrossRef]
52. Yulan, Z.; Tanguang, G.; Shichang, K.; Donghui, S.; Xi, L. Albedo reduction as an important driver for glacier melting in Tibetan Plateau and its surrounding areas. *Earth-Sci. Rev.* **2021**, *220*, 103735.
53. Yongpeng, G.; Xiaojun, Y.; Shiyin, L.; Miaomiao, Q.; Hongyu, D.; Juan, L.; Dahong, Z. Spatial-temporal variation of glacier resources in the Hexi interior from 1956 to 2017. *J. Glaciol. Geocryol.* **2019**, *41*, 1313–1325.
54. Wuchao, Y. Runoff Characteristics, Attribution Analysis and Futureprediction of the Cold and Arid Regions: A Case Study of Headwater of the Manas. Master Thesis, Northwest A&F University, XianYang, China, 2021.
55. Yanlong, G. Changes of Global Climate Landscape Pattern and lts Impact on Surface Water Resource in Qinghai-Tibet Plateau. Doctor Thesis, North China Electric Power University, Beijing, China, 2021.
56. Zongli, J.; Lei, W.; Zhen, Z.; Shiyin, L.; Yong, Z.; Zhiguang, T.; Junfeng, W.; Danni, H.; Shasha, Z. Surface elevation changes of Yengisogat Glacier between 2000 and 2014. *Arid. Land Geogr.* **2020**, *43*, 12–19.
57. Li, Z.; Shi, X.; Tang, Q.; Zhang, Y.; Gao, H.; Pan, X.; Déry, S.J.; Zhou, P. Partitioning the contributions of glacier melt and precipitation to the 1971–2010 runoff increases in a headwater basin of the Tarim River. *J. Hydrol.* **2020**, *583*, 124579. [CrossRef]
58. Megumi, W.; Aki, Y.; Yukiko, H.; Satoshi, W.; Akiko, S.; Shinjiro, K. Uncertainty from climate forcing of projections in glacier melt for high mountain asia. *J. Jpn. Soc. Civ. Eng. Ser. B1* **2018**, *74*.
59. Wang, J.; Chen, X.; Liu, J.; Hu, Q. Changes of Precipitation-Runoff Relationship Induced by Climate Variation in a Large Glaciated Basin of the Tibetan Plateau. *J. Geophys. Res. Atmos.* **2021**, *126*, e2020JD034367. [CrossRef]

water

MDPI

Article

Spatial and Temporal Variation Characteristics of Glacier Resources in Xinjiang over the Past 50 Years

Xueying Zhang [1,†], Lin Liu [1,2,†], Zhengyong Zhang [1,2,*], Ziwei Kang [1], Hao Tian [1], Tongxia Wang [1] and Hongjin Chen [1]

[1] College of Sciences, Shihezi University, Shihezi 832000, China; zxy970716@163.com (X.Z.); liulin779@163.com (L.L.); kangziwei0808@163.com (Z.K.); 20192018013@stu.shzu.edu.cn (H.T.); wtx0428@163.com (T.W.); chenhongjin@stu.shzu.edu.cn (H.C.)
[2] Key Laboratory of Oasis Town and Mountain-Basin System Ecology, Xinjiang Production and Construction Corps, Shihezi 832000, China
* Correspondence: zyz0815@163.com; Tel.: +86-180-4083-0081
† These authors contributed equally to this work.

Citation: Zhang, X.; Liu, L.; Zhang, Z.; Kang, Z.; Tian, H.; Wang, T.; Chen, H. Spatial and Temporal Variation Characteristics of Glacier Resources in Xinjiang over the Past 50 Years. *Water* **2022**, *14*, 1057. https://doi.org/10.3390/w14071057

Academic Editor: Maria Mimikou

Received: 17 February 2022
Accepted: 25 March 2022
Published: 28 March 2022

Abstract: Changes in glacier resources and their meltwater runoff contributions in Xinjiang are significant to the hydrological processes and water resources utilization. This study used the first and second Chinese Glacier Inventory, geomorphological and meteorological data. GIS spatial analysis technology was used to explore the characteristics of glacier change and its response to topography and climate change in Xinjiang in the last 50 years. The results show that there are currently 20,695 glaciers in Xinjiang with a total area of 22,742.55 km^2 and ice reserves of about 2229.17 km^3. Glaciers in Xinjiang are concentrated at 5100–6000 m. The Tianshan mountains have the largest number of glaciers. However, the Kunlun mountains have the largest glaciers and ice reserves. The scale of glaciers is significantly larger in the south than that in the north. The changes in glaciers in Xinjiang during the last 50 years are mainly receding and splitting, and their number, area, and ice reserves have decreased by 1359, 7080.12 km^2 and 482.65 km^3, respectively. Small glaciers are more sensitive to climate change. Glaciers are basically unchanged in regions above 6000 m. The glaciers on the south slope of mountains are more susceptible to climate change. The phenomenon of an increase in the number of glaciers but decreasing total area in the southern mountains is related to glacier extinction and splitting. Glacier development and formation are determined by the combination of topography and hydrothermal material conditions. The change of glacier areas in Xinjiang is jointly affected by climatic conditions (53.45%) and topographic conditions (16.55%), among which climatic conditions are more prominent.

Keywords: glacier resources; glacier inventory; climate change; topography; Xinjiang

1. Introduction

Glaciers are one of the crucial components of the cryosphere and are a dynamic and scarce natural resource. Mountain glaciers are widely distributed in middle and high latitudes, with high sensitivity and feedback to climate change. They are known as natural "recorders" and "early warners" [1]. Glacier meltwater has an obvious recharge and regulating function for many river runoffs and is known as an alpine "solid water reservoir". China is one of the countries with the most mountain glaciers, but it also features some of the world's desert areas and is considered a water-poor country. Glaciers and their meltwater are essential freshwater resources in the arid region of northwest China. They are the "lifeline" for the health of the regional ecological environment and the sustainable socio-economic development of the oasis region [2]. Xinjiang ranks second in China after Tibet in terms of total area and number of glacier resources and first in terms of ice reserves [1]. Glaciers play an essential role in the composition of water resources in Xinjiang, and their variation has a profound impact on Xinjiang and Central Asia.

Under the dual influence of climate change and human activities, the processes and effects of regional ecohydrology, resources and environment, and natural disasters caused by glacier structure and scale changes become more and more apparent [3]. Since the middle of the 20th century, the global mean surface temperature (GMST) has continued to rise, and the precipitation in the 30–60° N zone has increased significantly [4]. During this period, northwest China's average annual temperature rise reached 0.37 °C/10a. Under this climate background, glacier resources in Xinjiang have continued to shrink and become thinner [5]. The high mountains of Xinjiang, including Tian Shan, Altai Shan, Kunlun Shan, Karakorum Shan, and Pamir Plateau, provide a vast space for developing surviving glaciers. Due to the diversity of hydrothermal conditions and complex terrains in the southern, northern and central mountain systems (groups) with large latitude spans, glaciers vary in form and size.

After the Research Team of Alpine and Ice Utilization of the Chinese Academy of Sciences was established in 1958, glaciologists such as Shi, Y.F. began to investigate and conduct scientific research on glaciers in the Tianshan mountains [6]. So far, research on glacier changes in the Tianshan mountains has achieved remarkable results. Chen, H. [7], Zhao, Q. [8], Wang, S. [9] and Xing, W. [10] studied the temporal and spatial variation of the glacier area in the Tianshan mountains and its response to climate and showed that the glacier area in Tianshan mountains in China has shrunk by 11.5–18.4% in the most recent 50 years, and there is a great difference in the rate of decline in the eastern and western parts of the Tianshan mountains before and after 2000. Meanwhile, field observations and in-depth studies were made on the glacier area, mass balance, and glacier hydrometeorology of a single typical glacier in the Tianshan mountains, such as Tianshan No. 1 glacier [11], Hasilegen Glacier 51 [12], Qingbingtan Glacier 72 [13] and Hami Miaergou flat-topped glacier [14]. A large number of studies have also been carried out on the changes of glacier meltwater runoff in many inland river basins, such as the Manas river basin [15], the Aksu river basin [16] and the Kuitun river basin [17,18]. With the development and application of 3S technology, glacier research's spatial and temporal limitations by traditional means have been gradually broken. The "high spatial and temporal resolution" remote sensing monitoring of glaciers has entered a new era. Studies on glacier changes in Xinjiang also involve the Altai mountains in the north and the Pamir Plateau, Karakoram, Kunlun, and Altun mountains in the south (referred to as the Pakakuna mountain group). For example, Lv, H. [19], Huai, B. [20], Xu, C. [21], Lv, M. [22], Ke, L. [23], Yu, X. [24] and Liu, S. [25] have analyzed the glacier changes and their causes in the Alta mountains, Saurshmusi Island, the East Pamir-West Kunlun region and the Altun mountains. Comparatively, studies on glaciers in Xinjiang other than the Tianshan mountains are limited, especially the glaciers and their ice reserve changes in the Pakakuna mountain group. This is particularly important and urgent for the hydrological processes and water resources security of the Tarim basin, where glacial meltwater runoff recharge accounts for up to 43.3% [26].

In Xinjiang, 98% of surface water resources come from mountainous areas, and the "triadic" (rainfall, snow, glacier) flow-production model is a common feature of many watersheds in the region [27]. Glacier meltwater has a relatively stable replenishment and regulation effect on river runoff, and climate change has a significant impact on glacier ablation/accumulation. The process of glacier change and mass balance is of great importance to the basin's ecohydrology and water resources use. In the past, researchers have investigated a lot of glacier changes in most mountains and river basins in the study area. However, it is necessary to analyze the distribution and variation of glacier resources in mountain systems, watersheds, and prefectures and investigate further the response characteristics and sensitivity of glaciers to hydrothermal conditions and topography. This study analyzes the characteristics of glacier changes in Xinjiang over the past 50 years based on Chinese glacier inventory, combined with GIS spatial analysis techniques. The responses of glacier changes to topographic features and hydrothermal conditions are discussed to provide references for research and practice on ecological security and water resources utilization in Xinjiang.

2. Study Area

Xinjiang (34°22′~49°10′ N, 73°40′~96°23′ E) is located at the northwest border of China and is the largest administrative province in China (Figure 1). Located deep in the hinterland of the Eurasia continent, the high mountains and vast basins form a unique landform pattern of three mountains sandwiched by two basins. Coupled with the arid climate characteristics, the hydrological and water resources cycle process is more complex. The mountains are broad and have a wide range of elevations, providing a special topography for the development of mountain glaciers. Glacier resources in Xinjiang are mainly distributed in the Altai mountains in the north, the Tian Shan mountains in the central part of the province, and the Pamir, Karakorum, and Kunlun mountains in the south. It belongs to the Irtysh river watershed, Junggar basin, and Tarim basin, surrounded by high mountains. Xinjiang is located in the northwest desert area, where there is little rain and intense evaporation. Surrounded by high mountains where it is difficult for marine moisture to enter, it forms a distinct temperate continental climate. The average summer temperature can reach 18.24 °C, and the average annual precipitation is about 130 mm. However, the temperature and precipitation vary significantly from place to place, with the temperature in the south being higher than that in the north, and the precipitation is the opposite. The unique topographic advantages and climatic conditions provide conditions for the development of glaciers. Xinjiang's total surface water resources are about 7.93×10^{10} m^2, ranking only 12th in China, and glacial meltwater is the primary recharge source for many river runoffs [1].

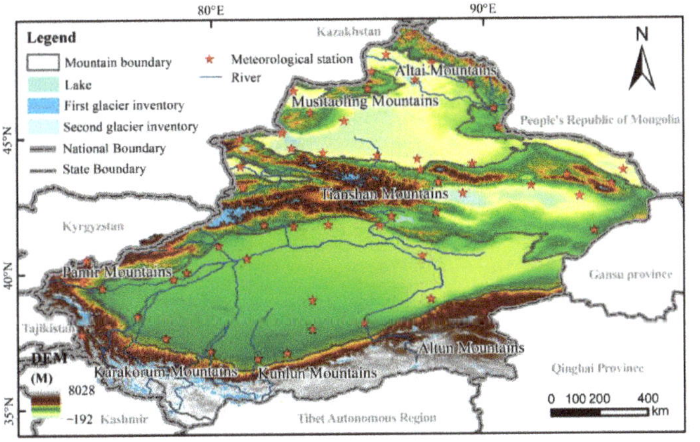

Figure 1. Location of the study area.

3. Data and Methods

The data in this study mainly include the Chinese Glacier Inventory, topographic and geomorphological data, and meteorological data (Table 1). The National Cryosphere Desert Data Center provided the data for the two periods of glacier inventory, and the cataloging methods are described in the literature [1,28]. For the first glacier inventory, topographic data such as elevation and orientation were extracted from single glaciers and assigned to each glacier; DEM and 1:1,000,000 geomorphological data were analyzed using zonal statistics of the elevation range and the topographic relief size of different mountain glaciers. Monthly precipitation and temperature data of the stations from 1961 to 2010 were selected as climate change indicators. Geostatistical analysis was carried out on the original data, and the ordinary kriging method with the lowest error was fixed for spatial interpolation. In order to facilitate comparative analysis, meteorological stations located in and around the mountains are divided into three regions: north, middle, and south. The annual or summer average temperature and precipitation increase or decrease in each mountain system (group) are calculated. The trends of their time series can identify the response of glacier changes to the climate.

Table 1. Data sources.

Category	Time	Resolution	Official Website (Accessed on 23 June 2021)
China Glacier Inventory	1960–2010	—	National Cryosphere Desert Data Center (https://www.crensed.ac.cn/portal/)
Digital Elevation Model (DEM)	—	30 m	Consultative Group on International Agricultural Research (CGIAR, http://srtm.csi.cgiar.org)
Digital geomorphological data	—	1:1,000,000	Website of the Data Center for Resource and Environmental Resource and Environment Science and Data Center (http://www.resdc.cn/)
Meteorological data	1961–2010	month	National Meteorological Science and Data Center (http://data.cma.cn/)

3.1. Glacier Area Change and Ice Reserves Estimation

Glacier area change is the most direct and specific means to assess changes in glacier resources. Due to the significant difference in the time interval between the first and second glacier inventory in Xinjiang, this study establishes the correspondence between the number of glaciers in the two glacier inventories by considering the number of omissions in the first glacier inventory and the number of increases and decreases in the second glacier inventory due to splitting or merging according to Li, L. [4]. Sun, M. [29] calculated the glacier area change rate and relative rates in different regions of the study area. Ice storage is an essential indicator for assessing glaciers and their changes and is a vital driving parameter for constructing glacier hydrological models [30]. Only a few glaciers in the world are currently available with relatively accurate thickness and volume data. Most ice reserves are still estimated mainly by indirect and fast empirical formulas, which are also commonly used to estimate glacier ice reserves in larger spatial scales [31]. Most of the current calculations of ice reserves use the volume–area empirical formula.

$$V = c \times A^{\gamma} \tag{1}$$

where: V is ice reserves (km^3); A is glacier area (km^2), c and gamma are empirical coefficients. In this paper, the coefficients proposed by Radić et al. [32], Grinsted [33], and Liu et al. [34] are used to calculate the glacier reserves in Xinjiang, and the average value of the three methods is used as the result.

3.2. Response Analysis of Glacier Change to the Topography and Climate

Glacier variability is not only directly affected by hydrothermal conditions but also obviously constrained by topographic conditions. Geodetectors can test both the spatial heterogeneity of a single variable and the possible causal relationship between two variables by trying the coupling of their spatial distribution [35]. To explore the contribution of topographic factors and meteorological factors to the change of the glacier area, Geodetector software was applied. For each factor, the data were divided into five levels using the natural breaks (jenks). The Geodetector model expression is:

$$q = 1 - \frac{\sum_{i=1}^{L} N_i \sigma_i^2}{N\sigma^2} \tag{2}$$

where L is the stratification of the independent and dependent variables, σ_i^2 and σ^2 are the variances of the independent variable topographic or meteorological factors and the dependent variable glacier area change. The q takes values in the range of [0, 1], and an enormous value of q indicates a more substantial explanatory power of the factors in glacier change. In the extreme case, a q value of 1 indicates that the factors control the spatial distribution of glacier change entirely, and a q value of 0 indicates no relationship between them.

In meteorological data, temperature and precipitation data processed by spatial interpolation are used to calculate air temperature and precipitation trend rates. The average temperature tendency, average precipitation tendency, summer temperature, and summer precipitation tendency are extracted for a single glacier. This paper uses the slope analysis method to analyze the spatial interannual variation trend of temperature and precipitation in Xinjiang. The calculation formula is as follows:

$$\theta = \frac{\sum_{i=1}^{n} a_i b_i - \frac{1}{n}\sum_{i=1}^{n} a_i \sum_{i=1}^{n} b_i}{\sum_{i=1}^{n} b_i{}^2 - \frac{1}{n}\left(\sum_{i=1}^{n} b_i\right)^2} \tag{3}$$

In the formula, θ is the interannual rate of change; n is the number of years from 1960 to 2010, and this study takes 50. The b is the time series, from 1960 to 2010 in turns 1–51. The a_i is the temperature or precipitation in the i-th year. $\theta < 0$ and $\theta > 0$ indicate increasing and decreasing with time, respectively, during the study period. The larger the positive value, the faster the temperature or precipitation rises, and the smaller the negative value, the quicker it decreases.

Terrain factors include elevation, slope, aspect, and topographic relief. Topographic relief is the difference between the highest and lowest altitude, an essential index for classifying geomorphic types. Based on DEM data, using the neighborhood analysis method in GIS, the topographic relief of a single glacier is extracted under a 3 × 3 rectangular analysis window.

4. Results

4.1. Status of Glacier Resources in Xinjiang

4.1.1. General Distribution Characteristics of Glaciers

According to the second glacier inventory of China, Xinjiang ranks second in China in terms of the number and area of glaciers and first in terms of ice reserves, accounting for 42.61%, 43.70%, and 47.97% of the country, respectively [1]. During 2005–2010, there were 20,695 glaciers in Xinjiang, with a total area of about 22,742.55 km^2 and ice reserves of about 2229.17 km^3. The average area is 1.10 km^2. From the statistical classification of glacier numbers and area (Figure 2), it can be seen that the number of glaciers is mostly less than 1 km^2 in area, with 16,996 glaciers accounting for 82.13% of the total number of glaciers. The area between 2–50 km^2 accounts for the largest proportion, accounting for 53.97% of the total glacier area in Xinjiang. The overall glacier area has a skew-normal distribution, and the number of glaciers shows a decreasing trend with increasing glacier area. There are only four glaciers in Xinjiang with an area larger than 200 km^2, with a total area of 1236.79 km^2, accounting for 5.98% of the total glacier area. The glacier with the largest area is the Yongsugeti glacier (code: 5Y654D53), located on the northern slope of Qogir in the Karakorum mountains, with an area of about 359.05 km^2 and an ice reserve of 105.75 km^3; it is also the largest valley glacier in China.

4.1.2. Characteristics of Glacier Distribution in Each Mountain System in Xinjiang

The absolute altitude of high mountains and the relative height difference above the glacial mass balance line are the main topographic conditions that determine the number and development scale of glaciers [36]. Glaciers in Xinjiang are distributed in seven mountain systems, including the Altai mountains, Tianshan mountains, and the Kunlun mountains from north to south (Table 2). Among them, the Tianshan mountains have the largest number of glaciers, with the Kunlun mountains ranking first in glacier area and ice reserves, accounting for 38.07% and 35.40% of the total glaciers in Xinjiang, respectively. There are 22 giant glaciers larger than 100 km^2 in China, 14 of which are located in Xinjiang, respectively in the Tian Shan (5), Pamir Plateau (1), Kunlun mountains (4), and Karakorum mountains (4) [1]. The glaciers in the Altai, Altun, and Musitaoling mountains in Xinjiang are relatively on a small scale. Glacier scale is the result of hydrothermal material and topography, and the high mountains provide a vast space for glacier development.

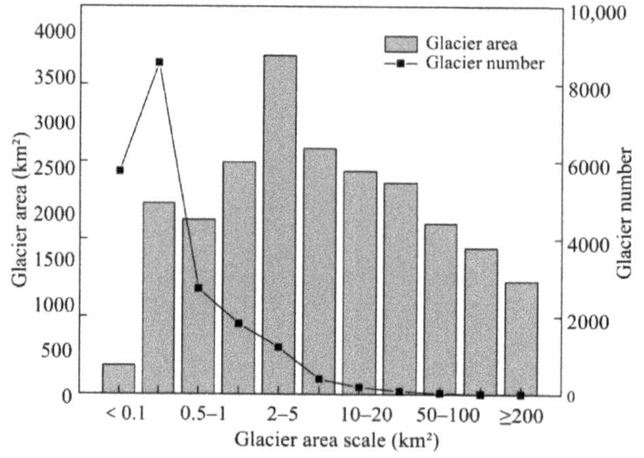

Figure 2. Number and area of glaciers in different scales in Xinjiang.

Table 2. The glacier resources in different mountains in Xinjiang.

Mountains	Quantities		Area		Ice Reserves	
	(Number)	(%)	(km^2)	(%)	(km^3)	(%)
Altai	273	1.32	178.79	0.79	10.61	0.48
Musitaoling	12	0.06	8.94	0.04	0.40	0.02
Tianshan	8017	38.74	7275.35	31.95	748.98	33.60
Pamirs Plateau	1179	5.70	1921.66	8.44	169.30	7.59
Karakorum	3414	16.50	4463.95	19.60	497.23	22.31
Kunlun	7379	35.66	8669.48	38.07	789.14	35.40
Altun	421	2.03	255.61	1.12	13.52	0.61

Glaciers in Xinjiang are developed in the area with an altitude above 2363 m. Most of the ridge areas above 7200 m are steep and unfavorable for glacier development, and the areas below 2600 m are not conducive to the accumulation of glaciers due to the influence of temperature. Therefore, glaciers are concentrated at 5100–6000 m, accounting for 52.67% of the total glacier area. The glacier with the lowest terminal elevation (2363 m) is located in the Burqin river basin, and its glacier number is 5A255F0041. The average elevation of the median glacier area in Xinjiang is 5785.8 m. Due to the large difference in mountain scale and altitude, the median glacier area appeared in different elevation intervals. The highest average elevation of glaciers in the Karakorum mountains is 5816 m, while the lowest in the Altai mountains is 3068 m. Relatively vast mountainous areas under a similar latitudinal zone are more suitable to provide the topographic conditions for the accumulation of glacier development. The Pakakuna mountain group is extremely high in altitude, with glaciers distributed at an average altitude of above 4000 m. The scale of glaciers in the Kunlun mountains is much larger than that in other mountainous areas below the same latitude. Although the Altai mountains are at a lower altitude, they are located at high latitudes and have abundant precipitation, promoting the development of glaciers.

4.1.3. Characteristics of Glacier Distribution in Each Water System in Xinjiang

According to the international glacier inventory standard, the distribution area of Xinjiang glaciers is firstly divided into endorheic and outflow regions and subdivided into 5 first-level and 11 second-level basins [1]. From the statistics of the second glacier inventory in China (Table 3), the number of glaciers in the East Asian endorheic region is the largest among the first-level basins (5Y), followed by the Ili river endorheic system (5X), with the least in the Irtysh river basin (5A). From the glacier extent in second-level basins, all glaciers in Xinjiang with an area greater than 100 km^2 are distributed in the Tarim endorheic basin (5Y6). In terms of glacier number, glacier area, and ice reserves, the Tarim basin ranks first, accounting for 61.80%, 78.25%, and 85.52% of the total glaciers in Xinjiang, respectively. The basin with the least distribution of glacier resources is the Kobdo river (5Y1), with only four glaciers and an average area of 4.85 km^2. The average glacier area in the Turpan Hami endorheic basin (5Y8) reaches 2.12 km^2, but the basin glaciers are only 0.79% of the total glacier area. The glaciers in the Tarim endorheic basin have a minimum average area of only 0.72 km^2. Limited glacial meltwater is particularly important for recharging local river runoff in Xinjiang, where precipitation is minimal.

Table 3. The glacier resources in different watersheds in Xinjiang.

Partition	Class I (II) Watersheds	Encoding	Quantities		Glacier Area		Ice Reserves	
			(Number)	(%)	(km^2)	(%)	(km^3)	(%)
Outflow region	Irtysh River	5A	279	1.35	186.09	0.82	10.96	0.49
	Indus River	5Q	696	3.36	367.18	1.62	18.71	0.84
Endorheic region	Ili River	5X	2121	10.25	1554.41	6.86	107.46	4.82
	Kobdo River	5Y1	4	0.02	0.82	0.00	0.02	0.00
	Qaidam Basin	5Y5	397	1.92	208.95	0.92	9.61	0.43
	Tarim Basin	5Y6	12,790	61.80	17,720.73	78.25	1906.36	85.52
	Junggar Basin	5Y7	3096	14.96	1737.35	7.67	100.66	4.52
	Turpan-Hami Basin	5Y8	378	1.83	178.11	0.79	8.37	0.38
	Qinghai-Tibet Plateau basin	5Z	934	4.51	788.88	3.48	67.03	3.01

4.1.4. Characteristics of Glacier Resources in Each City and Autonomous Prefecture in Xinjiang

The geographical location of Xinjiang is divided into the "South" and "North" by the Tianshan mountains, and the natural environments of these two places composed of climate and water resources are quite different. According to the statistics of glacier resources in Xinjiang (Table 4), the extent of the glaciers in southern Xinjiang is significantly larger than that in the northern area. Except for the city of Karamay, which has no glaciers, all other 13 cities and autonomous prefectures have some distribution. Hotan prefecture has the largest number, area, and ice reserves of glaciers in Xinjiang, and Turpan city has the smallest glacier extent. The order of the number, area and ice reserves of glaciers in other regions are basically the same. The number of glaciers in some regions such as Yili Prefecture, Bayingoleng Mongol Autonomous Prefecture, and Urumqi is significantly higher than the area and ice reserves ranking, indicating that the glaciers there are more fragmented. The opposite is true for the Aksu region, where the area of a single glacier is much larger. The ice reserves in southern Xinjiang are all ranked high, and the mountain glaciers contribute abundant glacial meltwater runoff to the local watersheds.

4.2. Changes in the Spatial and Temporal Distribution of Glacial Resources in Xinjiang

4.2.1. General Glacier Change Characteristics

By comparing the number and size of the glaciers inventory in the two periods, it is found that glaciers in Xinjiang have been shrinking sharply since 1960 (Figure 3). The first glacier inventory showed that there were 22,054 glaciers in Xinjiang, with a total area of about 29,822.67 km^2 and an average area of 1.35 km^2.

Table 4. The glacier resources in different mountains in Xinjiang.

Name of the Mountains	Quantities			Area			Ice Reserves		
	(Number)	(%)	Sorted	(km^2)	(%)	Sorted	(km^3)	(%)	Sorted
Aksu Area	1276	6.17	6	3542.06	15.57	3	539.27	24.19	2
Altai Area	284	1.37	11	187.07	0.82	10	10.98	0.49	9
Mongolia bayinguoleng autonomous prefecture	3952	19.10	2	2867.40	12.61	4	201.48	9.04	4
Bortala Mongol autonomous prefecture	408	1.97	9	219.87	0.97	9	10.53	0.47	10
Changji Hui autonomous prefecture	755	3.65	8	287.16	1.26	8	13.36	0.60	8
Hami city	205	0.99	12	123.71	0.54	11	6.05	0.27	11
Hotan prefecture	5654	27.32	1	6869.09	30.20	1	650.18	29.17	1
Kashí prefecture	3356	16.22	3	4871.47	21.42	2	534.49	23.98	3
Kizilsu Kirghiz autonomous prefecture	1582	7.64	5	1990.09	8.75	5	161.07	7.23	5
Tacheng prefecture	1152	5.57	7	769.05	3.38	7	46.81	2.10	7
Turpan City	69	0.33	13	16.86	0.07	13	0.57	0.03	13
Urumqi	270	1.30	10	64.92	0.29	12	2.00	0.09	12
Ili Kazakh autonomous prefecture	1732	8.37	4	933.79	4.11	6	52.37	2.35	6

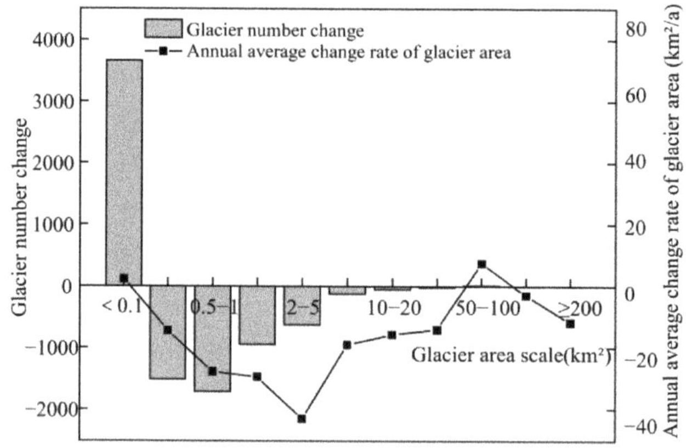

Figure 3. The glacier number change and area retreat rate in Xinjiang from 1960 to 2010.

From the perspective of the number of glaciers, the number of glaciers decreased by 1359 in 50 years, among which the number of glaciers with an area of 0.5–1.0 km^2 decreased the most, followed by 950 glaciers with an area of 1–2 km^2. The number of glaciers with an area less than 0.1 km^2 increased by 3656. In terms of glacier area, the glacier area decreased by 7080.12 km^2 during 50 years, with an area change rate of 141.60 km^2/a. Glaciers in the range of 1–2 km^2 and 2–5 km^2 have a greater reduction, with an area change rate of 25.96 km^2/a and 38.58 km^2/a. In addition, there is a partial increase in areal extent less than 0.1 km^2 and 50–100 km^2, with relative area change rates of 1.64%/a and 0.50%/a. The number and area of glaciers with an area of 0.1–5 km^2 decreased at a larger rate, while the number and area of glaciers smaller than 0.1 km^2 showed an increasing trend. From the perspective of glacier ice storage, the ice reserves in Xinjiang were reduced from 2711.81 km^3 to 2229.17 km^3 during the study period, with a deficit of about 482.65 km^3 and a deficit rate of 0.39%/a. The largest deficit rate was 0.99%/a for area 0.5–1 km^2, followed by 0.83%/a for area 2–5 km^2.

On the whole, the rate of larger glaciers breaking up to form small glaciers is significantly higher than the rate of ablation of small glaciers themselves. Therefore, the size of glaciers smaller than 0.1 km^2 was on the rise, thus showing that small glaciers are changing more intensely.

4.2.2. Characteristics of Glacier Orientation and Elevation Change

Apart from the prominent role of topographic differences in the movement of glaciers, they also have an important influence on the distribution and variation of glaciers—the higher the terrain and the lower the temperature, the more favorable the accumulation of glacier material. The statistical results show that the loss rate of glaciers below 5000 m above sea level is greater than 50%, and there is basically no change in glaciers in the area above 6000 m above sea level (Figure 4).

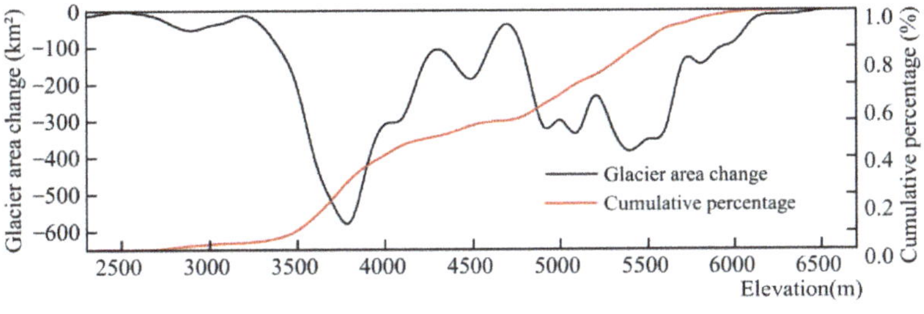

Figure 4. Changes in glacier area in different elevations of Xinjiang.

As the altitude rises, the glacier area is affected by temperature conditions. The statistics of glacier inventory data for the two periods by orientation (Figure 5) show that the change in the number of glaciers in each direction is consistent with the size of glaciers. The largest decrease in the number of glaciers was towards the north (377), followed by the northwest (310). The size of glaciers distributed in the northerly direction (N, NE, and NW) in Xinjiang is significantly larger than that in the southerly direction (SE, S, and SW). The number of glaciers in the northerly direction (shady slope) decreased by 826 glaciers, with 1743.24 km^2. The number of glaciers in the southerly direction (sunny slope) decreased by 281 glaciers, with 3105.98 km^2. The number and size of glaciers on the north slope are significantly larger than those on the south. For example, the larger glaciers on the north slope of the Kunlun mountains are more insensitive to climate change [37]. The result is that the glaciers on the north slope are less susceptible to climate warming and have more precipitation influenced by Atlantic moisture, which facilitates the accumulation of reliable water resources in the study area. The distribution of glaciers on the north and south slopes of the mountain system varies considerably due to hydrothermal conditions and topographic elements.

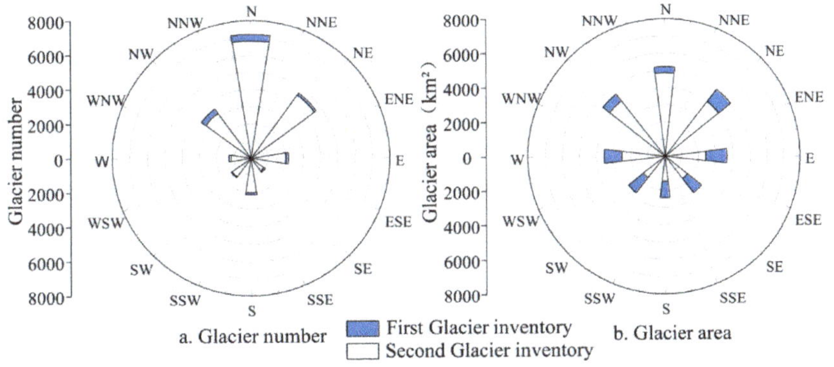

Figure 5. Based on the first and second Chinese Glacier Inventory of different orientations (**a**) and glacier area (**b**) in Xinjiang.

4.2.3. Characteristics of Changes in the Distribution of Glacial Resources in Mountain Systems

Mountain glaciers are mostly distributed in middle and high altitudes, and the zoning statistics on the glacier data of each mountain system in the two periods of the glacier inventory. The results showed that the number of glaciers in the Tianshan mountains had decreased the most, with 2206 glaciers, followed by 162 glaciers in the Altai mountains. The number of glaciers in the Pakakuna group has increased, with the largest increase in the Karakorum mountains, accounting for 16.08% of the glaciers in the study area. The glacier area retreated more severely. Among them, the glacier area in the Tianshan mountains decreased the most (62.16 km^2/a) and the least (0.16 km^2/a) in the Altun mountains. Considering the size of the initial area of each mountain area, the glacier area of Altai mountain decreases the most (1.37%/a), while the glacier area of Altun mountain remains the smallest (0.06%/a). The glacier reserves in each mountain system were in deficit, with the most severe deficit in the Karakorum mountains (170.15 km^3). Glaciers develop at very high altitudes without topographic conditions and low altitudes without climatic conditions. The central part is the optimal combination of climatic and topographic conditions, so the glacier area of each mountain system in Xinjiang has an approximate skew-normal distribution characteristic with altitude.

Zonal statistics were conducted for the glacier inventory of two periods under altitude at 100 m intervals (Figure 6) based on the DEM data. The altitude range of glacier development varied with the lapse rate of temperature in each mountainous system. The lowest altitude range of glacier area retreat in the Altai mountains is 2800–3200 m, accounting for 98.96% of the total retreat area in the mountains. The glaciers of the Pakakuna mountain group are concentrated at an altitude range of 4800–5500 m. The concentrated area of glaciers in the Tianshan mountains is between them, in which the area of 3400–4100 m glaciers retreats 2887.92 km^2, accounting for 94.53% of the total area. In general, the changes in glacier area and number in each mountain area are not consistent, with a retreating trend in the glacier area. Still, the number of glaciers in individual mountainous has increased slightly.

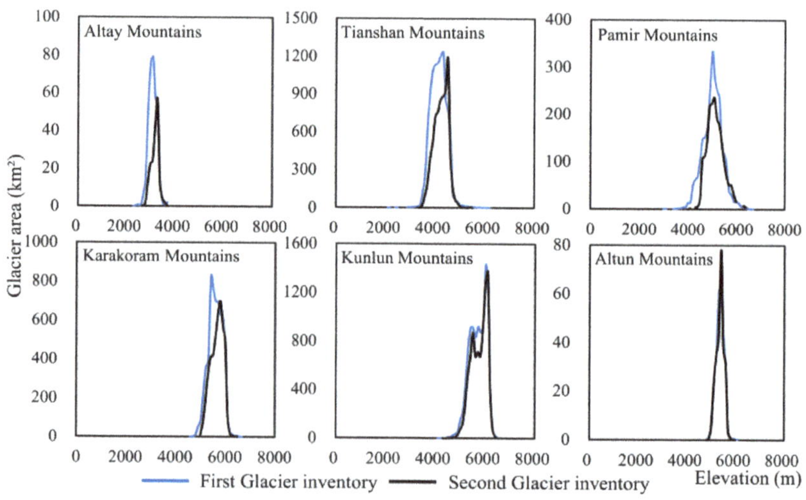

Figure 6. Hypsography of glacier area change in different mountains in Xinjiang.

4.2.4. Changes in the Distribution of Water-Based Glacier Resources

Glaciers have a critical ecological role in inland river basins in the arid zone, and glacier changes vary from basin to basin in Xinjiang. The largest decrease in the number of glaciers was 728 in the Junggar inland flow area, followed by 534 in the Ili river basin and

156 in the northern Erzis river basin, accounting for 35.86% of the number of glaciers in the basin. The glacier in each basin has decreased, with the largest glacier area and the most drastic change in the Tarim endorheic basin, with a change rate of 94.36 km^2/a. The second is the Junggar endorheic basin (18.26 km^2/a), and the lowest rate of area change (except for the Kobdo river) is the Qaidam basin (0.17 km^2/a) (Figure 7). Statistics show that the number of glaciers in the Qaidam basin, the Indus river basin, and the inland basins of the Tibetan Plateau has increased (in Xinjiang). Such results are probably due to local glacier splitting. According to Wang Yuan et al.'s [38] analysis of the glacier changes in the Goradanton region of the Tibetan Plateau, the number of glaciers in the region increased by 10 from 1964 to 2010, while the glacier area decreased by 45.75 km^2. There are individual glaciers in the region advancing, probably due to the glacier splitting phenomenon. Ice reserves have decreased in all basins, with about 3% in both the Ili river basin and the Junggar basin. In addition to the phenomena of extinction and splitting, the changes in the size of glaciers are also related to the reduction of glacier length and the phenomenon of jumping and over-covering of individual glaciers [39], which makes the trends of glacier number, area, and ice reserves not wholly consistent.

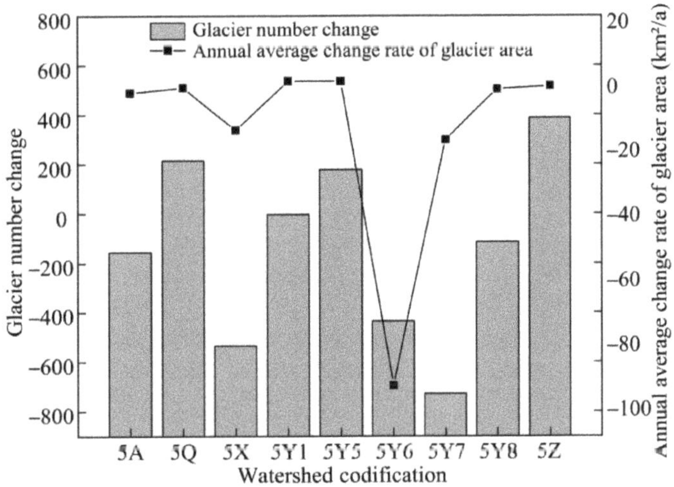

Figure 7. Changes in the quantity and area of glaciers in watersheds in Xinjiang from 1960–2010.

4.3. *Response of Glacial Change to Topographic Conditions and Climate Conditions*

4.3.1. Response of Glacial Change to Topographic Relief

The development of glaciers depends on high altitude and broad mountainous terrain conditions. Topographic relief refers to the difference between the altitude of the highest point and the altitude of the lowest point in a specific area, which can reflect the altitude change degree of the surface. Based on the spatial distribution data of 1:1,000,000 geomorphic types in China, this study investigates the differentiation regularity of glacier area changes in Xinjiang. Combined with the distribution of glacier terminus in the study area, the landform pattern of each mountain was classified by elevation and topographic relief. According to the elevation, it can be divided into three types: low-middle mountains (2000–4000 m), high mountains (4000–6000 m), and extremely high mountains (>6000 m); at the same time, the topographic relief is defined according to the relative elevation, including small relief (200–500 m), medium relief (500–1000 m), large relief (1000–2500 m) and extreme relief (>2500 m) (Figure 8).

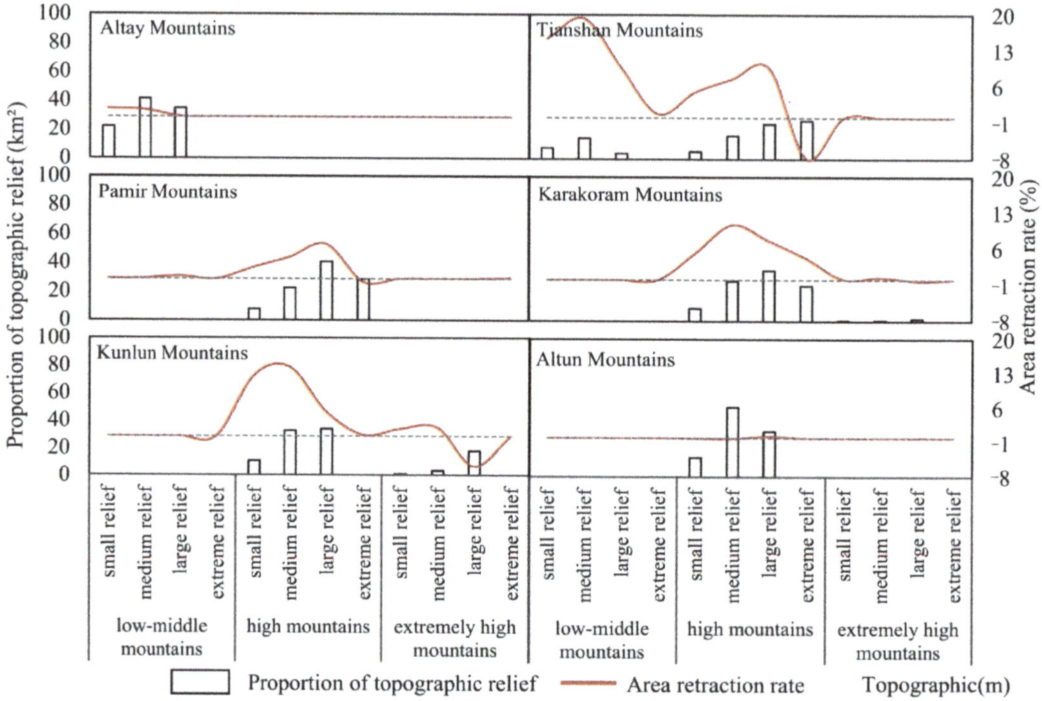

Figure 8. The proportion of topographic relief and glacier area retraction rate in each mountain of Xinjiang.

Combined with GIS spatial analysis, the current situation of glacier distribution in Xinjiang in 2010 and the shrinking rate of glacier area during the study period were statistically analyzed. There were apparent differences in glacier distribution and changes in different topographic reliefs of each mountain. In terms of the current distribution of glaciers, glaciers in the Altun mountains, Kunlun mountains, Karakorum mountains, and Pamir Plateau are mainly distributed in the high and extremely high mountains. Among them, the distribution of glaciers in the high mountains with medium and large topographic relief is the most concentrated, accounting for 83.93%, 66.42%, 62.94%, and 62.82% of the total area, respectively. Tianshan glaciers, known as the "wet island" of Xinjiang, are mainly distributed in the low-middle and high mountains. Relatively abundant precipitation is conducive to the development of glaciers at lower altitudes. At the lower altitude of the Altai mountains, 75.42% of the glaciers are distributed in the large and medium topographic relief of the low-middle mountains. According to the degree of topographic relief, the glaciers in large and medium relief mountains are the most widely distributed. Most of these landforms are distributed in the middle and upper part of the mountain, so this area has good hydrothermal conditions. On the contrary, the steep terrain of extremely high mountains and relief is not conducive to glacier development due to gravity. The higher the altitude, the stronger the solar radiation and wind, which is also not conducive to glacier accumulation. Only Pamir Plateau has less glacier distribution in extremely high mountains and relief.

Based on the changes of glaciers in various topographies in Xinjiang from 1960 to 2010, the retreat of glaciers with large and medium relief is the most significant. Glaciers in the extreme relief of the Tianshan mountains increased slightly. This is consistent with previous studies. Scholars suggest that the increase in precipitation in this region has led to more accumulation of glaciers than the melting of glaciers caused by the rise in temperature [2]. Large relief glaciers increased in extremely high mountain areas of the Kunlun mountains. The reason may be that the local topographic relief in the high-altitude areas is favorable for the accumulation of water vapor to form precipitation-rich areas so that the glaciers accumulate more than they ablate [40]. During the study period, the small increase in glaciers in the small and medium relief areas of the Karakorum mountains increased slightly, which may be caused by the migration of glaciers in large relief to relatively flat areas at lower altitudes under the action of gravity and stress [41]. The causes of glacier changes in climate warming and humidification in Xinjiang are diverse. The topographic relief and landform complexity of mountainous areas are closely related to the change processes of local glaciers, such as ablation/accumulation, advancing and retreating, and splitting.

4.3.2. Response of Glacial Change to the Climate

Glaciers are a product of climate, and their changes are extremely sensitive to water (precipitation) and thermal (temperature) conditions. Xinjiang has a large span from north to south. Differences in the sources and intra-regional circulation of water vapor in the major mountain systems cause a highly uneven spatial distribution of precipitation (Figure 9). The trend of glaciers scale in Xinjiang has been shrinking over the past 50 years, with varying glacier changes in each mountain. The greatest changes in the number and area of glaciers are in the Altai mountain system (including the Mustang Ridge), both of which have shrunk by 37.24% and 49.73%, respectively. It is due to the most pronounced increase in temperature (annual and summer half-year mean temperature) in this region (0.44 °C/10a, 0.32 °C/10a) and the slowest increase in precipitation (3.26 mm/10a) in summer half-year. The precipitation in the Altai mountain mainly increased after 2000, and the anomaly percentage of precipitation reached 30.92% in 2010. The variation trend of temperature and precipitation is consistent with the transition from warm-dry to warm-wet in Xinjiang, China, proposed by Shi, Y.F. [42].

In contrast, the glaciers change in the Pakhakuna mountain group in the south of Xinjiang is more unusual. The number of glaciers is increasing instead of decreasing, increasing by 10.4% during the study period. However, the total area of glaciers still decreased by 14.4%, and its retreat rate was the smallest. In recent years, scholars at home and abroad have begun to pay attention to the anomalies of glacier changes on Karakorum mountain and its surrounding areas, mainly in terms of anomalous temperature changes [43,44] and increased precipitation [45] to explain the "Karakorum anomaly". The change is that the southern part of Xinjiang is arid with little rainfall, and the annual mean temperature and precipitation changes (0.29 °C/10a, 4.89 mm/10a) are more prominent. In 2010, the percentage of precipitation anomaly reached 60.32%. The significant increase in temperature is the main reason for the area retreat of the Pakhakunga mountain group. In contrast, the increase in precipitation is the main reason for the minimal rate of glacier retreat in the region. Its degree of warming and humidification is lower than that of the Tianshan and Altai mountains. The increase in temperature accelerates the division of some mountain glaciers, leading to an increase in the number of glaciers. In addition, it is not enough to explain the abnormal reasons for glacier changes on Karakorum mountain and its surrounding areas only by temperature and precipitation. The enhancement of westerly circulation [46] and the alpine cold storage caused by extremely high terrain [47] are all related factors that affect the glacier changes. The number and area of glaciers in the Tianshan mountains decreased by 21.5% and 27.6%, respectively, and the greatest magnitude of change was between the northern and southern mountain systems.

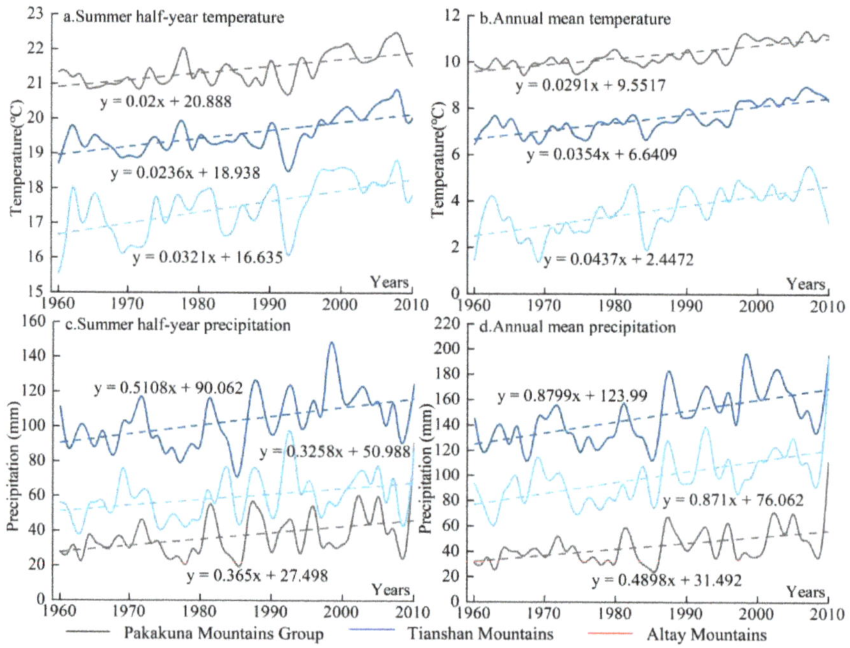

Figure 9. Changes in temperature and precipitation in Xinjiang during the whole year and summer from 1960 to 2010. Summer half-year temperature (**a**); Annual mean temperature (**b**); Summer half-year precipitation (**c**); Annual mean precipitation (**d**).

Kang, E. [48] and Liu, S. et al. [49] suggest that the glacier ablation caused by a 1 °C increase in temperature (summer temperature) needs to be compensated by a 40–47% increase in solid precipitation. Apparently, the temperature rising rate in each mountain is above 0.29 °C/10a, and the average temperature rising rate in summer is above 0.20 °C/10a. However, the precipitation reaches more than 40% only in individual years in the Pakakunya group of mountains, but not in the rest of the mountains. Although the precipitation has increased, it still cannot make up for the glacier ablation caused by the temperature rise. In the last 50 years, the temperature in Xinjiang's mountainous areas has increased by about 1.5–2.2 °C. However, the precipitation has not reached the corresponding increase, which is one of the reasons for the significant reduction of glacier resources in the study area.

5. Discussion

5.1. Comparison with Typical Mountain Glacier Changes in Western China

Under global warming, glaciers in western China are mainly retreating [50]. Due to differences in data sources, research periods, and initial values of glaciers in each mountain, it is impossible to compare them well. Therefore, the method proposed by Sun et al. [29] was used to calculate the relative change rate of glacier area and the change rate of glacier area for each mountainous area in western China over the last 50 years. It can be seen from Table 5 that the trend in the glacier area change rate and relative change rate is basically the same. In combination with published results, Altai mountains had the highest glacial retreat rate (1.37%/a), followed by Gangdisê mountains (1.09%/a). The warming rate of summer temperature on Altai mountain (0.32 °C/10a) is lower than that on Gangdisê mountain (0.37 °C/10a), but the annual average temperature changes greatly (0.44 °C/10a), which may be the main reason for the fastest glacier ablation on Altai mountain. The glacier area of Kangri Karpo (0.82%/a), Tianshan (0.71%/a), and Pamir (0.61%/a) decreased rapidly.

Table 5. Statistics of glacier changes in western China in recent decades.

Mountains	Study Period (Year)	Area Retreat Rate (%)	Relative Rate of Area Change (%/a)	Data Source	Author
Altai	1960–2010	−49.73	−1.37	TM/ETM+	our study
Tianshan	1960–2010	−29.93	−0.71	TM/ETM+	our study
Pamirs Plateau	1960–2010	−26.31	−0.61	TM/ETM+	our study
Karakorum	1960–2010	−24.55	−0.56	TM/ETM+	our study
Kunlun	1960–2010	−15.74	−0.34	TM/ETM+	our study
Altun	1960–2010	−3.08	−0.06	TM/ETM+	our study
Qilian	1960–2010	−20.88	−0.47	TM/ETM+	Sun [29]
Tanggula	1990–2015	−22.18	−1.00	TM/ETM+/OLI	Wang [51]
Qiangtang	1970–2000	−4.35	−0.15	TM/ETM+	Wang [52]
Gangdisê	1970–2016	−39.53	−1.09	TM/ETM+/OLI	Liu [53]
Himalaya	1990–2015	−10.99	−0.47	TM/ETM+/OLI	Ji [54]
Nyenchen Tanglha	1976–2011	−1.81	−0.05	MSS/TM/ETM+	Ji [55]
Anyemaqen	1966–2000	−17	−0.56	TM	Liu [56]
Kangri Karpo	1980–2015	−24.9	−0.82	TM/ETM+/OLI	Wu [57]
Gongga	1974–2010	−11.86	−0.35	MSS/TM/ETM+	Li [58]

The study shows that the glacier change on Gangri Gabu mountain dominated by the Indian monsoon is significantly larger than that on Tianshan mountain, dominated by the northern branch of the westerly belt. Overall, glaciers change most strongly on small-scale and low-altitude mountains. Oceanic glaciers in southeastern Tibet are less variable than continental glaciers in other regions.

5.2. Attribution Analysis of Glacier Area Change in Xinjiang

The survival, development, and scale change of glaciers depend on climate change and topography. To further explore the response of glacier change to climate and topography, this study used Geodetector to analyze the contribution of climate and topography to glacier area change (Table 6). Average elevation, slope, aspect, and relief were selected as topographic factors. Climate factors include annual precipitation tendency, annual temperature tendency, summer precipitation, and summer temperature tendency. The results show that the above factors have passed the aboriginality test. The summer temperature (23.93%) is the largest contributor to the glacier area change, followed by the average elevation (18.92%). This conclusion is also consistent with the findings of current scholars [59]. In the long timescale and large space range, glaciers are more affected by temperature. Glacier ablation mainly occurs in summer, so the influence of summer temperature on glacier change is stronger than that of annual average temperature (7.51%). Precipitation is crucial for glacier accumulation. The summer precipitation only greatly influences the warm and humid Tianshan mountains and the Altai mountains. Still, it plays a major role in the annual precipitation of the southern Pakakuna mountains on the accumulation of glaciers. In summary, the influence of annual precipitation (15.17%) on glacier change is stronger than that of summer precipitation (6.84%).

The distribution of glaciers is also affected by complex topographic conditions. Among them, the average elevation (18.92%), slope (13.92%), and topographic relief (9.80%) had a greater impact on glacier change, and the aspect had a relatively small impact (3.91%). It can be concluded that the glacier change is jointly affected by climatic conditions (53.45%) and topographic conditions (46.55%), among which climatic conditions are more prominent.

In addition, the results obtained by the Geodetector are mainly related to the classification and the selection of detection factors. Changes in glaciers are not only related to topography and climate, but also to other relevant factors that the authors have not considered. In the subsequent study, more factors should be considered to discuss the influence of glacier changes.

Table 6. Response of topographic and climate factors to glacier change.

	Average Elevation	Slope	Aspect	Terrain Relief	Annual Mean Precipitation	Annual Mean Temperature	Summer Half-Year Precipitation	Summer Half-Year Temperature
	X1	X2	X3	X4	X5	X6	X7	X8
Q Statistics	0.7720	0.5680	0.1593	0.4000	0.6189	0.3066	0.2790	0.9766
p	0.0000	0.0000	0.0000	0.0000	0.0000	0.0000	0.0000	0.0000
contribution	18.92%	13.92%	3.91%	9.80%	15.17%	7.51%	6.84%	23.93%

6. Conclusions

(1) At present, there are 20,695 glaciers in Xinjiang, with a total area of about 22,742.55 km^2 and ice reserves of about 2229.17 km^3. The average area is 1.10 km^2. The number of glaciers is mostly less than 1 km^2, and the area between 2–50 km^2 accounts for a larger proportion. The number of glaciers is decreasing with the increase of glacier area. Glaciers are developed in the area with an altitude above 2363 m, concentrated at the altitude of 5100–6000 m, accounting for 52.67% of the total glacier area. The Tianshan mountains have the largest number of glaciers, and the Kunlun mountains with higher altitudes have the largest glacier area and ice reserves. The glaciers extent in Tarim Basin is the largest, but the average area of glaciers is only 0.72 km^2. Except for the city of Karamay, which has no glaciers, all other 13 cities and autonomous prefectures have some distribution. The glaciers extent in southern Xinjiang is significantly larger than that in northern Xinjiang.

(2) Over 50 years, the number of glaciers in Xinjiang decreased by 1359, the area decreased by about 7080.12 km^2, and the ice reserves deficit was about 482.65 km^3. Small area glaciers change more intensely, among which the number and area of glaciers with an area of 0.1–5 km^2 are the most serious, while those with an area of less than 0.1 km^2 are on the rise. The glacier area of each mountain system is approximately in skew-normal distribution with altitude, and there is basically no change in glaciers area above 6000 m. The distribution of northerly glaciers is significantly more than southerly glaciers, and glaciers on the southern slopes of mountains are more sensitive to climate change. The phenomenon of an increasing number of glaciers but decreasing total area in the southern mountains is closely related to the change processes of local glaciers, such as ablation/accumulation, advancing and retreating, and splitting.

(3) Each mountain is divided into landform patterns based on elevation and topographic relief. The glaciers are most widely distributed and vary most significantly in the large and medium relief. Due to gravity, the extreme relief in the extremely high mountains is not conducive to the development of glaciers. In the last 50 years, the temperature in Xinjiang mountainous areas has increased by about 1.5–2.2 °C. However, the precipitation has not reached the corresponding increase, which is one of the reasons for the significant reduction of glacier resources in the study area.

(4) Compared with the glaciers in typical mountainous areas in western China, it is found that glaciers change most strongly in small-scale and low-altitude mountains. Oceanic glaciers in southeastern Tibet are less variable than continental glaciers in other regions. The change of glacier area in Xinjiang is jointly affected by climatic conditions (53.45%) and topographic conditions (46.55%), among which climatic conditions are more prominent.

Glaciers are dynamic and relatively independent natural resources. It is of great significance to the regional water cycle and water resource security and has unique and irreplaceable ecological service value for surface energy balance, climate comfort, biodiversity, ice, and snow culture. The proper application and adequate protection of glacial resources still has a long way to go. Based on the study of the physical and chemical properties of mountain glaciers, more research should be done on the influence of the surrounding environment on their changes at different spatial and temporal scales so as to control and slow down the rapid retreat of glaciers moderately.

Author Contributions: Conceptualization, Z.Z. and L.L.; Data curation, X.Z.; Investigation, T.W.; Methodology, Z.Z. and L.L.; Software, H.T.; Validation, H.C.; Visualization, Z.K.; Writing—original draft, X.Z.; Writing—review & editing, Z.Z. and L.L. All authors have read and agreed to the published version of the manuscript.

Funding: This research was supported by The National Natural Science Foundation of China (Grant No. 41761108, 41641003), The third Comprehensive Scientific investigation project in Xinjiang (Grant No. 2021xjkk08).

Data Availability Statement: Not applicable.

Acknowledgments: We acknowledge the research environment provided by Xinjiang Production and Construction Corps Key Laboratory of Oasis Town and Mountain-basin System Ecology.

Conflicts of Interest: The authors declare no conflict of interest.

References

1. Liu, S.; Yao, X.; Guo, W.; Xu, J.; Shangguan, D.; Wei, J.; Bao, W.; Wu, L.J. The contemporary glaciers in China based on the Second Chinese Glacier Inventory. *Acta Geogr. Sin.* **2015**, *70*, 3–16.
2. Zhang, Z.; Liu, L.; He, X.; Li, Z.; Wang, P. Evaluation on glaciers ecological services value in the Tianshan Mountains, Northwest China. *J. Geogr. Sci.* **2019**, *29*, 101–114. [CrossRef]
3. Bliss, A.; Hock, R.; Radić, V. Global response of glacier runoff to twenty-first century climate change. *J. Geophys. Res. Earth Surf.* **2014**, *119*, 717–730. [CrossRef]
4. Li, L.; Yao, X.; Liu, S.; Bu, Y.; Gong, P.; Li, X. Glacier changes along the Silk Road Economic Belt in China in the past 50 years. *J. Nat. Resour.* **2019**, *34*, 1506–1520.
5. Tang, X.-L.; Lv, X.; He, Y. Features of climate change and their effects on glacier snow melting in Xinjiang, China. *C. R. Geosci.* **2013**, *345*, 93–100. [CrossRef]
6. Yafeng, S.; Zichu, X.; Xiangsong, Z.; Maohuan, H. Twenty five years of glaciology in China: Retrospect and prospect. *Acta Geogr. Sin.* **1985**, *40*, 367–376.
7. Chen, H.; Chen, Y.; Li, W.; Li, Z. Quantifying the contributions of snow/glacier meltwater to river runoff in the Tianshan Mountains, Central Asia. *Glob. Planet. Chang.* **2019**, *174*, 47–57. [CrossRef]
8. Zhao, Q.; Zhang, S.; Ding, Y.J.; Wang, J.; Han, H.; Xu, J.; Zhao, C.; Guo, W.; Shangguan, D. Modeling Hydrologic Response to Climate Change and Shrinking Glaciers in the Highly Glacierized Kunma Like River Catchment, Central Tian Shan. *J. Hydrometeorol.* **2015**, *16*, 2383–2402. [CrossRef]
9. Wang, S.; Zhang, M.; Li, Z.; Wang, F.; Li, H.; Li, Y.; Huang, X. Glacier area variation and climate change in the Chinese Tianshan Mountains since 1960. *J. Geogr. Sci.* **2011**, *21*, 263–273. [CrossRef]
10. Xing, W.; Li, Z.; Zhang, H.; Zhang, M.; Liang, P.; Mu, J. Spatial-temporal variation of glacier resources in Chinese Tianshan Mountains since 1959. *Acta Geogr. Sin.* **2017**, *72*, 1594–1605.
11. Wu, L.; Li, H.; Wang, L. Application of a degree-day model for determination of mass balance of Urumqi Glacier No. 1, eastern Tianshan, China. *J. Earth Sci.* **2011**, *22*, 470–481. [CrossRef]
12. Zhang, H.; Li, Z.; Zhou, P.; Zhu, X.; Wang, L. Mass-balance observations and reconstruction for Haxilegen Glacier No.51, eastern Tien Shan, from 1999 to 2015. *J. Glaciol.* **2018**, *64*, 689–699. [CrossRef]
13. Wang, P.; Li, Z.; Li, H.; Wang, W.; Wang, F. Ice surface-elevation change and velocity of Qingbingtan glacier No.72 in the Tomor region, Tianshan Mountains, central Asia. *J. Mt. Sci.* **2011**, *8*, 855–864. [CrossRef]
14. Wang, P.; Li, Z.; Zhou, P.; Wang, W.; Jin, S.; Li, H.; Wang, F.; Yao, H.; Zhang, H.; Wang, L. Recent changes of two selected glaciers in Hami Prefecture of eastern Xinjiang and their impact on water resources. *Quat. Int.* **2015**, *358*, 146–152. [CrossRef]
15. Liu, Y.; Xinyu, L.; Liancheng, Z.; Yang, L.; Chunrong, J.; Ni, W.; Juan, Z. Quantifying rain, snow and glacier meltwater in river discharge during flood events in the Manas River Basin, China. *Nat. Hazards* **2021**, *108*, 1137–1158. [CrossRef]
16. Zhang, Q.; Chen, Y.; Zhi, L.I.; Yupeng, L.I.; Xiang, Y.; Bian, W. Glacier changes from 1975 to 2016 in the Aksu River Basin, Central Tianshan Mountains. *J. Geogr. Sci.* **2019**, *29*, 984–1000. [CrossRef]
17. Ziyabeike, A.; Gao, J. Impact of Climate Change on the Surface Runoff in the Kuytun River Basin on the North Slope of Tianshan Mountains. *J. Glaciol. Geocryol.* **2010**, *32*, 1186–1193.
18. Wang, P.; Li, Z.; Li, H.; Wang, W.; Wu, L.; Zhang, H.; Huai, B.; Wang, L. Recent Evolution in Extent, Thickness, and Velocity of Haxilegen Glacier No. 51, Kuytun River Basin, Eastern Tianshan Mountains. *Arctic Antarct. Alp. Res.* **2016**, *48*, 241–252. [CrossRef]
19. Lv, H.; Yang, T.; Tian, H.Z. Glacier changes in the northern Altay Mountains from 1980 to 2010. *J. Arid Land Resour. Environ.* **2012**, *26*, 10.
20. Baojuan, H.; Zhongqin, L.; Feiteng, W.; Wenbin, W.; Puyu, W.; Kaiming, L. Glacier volume estimation from ice-thickness data, applied to the Muz Taw glacier, Sawir Mountains, China. *Environ. Earth Sci.* **2015**, *74*, 1861–1870. [CrossRef]

21. Xu, C.; Li, Z.; Wang, F.; Mu, J. Spatio-Temporal Changes of Mass Balance in the Ablation Area of the Muz Taw Glacier, Sawir Mountains, from Multi-Temporal Terrestrial Geodetic Surveys. *Remote Sens.* **2021**, *13*, 1465. [CrossRef]
22. Lv, M.; Guo, H.; Lu, X.; Liu, G.; Yan, S.; Ruan, Z.; Ding, Y.; Quincey, D.J. Characterizing the behaviour of surge-and non-surge-type glaciers in the Kingata Mountains, eastern Pamir, from 1999 to 2016. *Cryosphere Discuss.* **2019**, *13*, 219–236. [CrossRef]
23. Ke, L.; Ding, X.; Song, C. Heterogeneous changes of glaciers over the western Kunlun Mountains based on ICESat and Landsat-8 derived glacier inventory. *Remote Sens. Environ.* **2015**, *168*, 13–23. [CrossRef]
24. Yu, X.; Lu, C. Alpine Glacier Change in the Eastern Altun Mountains of Northwest China during 1972-2010. *PLoS ONE* **2015**, *10*, e0117262. [CrossRef]
25. Liu, S.L.; Zhao, H.D.; Su, X.K.; Deng, L.; Dong, S.K.; Zhang, X. Spatio-temporal variability in rangeland conditions associated with climate change in the Altun Mountain National Nature Reserve on the Qinghai-Tibet Plateau over the past 15 years. *Rangel. J.* **2015**, *37*, 67–75. [CrossRef]
26. Li, Z.; Shi, X.; Tang, Q.; Zhang, Y.; Gao, H.; Pan, X.; Déry, S.J.; Zhou, P. Partitioning the contributions of glacier melt and precipitation to the 1971–2010 runoff increases in a headwater basin of the Tarim River. *J. Hydrol.* **2020**, *583*, 124579. [CrossRef]
27. Zhao, G.; Zhang, Z.; Liu, L.; Li, Z.; Wang, P.; Xu, L. Simulation and construction of the glacier mass balance in the Manas River Basin, Tianshan, China from 2000 to 2016. *J. Geogr. Sci.* **2020**, *30*, 988–1004. [CrossRef]
28. Guo, W.; Liu, S.; Xu, J.; Wu, L.; Shangguan, D.; Yao, X.; Wei, J.; Bao, W.; Yu, P.; Liu, Q.; et al. The second Chinese glacier inventory: Data, methods and results. *J. Glaciol.* **2015**, *61*, 357–372. [CrossRef]
29. Sun, M.; Liu, S.; Yao, X.; Guo, W.; Xu, J. Glacier changes in the Qilian Mountains in the past half-century: Based on the revised First and Second Chinese Glacier Inventory. *J. Geogr. Sci.* **2018**, *28*, 206–220. [CrossRef]
30. Hall, D.K.; Bayr, K.J.; Schöner, W.; Bindschadler, R.A.; Chien, J.Y. Consideration of the errors inherent in mapping historical glacier positions in Austria from the ground and space (1893–2001). *Remote Sens. Environ.* **2003**, *86*, 566–577. [CrossRef]
31. Ma, L.; Tian, L.; Pu, J.; Wang, P. Recent area and ice volume change of Kangwure Glacier in the middle of Himalayas. *Chin. Sci. Bull.* **2010**, *55*, 2088–2096. [CrossRef]
32. Radić, V.; Hock, R. Regional and global volumes of glaciers derived from statistical upscaling of glacier inventory data. *J. Geophys. Res. Atmos.* **2010**, *115*, F01010. [CrossRef]
33. Grinsted, A. An estimate of global glacier volume. *Cryosphere* **2013**, *7*, 141–151. [CrossRef]
34. Shiyin, L.; Wenxin, S.; Yongping, S.; Gang, L. Glacier changes since the Little Ice Age maximum in the western Qilian Shan, northwest China, and consequences of glacier runoff for water supply. *J. Glaciol.* **2003**, *49*, 117–124. [CrossRef]
35. Wang, J.; Xu, C. Geodetector: Principle and prospective. *Acta Geogr. Sin.* **2017**, *72*, 116–134.
36. Rabatel, A.; Letréguilly, A.; Dedieu, J.-P.; Eckert, N. Changes in glacier equilibrium-line altitude in the western Alps from 1984 to 2010: Evaluation by remote sensing and modeling of the morpho-topographic and climate controls. *Cryosphere* **2013**, *7*, 1455–1471. [CrossRef]
37. Zhang, J.; Fu, B.H.; Wang, L.M.; Maimaiti, A.; Ma, Y.X.; Yan, F. Glacier Changes in the West Kunlun Mountains Revealed by Landsat Data from 1994 to 2016. *IOP Conf. Ser. Earth Environ. Sci.* **2017**, *74*, 012002. [CrossRef]
38. Wang, Y.; Wu, L.-Z.; Xu, J.-L.; Liu, S.-Y. Variation and uncertainty analysis of the glaciers in the past 50 years in Geladandong of Tibetan Plateau. *J. Glaciol. Geocryol.* **2013**, *35*, 255–262.
39. Paul, F. A 60-year chronology of glacier surges in the central Karakoram from the analysis of satellite image time-series. *Geomorphology* **2020**, *352*, 106993. [CrossRef]
40. Jarosch, A.H.; Anslow, F.S.; Clarke, G.K.C. High-resolution precipitation and temperature downscaling for glacier models. *Clim. Dyn.* **2012**, *38*, 391–409. [CrossRef]
41. Gardelle, J.; Berthier, E.; Arnaud, Y. Slight mass gain of Karakoram glaciers in the early twenty-first century. *Nat. Geosci.* **2012**, *5*, 322–325. [CrossRef]
42. Shi, Y.; Shen, Y.; Li, D.; Zhang, G.; Ding, Y.; Hu, R.; Kang, E. Discussion on the present climate change from warm-dry to warm wet in northwest China. *Quat. Sci.* **2003**, *23*, 152–164.
43. Fowler, H.J.; Archer, D.R. Conflicting Signals of Climatic Change in the Upper Indus Basin. *J. Clim.* **2006**, *19*, 4276–4293. [CrossRef]
44. Zafar, M.U.; Ahmed, M.; Rao, M.P.; Buckley, B.M.; Khan, N.; Wahab, M.; Palmer, J. Karakorum temperature out of phase with hemispheric trends for the past five centuries. *Clim. Dyn.* **2016**, *46*, 1943–1952. [CrossRef]
45. Ridley, J.; Wiltshire, A.; Mathison, C. More frequent occurrence of westerly disturbances in Karakoram up to 2100. *Sci. Total Environ.* **2013**, *468–469*, S31–S35. [CrossRef]
46. Yao, T.D.; Thompson, L.; Yang, W.; Yu, W.S.; Gao, Y.; Guo, X.J.; Yang, X.X.; Duan, K.Q.; Zhao, H.B.; Xu, B.Q.; et al. Different glacier status with atmospheric circulations in Tibetan Plateau and surroundings. *Nat. Clim. Chang.* **2012**, *2*, 663–667. [CrossRef]
47. Hewitt, K. Glaciers of the Karakoram Himalaya. In *Encyclopedia of Snow, Ice and Glaciers*; Springer: Berlin/Heidelberg, Germany, 2014; pp. 429–436.
48. Kang, E.S. Energy balance characteristics and mass balance change calculation research of Cryosphere in High Asia. *J. Glaciol. Geocryol.* **1996**, *18*, 12–22.
49. Liu, S.; Xie, Z.; Wang, N.; Ye, B. Mass balance sensitivity to climate change: A case study of glacier No. 1 at urumqi riverhead, Tianshan Mountains, China. *Chin. Geogr. Sci.* **1999**, *9*, 134–140. [CrossRef]
50. Haeberli, W.; Cihlar, J.; Barry, R.G. Glacier monitoring within the Global Climate Observing System. *Ann. Glaciol.* **2000**, *31*, 241–246. [CrossRef]

51. Wang, C.; Yang, T.; Qin, J.; Yi, H. Remote sensing monitoring of glacier changes in the middle region of Tanggula Mountain. *Natl. Sci. Found. China* **2016**, *39*, 504–5012.
52. Wang, L.P.; Xie, Z.C.; Liu, S.Y.; Ding, L.F.; Yin, J.Q.; Wu, L.Z. Glacierized Area Variation and Its Response to Climate Change in Qangtang Plateau during 1970–2000. *J. Glaciol. Geocryol.* **2011**, *33*, 979–990.
53. Liu, J.; Yao, X.; Liu, S.; Guo, W.; Xu, J. Glacial changes in the Gangdisê Mountains from 1970 to 2016. *J. Geogr. Sci.* **2020**, *30*, 131–144. [CrossRef]
54. Ji, Q.; Dong, J.; Li, H.-R.; Qin, Y.; Liu, R.; Yang, T. Glacier variations in the Himalaya from 1990 to 2015 based on remote sensing. *Cryosphere Discuss.* **2020**, 1–28. [CrossRef]
55. Ji, Q.; Yang, T.; Tian, H.; He, Y.; Li, X. Relation between glacier retreat and climate change in the western Nyainqentanglha in the past 40 years. *J. Arid. Land Resour. Environ.* **2014**, *28*, 12–17.
56. Liu, S. Glacier fluctuations and the inferred climate changes in the A'Nyêmaqên Mountains in the source area of the Yellow River. *J. Glaciol. Geocryol.* **2002**, *24*, 701–707.
57. Wu, K.; Liu, S.; Jiang, Z.; Xu, J.; Wei, J.; Guo, W. Recent glacier mass balance and area changes in the Kangri Karpo Mountains from DEMs and glacier inventories. *Cryosphere* **2018**, *12*, 103–121. [CrossRef]
58. Li, X.; Yang, T.; Tian, H.; Ji, Q. Response of glacier in Gongga Mountain to climate change during the last 40 years. *Res. Soil Water Conserv.* **2013**, *20*, 125–129.
59. Li, L.-F.; Li, K.-M. Study on the Relationship Between Glacier Change and Topographic Factors in the Shiyang River Basin. *J. Glaciol. Geocryol.* **2019**, *41*, 10.

water

MDPI

Article

Ice Phenology in Eurasian Lakes over Spatial Location and Altitude

Matti Leppäranta [1,2] and Lijuan Wen [2,*]

1 Institute of Atmospheric and Earth Sciences, University of Helsinki, 00014 Helsinki, Finland; matti.lepparanta@helsinki.fi
2 Key Laboratory of Land Surface Process and Climate Change in Cold and Arid Regions, Northwest Institute of Eco-Environment and Resources, Chinese Academy of Sciences, Lanzhou 730000, China
* Correspondence: wlj@lzb.ac.cn

Abstract: Eurasian freezing lakes cover an almost 180° wide longitude sector between the latitudes 30° and 75° N, and their altitudes range from below the sea surface level up to 5 km elevation. Ice phenology varies widely in this region. However, these variations and their influence factors have been little studied. Analytic models are applied here to examine these variations supported by historical ice and weather data. These models are forced by a linear air–lake heat exchange formula based on local empirical fits. The weather brings latitude–longitude–altitude patterns to the large-scale lake ice phenology. Freezing and breakup dates are forced by the local air temperature and solar radiation, and their rates of change are also important. In addition, freezing depends on lake depth and breakup depends on accumulated ice thickness. Lake depth provides a lag and radiation balance provides a shift with respect to the air temperature in cooling of the lake, and breakup is dictated by spring warming conditions and ice thickness. Due to solar radiation forcing, the common degree-day approach is biased for modelling ice phenology, especially in low latitudes. Analytic models provide a first-order tool for climate sensitivity of ice seasons. The freezing date and breakup date both change by around five days per one-degree shift in air temperature away from the climatological ice margin; however, at this margin, the sensitivity is higher.

Keywords: ice phenology; Eurasia; analytical model; climate; altitude; spatial statistics

Citation: Leppäranta, M.; Wen, L. Ice Phenology in Eurasian Lakes over Spatial Location and Altitude. *Water* **2022**, *14*, 1037. https://doi.org/10.3390/w14071037

Academic Editor: Guido Paliaga

Received: 1 March 2022
Accepted: 23 March 2022
Published: 25 March 2022

1. Introduction

In the Eurasian continent, lakes freeze between the latitudes of 30–75° N. The extent is at most south over the high Tibetan Plateau, while freezing reaches only down to 50° N in western Europe due to the heating provided by the North Atlantic Current (Figure 1). The northern boundary of the continent is the coast of the Arctic Ocean at 70–75° N. Freezing and ice breakup bring major changes to the physics and ecology of lakes and to the human living conditions in lake districts [1]. The primary factors that drive the ice season are the air–lake interaction, radiation balance, and the characteristics of individual lakes [2]. Climate variations have a major impact on ice season characteristics; in other words, ice season characteristics are sensitive indicators of climate.

An ice phenology time-series consist of the dates of freezing and ice breakup and the number of ice days. There is a large set of these time-series, and many of them go back 100–200 years in history [3]. The thermal memory of lakes is too short in open water periods to show significant dependence between consecutive ice seasons. Therefore, the freezing date is independent of the previous ice season; however, in winter, time slows down, and there is a weak statistical connection between the freezing date and the following breakup date [4]. A connection through the ice season can also be seen between deep-water temperature and dissolved oxygen content [1]. During windy autumn seasons, forced deep mixing may continue until the water temperature is well below the temperature of maximum density. Then, after freeze-up, the oxygen storage in the cold lake is high enough to influence the oxygen content until spring turnover [5].

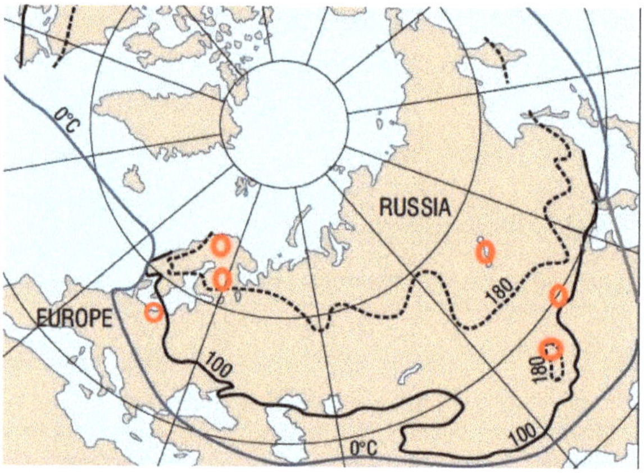

Figure 1. The Eurasian continent with contours for lake ice season of 100 and 180 days [6], together with the mean January 0 °C contour of air temperature. The study sites are marked by red circles.

Ice phenology time series predominantly show decrease in the length of the ice season in Eurasian lakes over the last 100 years [1,3,7–10]. The shortening is seen nearly equally in the dates of freezing and breakup. The trend comes from general climate warming in the 20th century in Eurasia and North America, where most of the lake ice investigations were made. Aperiodic variations and noise are superposed on the trend toward a milder ice climate. In the Tibetan Plateau, shortening of ice seasons is seen in satellite remote sensing data; however, some saline lakes show opposite behavior, because their salinity has been decreasing due to the increasing water volume that has influenced the cooling process [8,11–13].

In addition to the phenology, the thicknesses of ice and snow have been recorded in many lakes, and their annual maxima are a measure of the severity of ice seasons. In large lakes, the evolution of ice coverage has been monitored with its annual maximum as another characteristic of season severity [9]. A statistical approach is normally taken in analyses of lake ice time series [4,7,14,15].

The connection of the statistical characteristics of ice phenology to the physics of lakes needs to be understood, particularly to evaluate the impact of predicted future climate scenarios on the ice season in lakes. This can be approached by analytical modelling [9] that guides to understand the role of climatological forcing factors and lake properties, opens more physically-based views into the properties of ice time series, and reveals the climate sensitivity and future of winter conditions in Eurasian lakes.

Lake ice phenology is strongly related to weather conditions, as shown by field experiment outcomes [16,17] and numerical modelling [18,19]. Air temperature depends both on altitude and spatial location, while solar radiation depends primarily on latitude. The annual air temperature cycle can be similar in low-latitude high mountains and low-altitude polar regions, but they receive quite different solar radiation forcing. There is also a longitudinal temperature variation over Eurasia due to the variations in westerlies from the North Atlantic (the North Atlantic Oscillation, or NAO) and the continental climate impact that causes major differences in the heat balance, ice conditions, and primary production in Eurasian freezing lakes [2,20]. Latitude and altitude are expected to have different roles in ice phenology via radiation balances.

This paper examines the spatial characteristics of lake ice phenology in Eurasia. The study reference sites cover the Tibetan plateau and boreal and tundra zone in northern Europe. Analytic models are employed for the dates of freezing and breakup based on spatial location, altitude, and lake depth, and forced by solar radiation and air temperature.

The models are calibrated using the climatology of the study sites. Parameterized analytic models are used to examine the ice climatology of Eurasian lakes and the climate sensitivity of the lake ice season over the continent.

2. Materials and Methods

2.1. Lakes

A set of lakes was taken for the calibration of the climatological lake ice model (Table 1). They include lakes from tundra and boreal zones, and the Central Asian arid, cold climate zone. The latitude range is 34°, the longitude range is 94°, and the altitude range is 4.3 km in the sites. The data present the atmospheric and ice climatology obtained from the home laboratories of the authors, open data bases, and publications.

Table 1. Study lakes in the calibration of analytical models. The data are from local sources.

Lake	Latitude and Longitude	Altitude m	Area km^2	Depth m	Lake Type	Climate Zone
Kilpisjärvi	69° 03′ N 20° 50′ E	473	37	19.5	Oligotrophic	Arctic tundra
Kallavesi	62° 46′ N 27° 47′ E	82	478	9.7	Oligotrophic	Boreal
Pääjärvi	61° 04′ N 25° 08′ E	103	13	15.0	Meso-oligotr.	Boreal
Baikal	53° 18′ N 108° 00′ E	456	31,772	758	Oligotrophic	Boreal
Szczecin lagoon	53° 49′ N 14° 08′ E	0	687	4.6	Eutrophic	Boreal
Wuliansuhai	40° 56′ N 108° 52′ E	1019	233	1.1	Eutrophic	Cold arid
Qinghai	36° 53′ N 100° 11′ E	3260	4237	21.0	Oligotrophic	Tibetan
Gyaring	34° 56′ N 97° 17′ E	4294	526	8.9	Oligotrophic	Tibetan
Ngoring	34° 54′ N 97° 42′ E	4272	610	17.6	Oligotrophic	Tibetan

The date of freezing normally refers to the date when the lake froze or the date when the observation site was frozen. In the past, phenological observations were made from the shore and the observation area was limited, while current satellite data provide surface information for the whole lake. The breakup date is the last day when ice was observed. Sometimes, the beginning of the ice decay period is included in the analysis. It is defined as the day the thickness of ice starts to decrease after the annual maximum, or as the day the continuous ice cover of a lake experiences final breakage. The latter approach is useful when air–lake gas exchange questions are examined. However, field data have shown that in a fixed lake, the length of the decay period does not vary much but its starting time does [2,9,17]. Thus, the final breakup day and the start of the decay are highly correlated.

If a lake does not freeze every year, the statistical treatment of freezing and breakup dates needs care since standard statistics such as the mean and variance cannot be properly defined. The binary variable (1 for ice and 0 for no ice) is then taken into the ice phenology data to account for the statistics of ice occurrence, and fractiles of the phenological distributions are used for the statistics [6,21], e.g., median (50 % fractile) date $t_{0.5}$ can be used instead of average when the probability of freezing is at least 0.5; the median freezing date $t_{F0.5}$ is obtained from $P(t_F \leq t_{F0.5}) = 0.5$, where t_F is the annual freezing date. One lake included in this study was from the climatological ice margin where the probability of annual ice occurrence is between zero and one. In large lakes, although near-shore areas freeze annually, a complete freeze-over does not always take place. The maximum annual coverage is then recorded as another ice season characteristic [9].

2.2. Data

The wind, air temperature, and humidity were obtained from the Finnish Meteorological Institute and the China Meteorological Forcing Dataset for a detailed preparatory study of the heat budget. The freezing date and air temperature of lakes Kilpisjärvi, Kallavesi, and Pääjärvi were provided by the Finnish Meteorological Institute and Finnish Environmental Institute. In the climatology anaysis, air temperature (1980–2020) in Eurasia was otained from NCEP/NCAR Reanalysis 1, 2.5° × 2.5° latitude–longitude global grid.

2.3. Modelling Approach

Freezing of a lake, ice growth, and breakup are forced by the heat exchange with atmosphere–ice–water body–lake bottom interactions, and solar radiation. These interaction processes act on the boundary surfaces, while solar forcing provides heating throughout the ice and into the water body, and a part of the solar heat absorbed by the water is conducted to the ice bottom [15,22]. The basic analytical models are the slab model for lake cooling [23], freezing-degree-day model for ice growth [24], and positive-degree-day models for ice melting [25]. The connection of ice melting to air temperature is indirect, and models based on energy balance are more physically based [26].

Freezing date: Analytic modelling of autumn cooling is based on the slab model. The water temperature is assumed to be vertically uniform, and we have:

$$\frac{d(T_w H)}{dt} = \frac{Q_0 + Q_b}{\rho_w c_w} \tag{1}$$

where T_w is the mixed-layer temperature, H is the mixed-layer depth or lake depth (fixed), t is time, ρ_w and c_w are the density and specific heat of the water, respectively, and Q_0 and Q_b are the heat fluxes at the lake surface and bottom, respectively. The freezing date t_F is taken as the date when the mixed-layer temperature has reached the freezing point, $T_w(t_F) = T_f$.

Ice melting: Modelling the decay of lake ice is in principle straightforward, since in the melting season, ice loss is due to the net gain of energy from the external fluxes, and heat conduction is not significant. However, the space–time variability of the albedo and light attenuation coefficient of melting ice gives rise to parametrization problems for the solar flux. Ice melting progresses as:

$$\frac{dh}{dt} = -\frac{(Q_0 + Q_w)}{\rho_i L_f} \leq 0 \tag{2}$$

where h is ice thickness, Q_w is the heat flux from water to ice, ρ_i is ice density, and L_f is the latent heat of freezing. The breakup date t_B is taken from $h(t_B) = 0$ after the thickness has reached its annual maximum $h = h_{max}$. To predict the breakup date, this thickness needs to be known. It can be estimated as $h_{max} = \sqrt{a^2 S + b^2} - b$, where $a \approx$ 2–3 cm $(°\text{C d})^{-1}$ and $b \approx 10$ cm are the model parameters; S is the cumulative temperature of freezing-degree-days [24,27]. Melting of ice also depends on the ice structure, in particular on the fractions of clear congelation ice and opaque snow-ice. Light transfer properties largely determine how much melting takes place on the top and bottom surfaces and in the interior. However, the net heat flux can be taken as the first order forcing factor.

Forcing: These models are forced by solar radiation, atmospheric heat flux, and heat flux from below. The surface+internal forcing is expressed in linear form, which is convenient in analytic analysis (see Appendix A):

$$Q_0 = K_0 + K_1(T_a - T_0) \tag{3}$$

where K_0 and K_1 are the time-dependent model parameters, and T_a and T_0 are the air temperature and surface temperature, respectively. The parameter K_0 is governed by solar radiation, while K_1 builds up from turbulent heat exchange. These parameters were estimated for the control sites in Finland and Tibet from climate data (Figure 2). The forcing from below the mixed layer (or lake bottom) can be formulated as $Q_b = K_b(T_b - T_w)$, where K_b is the heat transfer coefficient and T_b is the temperature below the mixed layer (or lake bottom).

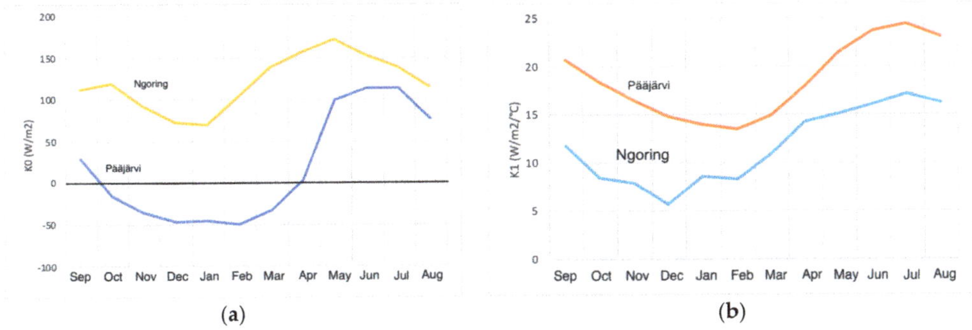

Figure 2. The parameters K_0 (**a**) and K_1 (**b**) of the linear surface heat transfer formula $Q_0 = K_0 + K_1(T_a - T_0)$ for Lake Pääjärvi (based on climate data from Jokioinen, 60°49′ N 23°30′° E) and Ngoring Lake (34°54′ N 97°42′ E).

The parameter K_0 has a strong annual cycle from solar radiation. The level is higher and amplitude lower in the Tibetan site, compared to the sites in Jokioinen, southern Finland. The annual averages are 120.5 W m^{-2} in Ngoring (range 150 W m^{-2} in monthly means) and 17.8 W m^{-2} (range 170 W m^{-2} in monthly means) in Jokioinen, which represents Lake Pääjärvi. The parameter K_1 is rather stable and the level is lower in Tibet than in Finland, and the annual cycle is mostly due to the temperature dependence of the saturation water vapour pressure. The annual averages are 11.7 W m^{-2} °C^{-1} in Ngoring and 18.6 W m^{-2} °C^{-1} in Jokioinen.

3. Geography of Ice Phenology

The study lakes are located within latitudes 34–69° N, longitudes 14–109° E, and altitudes 0.1–4.3 km (Table 1). In all cases, the mean air temperature is below the freezing point in December–March. The northern lakes freeze in November and the Tibetan lakes freeze in December. The air temperature is quite low in Tibet in December (Table 2), which shows that solar radiation places a long delay on the freezing, compared to the north. The very shallow Lake Wuliansuhai in Inner Mongolia freezes early, and the very deep Lake Baikal freezes only in January. Close to the climatological margin of freezing lakes, the freezing date and breakup date have a large variability [15,28].

Ice formation, growth, and melting depend on the latitude Φ, longitude Λ, and altitude Z via air temperature and solar radiation. Air temperature decreases upward by the adiabatic lapse rate $\Gamma \approx 6$ °C km^{-1}, while, according to climatology, latitude-wise and longitude-wise cooling in Eurasian winter are about 1.2 °C deg lat^{-1} and 0.3 °C deg lon^{-1}, respectively. Thus, 1 km increase in altitude corresponds to 5 degrees increase in latitude and 20 degrees in longitude. The location difference between Ngoring and Kilpisjärvi is 34 degrees in latitude, 77° in longitude, and 3.8 km in altitude, and therefore their temperatures should be close. In autumn and winter, Ngoring is colder but in spring and summer, the graphs are close (Figure 3). A similar condition exists between Qinghai and Kallavesi.

Comparing Kilpisjärvi with the Tibetan lakes, it was seen that the December air temperature is lower than in Qinghai but higher than in the other sites. The freezing dates then show that the latitudinal, i.e., solar radiation, effect accounts for about one month between the Tibetan plateau and the European Arctic tundra. For similar air temperature evolution, freezing takes place about one month later in the Tibetan sites, as compared to the Arctic (Figure 3). These Tibetan lakes freeze when the air temperature is 10–15 °C below the freezing point. A corresponding difference appears in the breakup date, and the breakup takes place in the Tibetan plateau at air temperatures below the freezing point. Figure 3 clearly illustrates the limitations of the normal freezing-degree-days and positive-degree-days formulae for the freezing and breakup dates, suggesting that they need to be modified in respect to the radiation balance. The Szczecin lagoon is shown as the reference, where the annual probability of ice occurrence is lower than one (0.78).

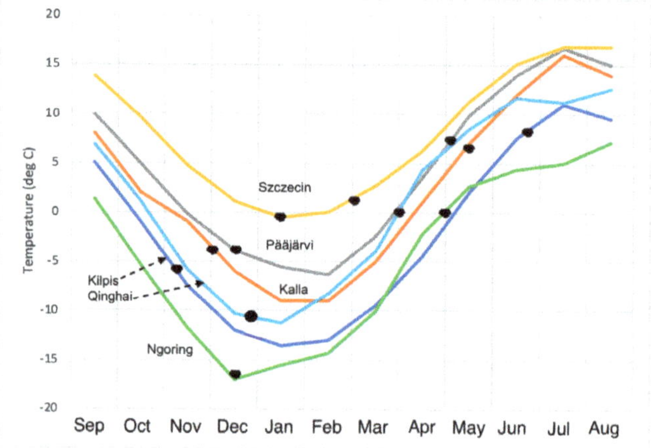

Figure 3. Annual average temperature cycle in the lakes under study. The black dots show the freezing and breakup dates.

In November–December, solar radiation is nearly zero in Kilpisjärvi, but in Ngoring the daily mean is around 150 W m^{-2} according to observations; thus, Ngoring is able to resist freezing until the air temperature is down to about −10 °C. The daily solar flux is above 150 W m^{-2} all winter and its penetration into ice is strong due to the absence of snow in dry climates, e.g., [29].

Table 2. Ice phenology in the selected Eurasian lakes. Notation ± nn refers to the standard deviation in days (d). The ice phenology data are based on the periods of 1960–2020 (Finland), 1991–2020 (China), 1896–2000 (Szczezin logoon), and 1960–2000 (Baikal), and the air temperature data are for the period of 1981–2010.

Lake	Freezing Date	December Mean Air Temperature	April Mean Air Temperature	Breakup Date
Kilpisjärvi	8 Nov. ± 8 d	−12 °C	−5 °C	18 Jun. ± 7 d
Kallavesi	30 Nov. ± 14 d	−6 °C	+1 °C	8 May ± 8 d
Pääjärvi	13 Dec. ± 16 d	−5 °C	+4 °C	5 May ± 6 d
Szczecin lagoon [21,30]	15 Jan.	−0.5 °C	+6 °C	1 Mar.
Wuliansuhai [31]	7 Dec.	−10 °C	+4 °C	31 Mar.
Baikal [32]	10 Jan.	−18 °C	+1 °C	1 May.
Gyaring [33]	30 Nov. ± 21 d	−18 °C	−2 °C	21 April ± 17 d
Ngoring [33]	15 Dec. ± 10 d	−17 °C	−2 °C	1 May ± 8 d
Qinghai [33]	29 Dec. ± 4 d	−10 °C	+4 °C	1 April ± 7 d

Next, analytic solutions were derived for the first-order explanations of the characteristics of the field data on ice phenology. Forcing of the system is taken care of by the linearized atmospheric heat flux (Equation (3)).

4. Analytic Model

4.1. Freezing Date

The slab model (1) with the surface heat flux (3) is first written as:

$$\frac{dT_w}{dt} = \lambda_a(T_a + T_R - T_w) + \lambda_b(T_b - T_w); \lambda_a = \frac{K_1}{\rho_w c_w H}, \ \lambda_b = \frac{K_b}{\rho_w c_w H}, \ T_R = \frac{K_0}{K_1} \quad (4)$$

Here, λ_a and λ_b are the relaxation times of the mixed layer for the surface and bottom forcing, and T_R (Figure 4) is a temperature shift term originating from the radiation balance. This equation can be directly integrated:

$$T_w(t) = \int_{-\infty}^{t} e^{-\lambda(t-\tau)}[\lambda_a(T_a + T_R) + \lambda_b T_b]d\tau \quad (5)$$

where $\lambda = \lambda_a + \lambda_b$, and $\tau_L = \lambda^{-1} \propto H$ is the response time of the lake with the proportionality coefficient of about 2 d m^{-1}.

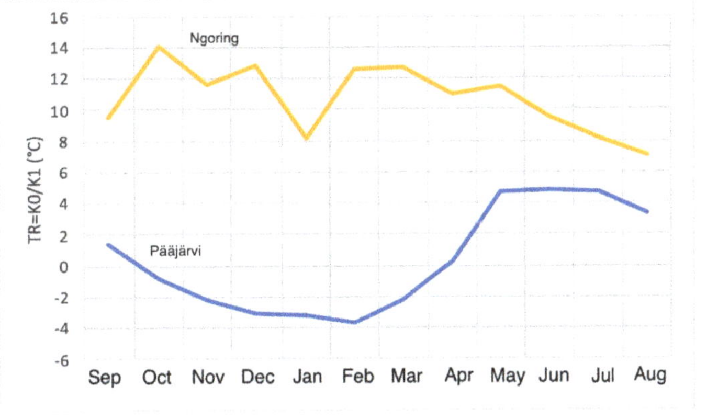

Figure 4. The temperature parameter T_R in Lake Pääjärvi and Ngoring Lake. This temperature indicates how much solar radiation compensates for air temperature forcing in the annual cycle of the lakes. Data from Figure 2.

The fundamental question in ice phenology is whether a lake is ice-free or freezes over. Take a parabolic form of the winter temperature evaluation, expressed as $T_a(t) = ct^2 + \hat{T}_a$, where $\hat{T}_a$ is the minimum air temperature in winter and $c > 0$ is a shape factor. Freezing takes place in late fall when the radiation balance is rather stable, and therefore the term T_R is taken as a constant (Figure 4). The lake bottom temperature is fixed. The condition of lake freezing is:

$$c\tau_L^2 < -\left\{\hat{T}_a - \left[T_f - T_R - \frac{\lambda_b}{\lambda_a}\left(T_b - T_f\right)\right]\right\} = -(\hat{T}_a - T_c) \quad (6)$$

where we introduced the temperature $T_c = T_f - T_R - \frac{\lambda_b}{\lambda_a}\left(T_b - T_f\right)$ as the critical temperature, which provides a necessary condition for any lake to freeze as $\hat{T}_a < T_c$. Then from $T_a(t) = T_c$ the shape factor c is obtained as $c\tau_f^2 = 4(T_c - \hat{T}_a)$, where τ_f is the length of the 'cold period' $T_a < T_c$. The condition (6) becomes simply $\tau_f > 2\tau_L$, i.e., for a lake to freeze, the cold period must be at least twice the response time of the lake. The temperature condition is external, but the required length of the cold period is lake-specific since $\tau_L \propto H$. The heat flux from the sediments is generally much smaller than the surface heat flux in the cooling season, but geothermal lakes may stay ice-free by condition (6).

Assume that the heat flux from the bottom is negligible. Equation (6) shows that a necessary condition for freezing of any lake is $\hat{T}_a - T_f < -T_R$; e.g., $T_R \approx -3$ °C in Lake Pääjärvi and $T_R \approx 12$ °C in Ngoring Lake (Figure 4). Thus, in the north, radiational and evaporation losses can freeze the surface even when the air temperature is above the freezing point, but in low latitudes, solar radiation can compensate for the losses down to quite low temperatures. For a lake of finite depth, the minimum air temperature must be still lower so that the length of the cold period is longer than twice the lake response time. e.g., if $T_R = 0$ and $\lambda_b = 0$, for a 5 m deep lake, the air temperature must be below the freezing point at least 10 days. In Szczecin Lagoon, Poland (53.8° N, 14.1° E, mean depth 4 m), the probability of ice occurrence is 0.78, and therefore in an average winter, the basin is just about to freeze. We have $\hat{T}_a \approx 0$ °C, $T_R \sim -1$ °C and $\tau_L \approx 8$ d. This means that a period of 16 days colder than +1 °C is needed to create an ice cover that corresponds to the observed ice climatology.

Assuming freezing takes place, to examine the freezing date, a linear cooling law $T_a(t) = T_f - \alpha t$ is taken, where α is the rate of cooling and $t = 0$ refers to the time the decreasing air temperature passes the freezing point, $T_a(0) = T_f$. The water temperature is obtained from Equation (4), and thereafter the freezing time is obtained from $T\left(t_f\right) = T_f$:

$$t_F = \tau_L + \frac{1}{\alpha}\left[T_R + \frac{\lambda_b}{\lambda_a}\left(T_b - T_f\right)\right] \tag{7}$$

Thus, the solution contains delay terms due to the lake response time, heat flux from bottom, and radiation balance. The local climate impact is included in α, the time zero, and T_R, while the response time and bottom heating are lake specific.

Assuming that the bottom heat flux is small, the freezing date is delayed behind $t = 0$ by the lake response time plus the lag $\Delta t = \alpha^{-1} T_R$, $t_F = \tau_L + \Delta t$, from the air temperature. It is interesting to compare the cases of Lake Kilpisjärvi in Arctic tundra and Ngoring Lake in Tibetan tundra, since their temperature cycles and depths are close (Figure 3). In Kilpisjärvi, we have $\Delta t = -15$ days; with $\tau_L \approx 2H$ d m^{-1}, the estimated response time is $\tau_L = 39$ days, and thus the predicted freezing time is $t_F = 24$ days, 6 days less than in the true climatology. In Ngoring Lake, $\Delta t = 50$ days, the estimated response time is 35 days, and the predicted freezing date is 85 days, 5 days more than in the true climatology. Thus, the analytic model provides an acceptable estimator of the freezing date, and it is feasible in analysing lake ice climatology across latitude and altitude. For the other lakes, corresponding results are obtained. e.g., in Szczecin lagoon, $\Delta t = -5$ d and $\tau_L = 8$ d, i.e., the basin freezes soon after the air temperature has gone to the cold level. The lake response time is a lake characteristic, while the lag due to forcing depends on the local climate.

4.2. Breakup Date

Analytic models of ice melting are based on the heat gain by the ice that is used to decrease the ice volume, and 'ice thickness' represents the volume of ice per unit area. This approach corresponds to the physical picture where significant melting takes place at both boundaries and in the ice interior.

Using the linear surface heat flux (Equation (3)), the thickness of ice during the melting period is:

$$h_{max} - h(t) = \frac{1}{\rho_i L_f}\int_{t_M}^{t}\left[K_0 + K_1\left(T_a - T_f\right) + Q_w\right]d\tau \tag{8}$$

where $h_{max} = h(t_M)$ is the ice thickness in the beginning of melting. In the melting period, net solar radiation increases fast and therefore we take $K_0 = K_0(t)$ and thus $T_R = T_R(t)$ (see Figures 2 and 4). At the start-up time t_M, the surface heat flux turns positive after the maximum ice thickness has been reached. This time is approximated from the solution $t = t_M$ of the equation $K_0(t) + K_1\left[T_a(t) - T_f\right] = 0$:

$$T_a(t_M) + T_R(t_M) = T_f \tag{9}$$

This is a delicate formula, since in spring both air temperature and solar radiation increase, and the temperature T_R depends largely on the latitude and altitude (Figure 4). The date of breakup, t_B is obtained from the implicit equation $h(t_B) = 0$. Since melting adds on the incoming heat fluxes, the heat flux from water can be absorbed to the parameter K_0, as is clear from Equation (8).

The K_1 term in Equation (8) provides a positive-degree-day formula. For $K_1 = 15$ W m^{-2} °C^{-1}, we have the degree-day coefficient $\frac{K_1}{\rho_i L_f} = 0.42$ cm (°C d)$^{-1}$. When the melting period has begun, the air temperature and the K_0-term are assumed to increase linearly by the rates β and γ, respectively. Ignoring Q_w (or absorbing to K_0), we have an equation for the breakup date:

$$\frac{1}{2}(\gamma + K_1\beta)(t_B - t_M)^2 = \rho L_f h_{max} \tag{10}$$

The square of the period $t_B - t_M$ illustrates the rather low sensitivity of the breakup date to ice thickness and the model parameters.

The parameters of Equation (10) for our study sites were obtained from the data shown in Figures 2 and 3. For Ngoring lake, $\gamma \approx 1$ W m^{-2} d^{-1}, for Kallavesi $\gamma \approx 1.5$ W m^{-2} d^{-1}, and for both $K_1 \approx 15$ W m^{-2} °C^{-1} and $\beta \approx 0.17$ °C d^{-1}. An approximation for the date of ice breakup is:

$$t_B = t_M + \sqrt{\frac{2\rho L_f}{\gamma + K_1\beta} h_{max}} \approx t_M + \left(\sqrt{A\Delta h_{max}}\right)\mathrm{d} \tag{11}$$

where $A \approx$ 15–20 cm^{-1} Thus, in the first-order approximation, the length of the melting period depends on the initial ice thickness, and for half-meter ice, it is about 30 days. Figure 3 does show that the melt rates overlap, but the breakup depends on the start-up time and initial ice thickness. The maximum ice thickness is typically 70 cm in Ngoring Lake and 90 cm in Lake Kilpisjärvi. Furthermore, the estimated time t_M is 15 March for Ngoring Lake, 31 March for Lake Kallavesi, and 10 May for Lake Kilpisjärvi, and the observed mean breakup dates are for these sites, respectively, 30 April, 8 May, and 18 June. In the observations, melting took 35–45 days from the start in these sites, while the start-up dates ranged over 65 days (Table 2, Figure 3). In the Tibetan lakes, at the time of ice breakup, the air temperature had just reached the freezing point. At the time of start-up, in Kallavesi, the temperature is at the freezing point, but in Kilpisjärvi, it is about 2 °C higher than the freezing point.

5. Discussion and Conclusions

The ice phenology time series from 10 Eurasian sites illustrated the roles of local air temperature (determined by altitude, latitude, longitude) and solar radiation (determined by latitude) to determine the statistics of ice occurrence, freezing date, and breakup date. The role of solar radiation shows up strongly in Eurasia, where the latitudinal extent of freezing lakes goes from 35° N to 70° N. In the Eurasian continent, winters also become more severe toward the east due to continental vs. maritime climate (Figure 5).

In autumn, surface temperature of a lake follows the air temperature with a lag depending in the lake depth and with a shift depending on the net solar radiation. Ice phenology depends on the lake morphology and weather. For freezing, the air temperature must go down enough and stay there long enough for freezing to occur. The critical temperature is close to the freezing point at 60° N but less by 10 °C in the Tibetan plateau at 35° N. After freeze-up, ice thickness follows the freezing-degree-days. The melting period begins when the surface heat balance turns positive in late winter. This date depends on the solar radiation and air temperature. Again, at 60° N, the air temperature is then close to the freezing point, but in Tibetan plateau it is still around 10 degrees lower. In low latitudes, solar radiation has a dominant role in the ice season heat balance, and therefore statistical degree-day models do not work there. Rather, the level of solar radiation determines how low the air temperature must go before ice formation is possible. Under similar temperature conditions, ice season is longer in a high latitude than in high altitude.

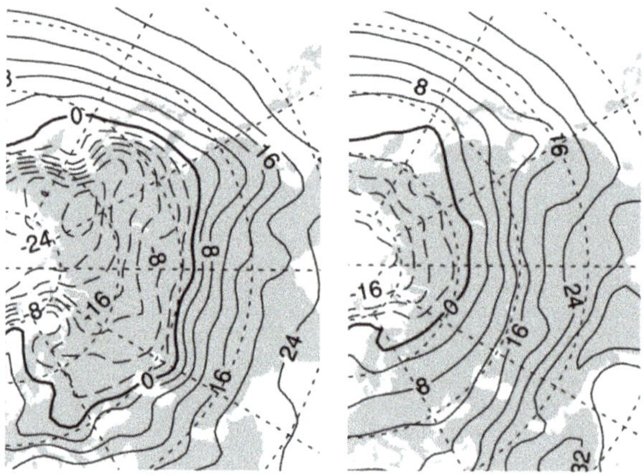

(a) (b)

Figure 5. Mean air temperature (1980–2020) in Eurasia in December (**a**) and April (**b**).

The dependence of the freezing and breakup dates on the latitude in East Europe was already presented [1]. Lake sites were at lower levels than 0.5 km above the sea level, and there were clear trends for both dates (Figure 6). Here, we added our sites from the Asian side of the continent. In addition, Lake El'gygytgyn (67°30′ N 172° E) from Northeast Siberia has been added in the plot [34]. This lake is a deep crater lake with an area of 14 km^2. In principle, at high altitudes, the trend lines are shifted, as suggested by our new data points, but the eastward winter cooling effects are also included. The Lake Baikal breakup date fits in the picture but freezing date is quite late, because the large mixed-layer depth in Baikal delays the freezing date but does not affect the breakup.

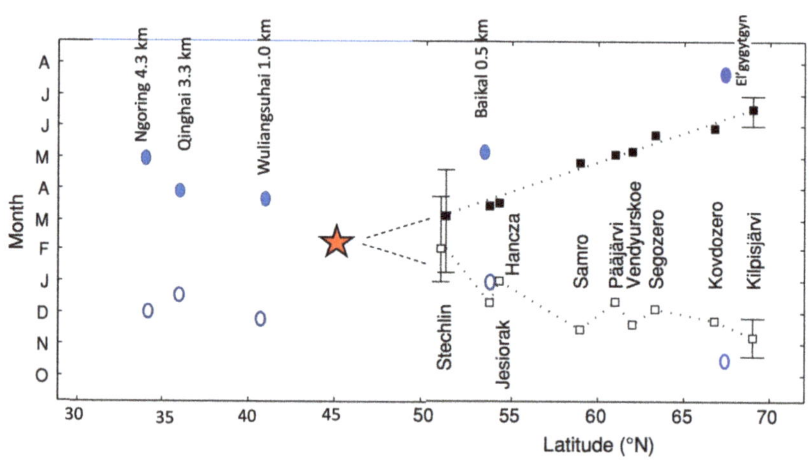

Figure 6. Long-term averages of the freezing date and breakup date in several Eurasian lakes. Original plot with white squares and black squares from Kirillin et al. [1]. Blue ovals have been added in this work for Asian lakes.

Analytic models were presented for the freezing date and breakup date forced by a linearized surface heat budget, heat flux from below the mixed layer, the mixed layer depth, and freezing point of lake water. The linearized heat budget was expressed as $Q_0 = K_0 + K_1(T_a - T_0)$ including two climate parameters $K_0 = K_0(t)$ and $K_1 \approx const.$ (see Appendix A). These form a temperature shift parameter $T_R = K_0/K_1$. Together with K_0, the local air temperature cycle defines the timing of the freeze-up and melting periods. The former depends also on the lake depth and the latter on the ice thickness.

The analytic model can be employed to examine the climate change influence. For a change ΔT_a in air temperature, the first-order approximation is that the parameters K_0 and K_1 do not change, but the climate impact comes from the shift of the annual temperature cycle. First, the condition of freeze-up, $\tau_f > 2\tau_L$ or the cold period, must be at least twice the memory of the lake, and becomes tighter with $\Delta T_a > 0$. Following the model of Equation (7), the freezing date would be changed by $\alpha^{-1}\Delta T_a$, where α is the atmospheric cooling rate in fall. e.g., in northeastern Europe, $\alpha \sim 5$ °C month^{-1}, and the freezing date would change by 6 days for $\Delta T_a = 1$ °C. The breakup date would be changed by $\beta^{-1}\Delta T_a$, where β is atmospheric warming date in spring, $\beta \sim \alpha$. A secondary effect to the breakup date comes from less ice to melt (Equation (10)), since temperature change implies change in freezing degree-days S, and consequently ice thickness changes as $\delta h \propto \delta S/h$. The length of the melting period is proportional to $\sqrt{h_{max}}$, and therefore the sensitivity of the breakup date to ice thickness is large only for thin ice.

The quality of ice cover can be categorized into weak and stable, where the latter refers to landfast ice cover thick enough for traffic and various human activities. The transition lies between 20–30 cm thickness, apart from the soft and porous ice cover in the melting season. In addition to ice phenology, the climatology of ice quality and thickness are also key characteristics for the society, which will be examined in a future study based on modelling and remote sensing.

Many ice phenology time series papers have shown that the climate warming in recent decades has had similar cuts of some days in the ice season in its both ends, which can now be understood from the analytic model [9,15]. Even though the model looks crude, the climate change prediction is a small differential, and one would expect these predictions to be robust.

Author Contributions: M.L. put together the data, performed the data analysis and wrote the original draft. L.W. contributed to the study conception and design, collected the Chinese data, and took part in the manuscript write-up. All authors have read and agreed to the published version of the manuscript.

Funding: The work was supported by the National Key Research and Development Program of China (2019YFE0197600), CAS "Light of West China" Program (E129030101, Y929641001), PIFI Fellowship of CAS 2021, the Academy of Finland project 333899, the National Natural Science Foundation of China (11930759, 41975081), and the Opening Research Foundation of the Key Laboratory of Land Surface Process and Climate Change in Cold and Arid Regions, Chinese Academy of Sciences (LPCC2019005).

Institutional Review Board Statement: Not applicable.

Informed Consent Statement: Not applicable.

Data Availability Statement: The climate data from Jokioinen can be obtained by contacting M.L. The China meteorological forcing dataset can be downloaded from http://data.tpdc.ac.cn (accessed on 1 December 2021); NCEP/NCAR Reanalysis data can be downloaded from https://psl.noaa.gov/data/gridded/data.ncep.reanalysis.pressure.html (accessed on 1 December 2021).

Acknowledgments: Two anonymous referees are thanked for their constructive comments.

Conflicts of Interest: The authors declare no conflict of interest.

Appendix A

Linearized Atmospheric Heat Flux

Linear solar and atmospheric heat flux is expressed as $Q_a = K_0 + K_1(T_a - T_0)$, where T_a and T_0 are the air temperature and surface temperature, respectively, and the coefficients K_0 and K_1 do not depend explicitly on T_0. It is obtained as follows: The common formulations for the terrestrial radiation and turbulent fluxes are based on the standard weather variables: air temperature, relative humidity R, wind speed U_a, and cloudiness N. The net solar radiation is $(1-\alpha)Q_s$, where Q_s is incident solar radiation and α is albedo.

The incoming terrestrial radiation from the atmosphere is a purely external factor, while the sensible heat flux is proportional to the temperature difference $T_a - T_0$ and thus the proportionality coefficient goes to K_1.

The outgoing terrestrial radiation and the latent heat flux contribute to both coefficients. The former is split into two parts by using the binomial formula, and then the net terrestrial radiation is

$$Q_{nL} = \varepsilon_0\sigma\left(\varepsilon_a T_a^4 - T_0^4\right) \approx -\varepsilon_0\sigma(1-\varepsilon_a)T_a^4 + 4\varepsilon_0\sigma T_a^3(T_a - T_0) \tag{A1}$$

where ε_0 and $\varepsilon_a = \varepsilon_a(T_a, R, N)$ are the emissivities of the surface and atmosphere, respectively, and σ is the Stefan–Boltzmann constant. The truncated formula is accurate to about 1% when $|T_a - T_0| < 10$ °C and $T_a \sim 0$ °C. The latent heat exchange is split using the Clausius-Clapeyron equation for the saturation water vapour pressure e_s and assuming that humidity is at the saturation level at the surface, $e_o = e_s(T_0)$. The latent heat flux is:

$$Q_e = -\rho_a L_E C_E \frac{0.622 e_s(T_a)}{p_a}\left[(1-R) - \frac{L_E}{R_v T_a^2}(T_a - T_0)\right]U_a \tag{A2}$$

where ρ_a is air density, L_E is latent heat of evaporation (water surface) or sublimation (ice surface), C_E is turbulent transfer coefficient of moisture, p_a is air pressure, and R_v is the gas constant of dry air. Combining all together, the coefficients K_0 and K_1 become

$$K_0 = (1-\alpha)Q_s - \varepsilon_0\sigma(1-\varepsilon_a)T_a^4 - \rho_a L_E C_E \frac{0.622 e_s(T_a)}{p_a}(1-R)U_a \tag{A3}$$

$$K_1 = 4\varepsilon_0\sigma T_a^3 + \rho_a c_a C_H U_a + \rho_a L_E C_E \frac{0.622 e_s(T_a)}{p_a}\frac{L_E}{R_v T_a^2}U_a \tag{A4}$$

These expressions look complicated, but the resulting climatological values are rather smooth with seasonal variation coming from solar radiation and latent heat flux.

References

1. Kirillin, G.; Leppäranta, M.; Terzhevik, A.; Granin, N.; Bernhardt, J.; Engelhardt, C.; Efremova, T.; Golosov, S.; Palshin, N.; Sherstyankin, P.; et al. Physics of seasonally ice-covered lakes: A review. *Aquat. Sci.* **2012**, *74*, 659–682. [CrossRef]
2. Leppäranta, M. *Freezing of Lakes and Evolution of Their Ice Cover*, 2nd ed.; Springer: Berlin/Heidelberg, Germany, 2022.
3. Korhonen, J. Long-term changes in lake ice cover in Finland. *Nord. Hydrol.* **2006**, *37*, 347–363. [CrossRef]
4. Leppäranta, M. Interpretation of statistics of lake ice time series for climate variability. *Nord. Hydrol.* **2014**, *45*, 673–683. [CrossRef]
5. Salonen, K.; Leppäranta, M.; Viljanen, M.; Gulati, R.D. Perspectives in winter limnology: Closing the annual cycle of freezing lakes. *Aquat. Ecol.* **2009**, *43*, 609–616. [CrossRef]
6. Bates, R.E.; Bilello, M.A. *Defining the Cold Regions of the Northern Hemisphere*; Technical Report 178; US Army Cold Regions Research and Engineering Laboratory: Hanover, NH, USA, 1966.
7. Magnuson, J.J.; Robertson, D.M.; Benson, B.J.; Wynne, R.H.; Livingstone, D.M.; Arai, T.; Assel, R.A.; Barry, R.G.; Card, V.; Kuusisto, E. Historical trends in lake and river ice cover in the Northern Hemisphere. *Science* **2000**, *289*, 1743–1746. [CrossRef]
8. Yao, X.; Long, L.I.; Zhao, J.; Sun, M.; Jing, L.I.; Gong, P.; Lina, A.N. Spatial-temporal variations of lake ice phenology in the Hoh Xil region from 2000 to 2011. *J. Geogr. Sci.* **2015**, *31*, 70–82. [CrossRef]
9. Karetnikov, S.; Leppäranta, M.; Montonen, A. A time series of over 100years of ice seasons on Lake Ladoga. *J. Great Lakes Res.* **2017**, *43*, 979–988. [CrossRef]
10. Sharma, S.; Blagrave, K.; Magnuson, J.J.; O'Reilly, C.M.; Oliver, S.; Batt, R.D.; Magee, M.R.; Straile, D.; Weyhenmeyer, G.A.; Winslow, L. Widespread loss of lake ice around the Northern Hemisphere in a warming world. *Nat. Clim. Chang.* **2019**, *9*, 227–231. [CrossRef]
11. Kropacek, J.; Maussion, F.; Chen, F.; Hoerz, S.; Hochschild, V. Analysis of ice phenology of lakes on the Tibetan Plateau from MODIS data. *Cryosphere* **2013**, *7*, 287–301. [CrossRef]
12. Cai, Y.; Ke, C.Q.; Duan, Z. Monitoring ice variations in Qinghai Lake from 1979 to 2016 using passive microwave remote sensing data. *Sci. Total Environ.* **2017**, *607–608*, 120–131. [CrossRef]
13. Cai, Y.; Ke, C.Q.; Li, X.; Zhang, G.; Duan, Z.; Lee, H. Variations of lake ice phenology on the Tibetan Plateau from 2001 to 2017 based on MODIS data. *J. Geophys. Res.* **2019**, *124*, 825–843. [CrossRef]
14. Mishra, V.; Cherkauer, K.A.; Bowling, L.C.; Huber, M. Lake ice phenology of small lakes: Impacts of climate variability in the Great Lakes region. *Glob. Planet. Chang.* **2011**, *76*, 166–185. [CrossRef]

15. Bernhardt, J.; Engelhardt, C.; Kirillin, G.; Matschullat, J. Lake ice phenology in Berlin-Brandenburg from 1947–2007: Observations and model hindcasts. *Clim. Chang.* **2012**, *112*, 791–817. [CrossRef]
16. Wang, C.; Shirasawa, K.; Leppäranta, M.; Ishikawa, M.; Huttunen, O.; Takatsuka, T. Solar radiation and ice heat budget during winter 2002–2003 in Lake Pääjärvi, Finland. *Verh. Int. Verein Limnol.* **2005**, *29*, 414–417. [CrossRef]
17. Leppäranta, M.; Lindgren, E.; Shirasawa, K. The heat budget of Lake Kilpisjrvi in the Arctic tundra. *Hydrol. Res.* **2016**, *48*, 969–980. [CrossRef]
18. Mironov, D.; Ritter, B.; Schulz, J.-P.; Buchhold, M.; Lange, M.; Machulskaya, E. Parameterisation of sea and lake ice in numerical weather prediction models of the German Weather Service. *Tellus A* **2012**, *64*, 17330. [CrossRef]
19. Huang, W.F.; Cheng, B.; Zhang, J.R.; Zhang, Z.; Vihma, T.; Li, Z.J.; Niu, F.J. Modeling experiments on seasonal lake ice mass and energy balance in the Qinghai-Tibet Plateau: A case study. *Hydrol. Earth Syst. Sci.* **2019**, *23*, 2173–2186. [CrossRef]
20. Song, S.; Li, C.Y.; Shi, X.H.; Zhao, S.N.; Tian, W.D.; Li, Z.J.; Bai, Y.L.; Cao, X.W.; Wang, Q.K.; Huotari, J.; et al. Under-ice metabolism in a shallow lake in a cold and arid climate. *Freshw. Biol.* **2019**, *64*, 1710–1720. [CrossRef]
21. Jevrejeva, S.; Drabkin, V.; Kostjukov, J.; Lebedev, A.; Leppäranta, M.; Mironov, Y.U.; Schmelzer, N.; Sztobryn, M. Baltic Sea ice seasons in the twentieth century. *Clim. Res.* **2004**, *25*, 217–227. [CrossRef]
22. Leppäranta, M.; Lindgren, E.; Wen, L.; Kirillin, G. Ice cover decay and heat balance in Lake Kilpisjrvi in Arctic tundra. *J. Limnol.* **2019**, *78*, 163–175. [CrossRef]
23. Rodhe, B. On the relation between air temperature and ice formation in the Baltic. *Geogr. Ann.* **1952**, *34*, 175–202.
24. Barnes, H.T. *Ice Engineering*; Montreal Renouf Publishing Co.: Montreal, QC, Canada, 1928; pp. 1–364.
25. Mueller, D.R.; Hove, P.V.; Antoniades, D.; Jeffries, M.O.; Vincent, W.F. High Arctic lakes as sentinel ecosystems: Cascading regime shifts in climate, ice cover, and mixing. *Limnol. Oceanogr.* **2009**, *54*, 2371–2385. [CrossRef]
26. Zhang, T.; Jeffries, M.O. Modeling interdecadal variations of lake-ice thickness and sensitivity to climatic change in northernmost Alaska. *Ann. Glaciol.* **2000**, *31*, 339–347. [CrossRef]
27. Ashton, G.D. *Dynamics of Snow and Ice Masses*; Academic Press: New York, NY, USA, 1980; pp. 261–304.
28. Barańczuk, K.; Barańczuk, J. The ice regime of Lake Ostrzyckie (Kashubian Lakeland, northern Poland). *Limnol. Rev.* **2019**, *19*, 105–112. [CrossRef]
29. Huang, W.; Zhang, Z.; Li, Z.; Leppäranta, M.; Arvola, L.; Song, S.; Lin, Z. Under-ice dissolved oxygen and metabolism dynamics in a shallow lake: The critical role of ice and snow. *Water Resour. Res.* **2021**, *57*, e2020WR027990. [CrossRef]
30. Mietus, M. *The Climate of the Baltic Sea Basin*; WMO/TD-No. 933; MMROA Report-No. 41; World Meteorological Organization: Geneva, Switzerland, 1998.
31. Yang, F.; Li, C.Y.; Shi, X.H.; Zhao, S.N.; Hao, Y.Z. Impact of seasonal ice structure characteristics on ice cover impurity distributions in Lake Ulansuhai. *J. Lake Sci.* **2016**, *28*, 455–462.
32. Kouraev, A.V.; Semovski, S.V.; Shimaraev, M.N.; Mognard, N.M.; Legrésy, B.; Rémy, F. The ice regime of Lake Baikal from historical and satellite data: Relationship to air temperature, dynamical, and other factors. *Limnol. Oceanogr.* **2007**, *52*, 1268–1286. [CrossRef]
33. Qiu, Y.; Guo, H.; Ruan, Y.; Fu, X.; Shi, L.; Tian, B. *A Dataset of Microwave Brightness Temperature and Freeze-Thaw for Medium-to-Large Lakes over the High Asia Region 2002–2016*; Science Data Bank: Beijing, China, 2017. [CrossRef]
34. Nolan, M.; Liston, G.; Prokein, P.; Brigham-Grette, J.; Sharpton, V.L.; Huntzinger, R. Analysis of lake ice dynamics and morphology on Lake El'gygytgyn, NE Siberia, using synthetic aperture radar (SAR) and Landsat. *J. Geophys. Res.* **2002**, *107*, ALT 3-1–ALT 3-12. [CrossRef]

water

Article

Monitoring Ice Phenology in Lake Wetlands Based on Optical Satellite Data: A Case Study of Wuliangsu Lake

Puzhen Huo [1], Peng Lu [1,*], Bin Cheng [2], Limin Zhang [1], Qingkai Wang [1] and Zhijun Li [1,*]

[1] State Key Laboratory of Coastal and Offshore Engineering, Dalian University of Technology, Dalian 116024, China
[2] Finnish Meteorological Institute, 00101 Helsinki, Finland
* Correspondence: lupeng@dlut.edu.cn (P.L.); lizhijun@dlut.edu.cn (Z.L.)

Abstract: It is challenging to obtain the ice phenology for a lake covered with a vast area of aquatic (shallow lake wetlands) using optical satellite data because possible clouds above the lake could contaminate the result. We developed a new method to tackle this challenge. Our target was Wuliangsu Lake, a large (330 km^2) and shallow (1.6 m average depth) lake wetland in the Inner Mongolia Plateau. We used Landsat and Sentinel-2 imageries to extract the lake water boundary. The MOD09GQ/MYD09GQ dataset, having the highest spatial resolution among MODIS reflectivity products, was first selected to differentiate water and ice pixels. Then, we used the reflectivity state parameters containing cloud information in the dataset to filter out the cloud pixels. The ice phenology characteristics, such as freeze-up, break-up dates, and ice cover duration (ICD) between 2013 and 2022 were obtained. We further applied the air temperature correction technique to remove the outliers. The average of ICD in Wuliangsu Lake was about 127 ± 6 days. The freeze-up start and break-up end occurred on 17 November ± 5 days and 25 March ± 4 days, respectively. The remote sensing results agree well with the field observation, with a mean absolute error of 2 days. The algorithm can effectively remove the influence of aquatic plants and clouds on lake ice identification, thereby satisfying the needs of daily monitoring and ice phenology research in the lake wetlands.

Keywords: lake; wetlands; ice phenology; optical satellite data; cloud; Wuliangsu Lake

Citation: Huo, P.; Lu, P.; Cheng, B.; Zhang, L.; Wang, Q.; Li, Z. Monitoring Ice Phenology in Lake Wetlands Based on Optical Satellite Data: A Case Study of Wuliangsu Lake. *Water* **2022**, *14*, 3307. https://doi.org/10.3390/w14203307

Academic Editor: Riccardo Scalenghe

Received: 4 September 2022
Accepted: 18 October 2022
Published: 19 October 2022

1. Introduction

As a critical ecosystem in the world, the lake wetlands are uniquely crucial to water supply, water purification, water storage, biodiversity maintenance, and local climate adaptation. Lakes are essential to ensuring global water ecological security patterns [1,2]. Compared with tropical regions, lakes in middle and high latitudes freeze in winter. Ice serves as a critical medium for the mass and energy exchange between air and water, further maintaining the local climate and environmental stability [3,4]. Especially in arid regions, ice-covered period lakes affect local ecological and economic development and impact human activities. For example, lake freezing reduces evaporation from lakes, limits greenhouse gas production in the water bodies due to biological and chemical processes, and increases recreational activities (such as ice skating and winter swimming) [5–7]. Ice phenology studies in China are relatively rare and mainly focused on the Qinghai-Tibet Plateau compared to the abundant publications on boreal lakes [8,9]. While few studies on lake wetlands in the Inner Mongolia Plateau exist, they are entirely different from most of the aforementioned areas. Taking Wuliangsu Lake as an example, it has extensive aquatic plants, characteristic of a shallow lake. High incident solar radiation and low precipitation make the lake more sensitive to climate fluctuations, further affecting the ice phenology characteristics (such as ice cover duration, freeze-up start, and break-up end dates). These ice phenology data are good indicators of local climate change and can be widely employed in various climate studies [10,11].

Accurate records of lake ice phenology are essential to better understanding the relationship between climate change and the local environment. Field observations provide a certain amount of ice phenology data but have many limitations due to objective factors, such as logistical support difficulties and poor temporal continuity. In contrast, satellite remote sensing can cover much broader spatial and temporal ranges; therefore, it is used in lake ice phenology studies [12,13].

Active microwave sensors provide high spatial resolution (20–100 m). However, their lower temporal resolution (>3 days) limits their ability to monitor surface cover types of variations of lakes. For some shallow lakes with a short opening/closing period of ice cover, many errors would be introduced when extracting ice phenology dates [14]. Similarly, passive microwave sensors are characterized by high temporal resolution (mostly twice a day or better). Still, their lower spatial resolution (2–25 km) is only adapted to detecting large water bodies [15–17]. During pixels classification, significant errors also exist in investigating wetlands with narrow lake shoreline shapes and lush aquatic plants. In contrast, optical sensors, such as the Moderate Resolution Imaging Spectroradiometer (MODIS), are available for daily monitoring of surface cover types of lake wetland with temporal resolution (1–8 days) and spatial resolution (250–1000 m). MODIS surface reflectance products have been used to characterize lake ice phenology in harsh environments such as the Arctic and the Qinghai-Tibet Plateau [18–20]. Elsewhere, Williamson et al. [21] combined Sentinel-2 and Landsat-8 optical imageries to calculate the area, depth, and volume of supraglacial lakes in Greenland, and identified lakes that drain rapidly ($\leq$ 4 days) using an automated algorithm. In addition, MODIS daily temperature products, and MODIS surface reflectance products combined with random forests have similarly demonstrated robust accuracy in lake ice phenology studies [22,23].

Although optical sensors have been successfully used to study lake evolution, their application in lake wetland ice phenology is still complicated. For example, determining the lake wetland boundary is the first difficulty because there are plenty of reeds and other aquatic plants on the lake water boundary. Although lake boundaries from databases such as HydroLAKES [24] have been widely used [13,25], it is cumbersome to manually distinguish them one by one due to the large data. Moreover, there is a significant interannual variation in the growth of reeds. A fixed lake water boundary would cause errors in the statistics of lake water areas, thereby increasing the uncertainty of pixel classification. Usually, the opening/closing period of ice cover is transient, and more pixel classification errors would miss capturing ice phenology dates. Therefore, to obtain accurate ice phenology in the wetland of a shallow lake, the influence of aquatic plants must be removed before accurately identifying the water/ice extent.

Moreover, clouds inevitably affect optical sensors, leading to the misjudgment of lake surface cover types. Tai et al. [26] used MODIS and Landsat products to calculate the ice phenology of Lake Selincuo over the past 20 years. The impact of clouds was removed by visual interpretation and culling of cloud-covered pixels, which would lose some of the pixels, leading to errors in the extracted ice phenology data.

To solve the existing problems in studying lake wetland ice phenology, this paper proposes a new algorithm. First of all, Landsat and Sentinel-2 products are used to separate the extent of aquatic plants, thus accurately extracting the lake water boundary. Secondly, we used the dynamic threshold method to cyclically extract water and ice pixels from the MODIS reflectance product to determine the optimal threshold that distinguishes water and ice. Thirdly, we combined four MODIS datasets to remove and re-fill cloud pixels. Finally, we removed outliers in the time series of water/ice coverage through air temperature correction. Wuliangsu Lake of Inner Mongolia was taken as an example in this study, the algorithm's reliability was verified by comparison with field observations for future studies on the interaction between lake wetlands and the atmosphere, and the variations and trends of ice phenology in the Wuliangsu Lake were obtained according to the algorithm for the last ten years.

2. Study Area and Data

2.1. Study Area

Wuliangsu Lake (40°46′–41°7′ N, 108°41′–108°8′ E), located in Bayannur City, Inner Mongolia, China (Figure 1), is one of the eight major freshwater lakes in China. It is the largest lake wetland in the world at the same latitude. The lake has an elevation, total area, and water area of about 1018 m, 330 km^2, and 130 km^2, respectively. The water depth is 1.2–2.7 m, averaging 1.6 m. Wuliangsu Lake is a semi-dry and humid mid-temperate climate region, with an annual mean air temperature of 7.5 °C and relative humidity of 49.5% [27,28]. The ice-covered period usually lasts from mid-to-late November to late March. It is a lake of interest to climate and ecological researchers because of its typical shallow lake characteristics, lush aquatic plants, nearly 200 species of birds, and over 20 species of fish thriving there [27,29,30]. However, there is still a lack of long-term interannual variations in lake ice phenology, and this paper focuses on ice phenology from 2013 to 2022. As with Cai et al. [20], the range of each year is defined from 1 August to 30 July. For example, 2021 denotes 1 August 2020–30 July 2021.

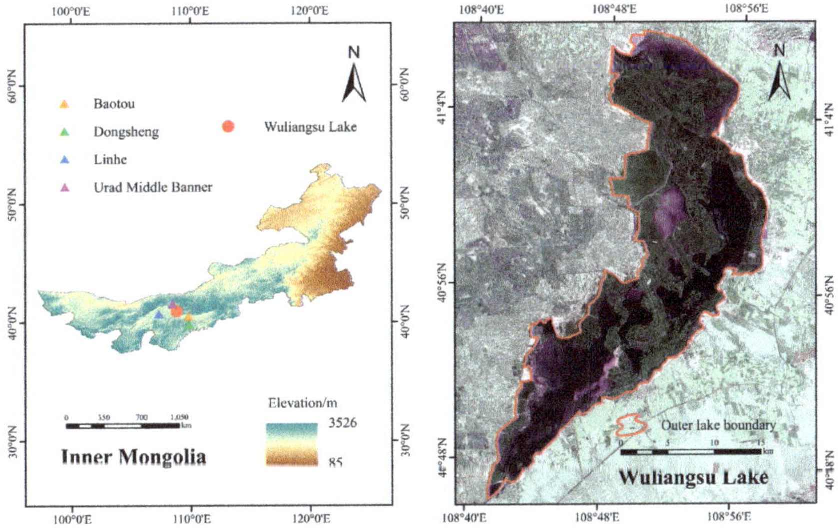

Figure 1. Location of the Wuliangsu Lake (the image on the right is the Sentinel-2's imagery on 2 November 2020, the outer lake boundary contains lake water and extensive aquatic plants).

2.2. Optical Satellite Data

2.2.1. MODIS Data

The National Aeronautics and Space Administration (NASA) carried the same MODIS sensor in the Terra and Aqua satellites launched on 18 December 1999 and 4 March 2002, respectively. The Terra and Aqua satellites crossed the equator at 10:30 am and 1:30 pm local time, respectively, resulting in different satellite altitude angles, relative azimuths, and cloudiness. The National Aeronautics and Space Administration Land Surface Distributed Data Center (https://ladsweb.modaps.eosdis.nasa.gov/ (accessed on 6 May 2022)) provides 2013–2022 reflectance products for the h26v04 region, which contains the complete Wuliangsu Lake. Downloaded HDF files were converted to a WGS84 geographic coordinate system and visualized in GEOTIF format. In the MODIS product, the MOD09GQ and MYD09GQ datasets provide information in the red band with 250 m spatial resolution, while the MOD09GA and MYD09GA datasets provide cloud information in the state assessment state_1km_1 parameter.

The MOD09GQ dataset was chosen because it has the highest spatial resolution among all MODIS reflectivity products, satisfying the requirements of dealing with the abundant reeds in Wuliangsu Lake and the long and narrow shape of the lake. Secondly, the temporal resolution of 1 day can satisfy the daily calculation of water/ice coverage and significantly improve variation continuity in lake surface cover types if the MYD09GQ dataset is combined to make up the vacant data. Finally, the classification by threshold of the red band in the MOD09GQ dataset can accurately extract the water/ice extent of the lake wetland.

2.2.2. Landsat and Sentinel-2 Data

To obtain the accurate lake water boundary and water/ice area, optical satellite products with higher spatial resolution than MODIS products are required. In total, 279 Landsat-7, Landsat-8, and Sentinel-2 optical imageries were downloaded during 2013–2022, and each imagery was screened individually to ensure no or minimal cloud ($\leq$30%). The spatial resolutions of Landsat and Sentinel-2 imageries were 30 m and 10 m, respectively. In addition, the onboard scan line corrector of Landsat-7 failed on May 31, 2003, resulting in missing data strips in subsequent images. Mask function and statistics function are used to make up for missing values through ArcGIS.

2.3. Meteorological Data

To accurately match the meteorological data of the ice-covered period, we selected the meteorological station data from the Dataset of Daily Climate Data from Chinese Surface Stations for Global Exchange. The data were released by the National Meteorological Information Center (NMIC) of the China Meteorological Administration (CMA) (http://data.cma.cn (accessed on 6 May 2022)). Four proximate meteorological stations measured these data at Baotou, Dongsheng, Linhe, and Urad Middle Banner (Figure 1). The daily mean, maximum, and minimum air temperatures of Wuliangsu Lake were obtained using the daily meteorological data according to the inverse distance weighted method. To verify the calculation, field observations from 2016 to 2018 were employed [27], using a PTWD sensor with a −40–80 °C measurement range and a ±0.04 °C precision. Furthermore, the comparison between them (Figure 2) shows that the mean absolute error (*MAE*) was 0.93°C, and the correlation coefficient (*R*) equaled = 0.98 for the daily mean air temperature. It indicates that the data can represent the natural meteorological conditions in Wuliangsu Lake.

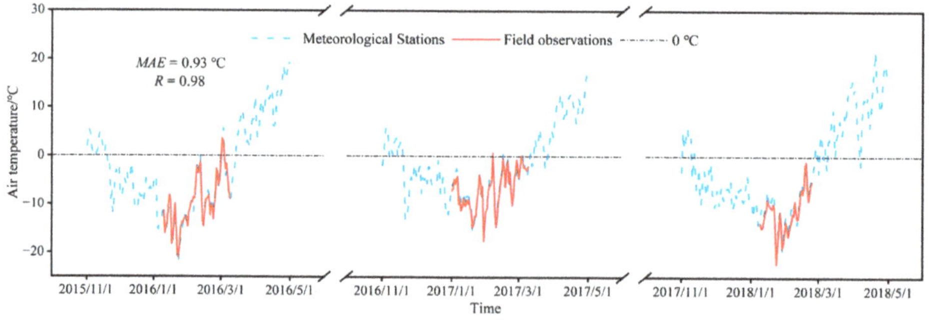

Figure 2. Comparison of daily mean air temperature between meteorological stations and field observations from 2016 to 2018. (*MAE* is the mean absolute error, *R* is the correlation coefficient).

3. Methods

3.1. Lake Ice Phenology Extraction

The algorithm for obtaining the lake wetland ice phenology can be divided into five steps (Figure 3). Step 1: The lake water boundary in Landsat and Sentinel-2 imageries are extracted. Step 2: The water/ice pixels within the lakes observed in MODIS, Landsat, and Sentinel-2 imageries are extracted. Step 3: The extracted results of MODIS products are processed to fill the cloud-covered pixels. Step 4: MODIS imagery classification results are compared with those of Landsat and Sentinel-2, with a higher spatial resolution to determine the optimal threshold for distinguishing water/ice pixels. Step 5: The ice phenology data are obtained from the time series of cloud/water/ice coverage after air temperature correction.

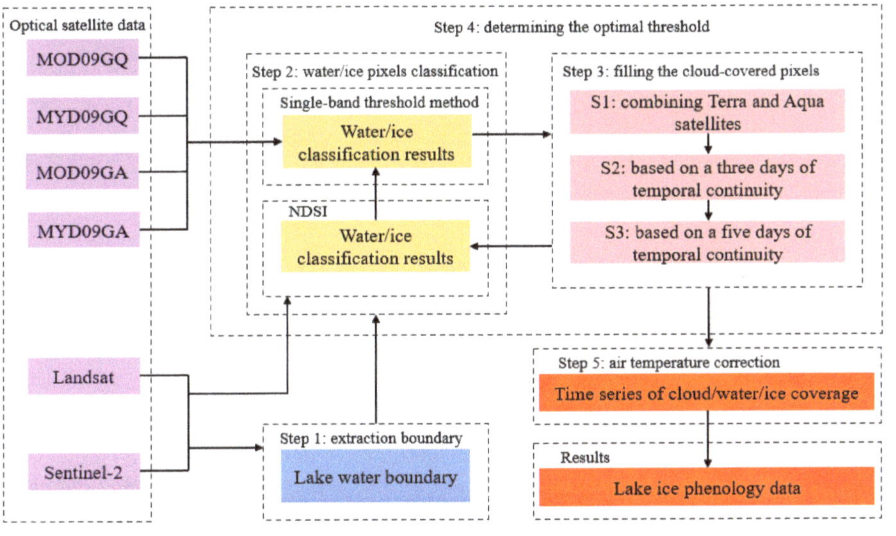

Figure 3. Algorithm flow chart.

3.1.1. Extraction of Lake Water Boundary

Considering the interannual variation in aquatic plant extent, the lake water boundaries must be extracted for each year based on Landsat and Sentinel-2 imageries. We employed the Normalized Difference Water Index (NDWI) formula proposed by McFeeters [31]. It takes advantage of the fact that soil and vegetation usually have high near-infrared (NIR) reflectance, strongly absorbed by water. The green band improves the reflectance of water and can meet the need to distinguish reeds from water bodies.

Using Equation (1):

$$\mathrm{NDWI} = \frac{\text{Band 2} - \text{Band 4}}{\text{Band 2} + \text{Band 4}} \tag{1}$$

where Band 2 and Band 4 represent the reflectance of the green and NIR bands, respectively, in Landsat and Sentinel-2 products; the NDWI values range from -1 to 1; and a positive value represents a lake water extent that should be checked visually to minimize the influence of lakeshore pixels on water/ice discrimination. A comparison of the lake water boundary between the HydroLAKES dataset and the result of Equation (1) is shown in Figure 4. We observed that Equation (1) could effectively annul the influence of wetland aquatic plants on the lake water boundary and validly retain the valuable pixels in the lake water with higher accuracy. In contrast, the lake water boundary of the HydroLAKES dataset contains many aquatic plants. It is also less accurate at the intersection with land, leading to incorrect pixel statistics.

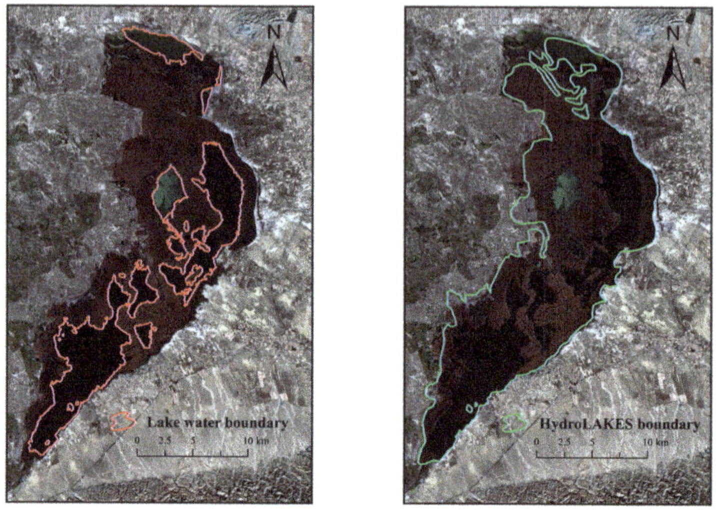

Figure 4. Comparison of the lake water boundary between the HydroLAKES dataset and the result of the NDWI formula in Wuliangsu Lake (the Landsat-8 image on 7 November 2013 was the selected example).

3.1.2. Extraction of Water and Ice Pixels

First of all, within the boundary of lake water, the single-band threshold was employed to extract water/ice pixels. Meanwhile, the reflectance band quality parameter QC_250M_1 in the MOD09GQ dataset can eliminate error pixels and categorize them into clouds, which can effectively use the information in the MOD09GQ dataset and ensure the accuracy of the threshold. The formula is as follows:

$$Pixels = \begin{cases} \text{ice}, & \text{Band 1} > a \\ \text{water}, & \text{Band 1} < a \end{cases} \quad (2)$$

where Band 1 represents the reflectance of the red band in the MOD09GQ dataset and a represents the threshold to distinguish between water and ice.

Then, we employed the Normalized Difference Snow Index (NDSI) formula for Landsat and Sentinel-2 products [32]. The formula is as follows:

$$\text{NDSI} = \frac{\text{Band 2} - \text{Band 5}}{\text{Band 2} + \text{Band 5}} \quad (3)$$

where Band 2 and Band 5 represent the reflectance of the green and shortwave infrared bands, respectively, in Landsat and Sentinel-2 products. NDSI of = 0.4 was selected as the threshold [20]. The vector boundary of water/ice was extracted with manual visual judgment, and the water/ice coverage in Landsat/Sentinel-2 imageries could be calculated.

Finally, the dynamic threshold method was employed to constantly modify the threshold a to change the water/ice classification results of the MOD09GQ dataset, and compared with the water/ice coverage area of Landsat and Sentinel-2 products with high spatial resolution to find the optimal threshold a.

3.1.3. Cloud Removal by Gap Filling

It is crucial to remove cloud pixels, which always affect optical imageries. The reflectivity band state parameter state_1 km_1 of MOD09GA and MYD09GA datasets, a 16-bit binary data with a spatial resolution of 1 km, was employed here. To retain the valuable pixels in the water bodies as much as possible, the cloud mixed and cloud shadow pixels were not considered in this paper. The specific method is similar to that of Zhang et al. [13]. The state_1km_1 value was first interpolated into a 250 m resolution image. Then, the cloud pixel was extracted by coding to synthesize a mask used to remove the cloud pixels in the daily corresponding Band 1 image. Zhang et al. [13] verified the method with 296 lakes, achieving excellent results. However, the method only removed the cloud pixels and did not fill the gaps under the removed clouds, thereby losing some relevant information such as extracting ice phenology. Therefore, a further filling process, i.e., cloud-covered pixel filling, was performed after cloud identification.

To improve the cloud gap filling accuracy, MOD09GQ from Terra satellite and MYD09GQ from the Aqua satellite were combined. For the same day, if the two satellite products were different in pixel classifications, the results of the Aqua satellite were reliable. The comparison with Landsat and Sentinel-2 products showed that the accuracy of Aqua was 71.4%, higher than 28.6% of Terra. The first step to fill the gaps (S1 in Figure 3) is shown in Equation (4), where P denotes pixels and output is the pixel type (water/ice) of the output satellite. It is divided into two cases: (1) if Aqua data are identified as water (ice) at A pixel, A pixel is defined as water (ice); (2) if Aqua data are identified as cloud at A pixel and Terra satellite is identified as water (ice) at A pixel, A pixel is defined as water (ice).

$$P_{(A,\text{output})} = \begin{cases} \text{water/ice, if } P_{(A,\text{Aqua})} = \text{water/ice} \\ \text{water/ice, if } P_{(A,\text{Aqua})} = \text{cloud AND } P_{(A,\text{Terra})} = \text{water/ice} \end{cases} \quad (4)$$

Equation (4) can fill some but not all the cloud pixels' gaps because the MOD09GQ and MYD09GQ datasets have missing values. Therefore, the second step uses the temporal continuity principle (S2 in Figure 3) to associate the type of A pixel on day t with the same pixel on days $t-1$ and $t+1$ [33]. The detail is if the A pixel on day t is a cloud pixel and is discriminated as water (ice) on days $t-1$ and $t+1$, then the pixel on day t is defined as water (ice). The formula is as follows:

$$P_{(A,t)} = \text{water/ice, if } P_{(A,t-1)} = P_{(A,t+1)} = \text{water/ice} \quad (5)$$

Some cloud pixels that exist after applying Equation (5) are further calculated based on the five days of temporal continuity [33] (S3 in Figure 3). As shown in Equation (6), there are two cases: (1) if the A pixel is cloud on days $t-1$ and t, and is water (ice) on days $t-2$ and $t+1$, then the A pixel is defined as water (ice) on days $t-1$ and t; (2) if the A pixel is cloud on days t and $t+1$, and is water (ice) on days $t-1$ and $t+2$, then the A pixel is defined as water (ice) on days t and $t+1$.

$$\begin{cases} P_{(A,t)} = P_{(A,t-1)} = \text{water/ice, if } P_{(A,t-2)} = P_{(A,t+1)} = \text{water/ice} \\ P_{(A,t)} = P_{(A,t+1)} = \text{water/ice, if } P_{(A,t+2)} = P_{(A,t-1)} = \text{water/ice} \end{cases} \quad (6)$$

3.1.4. Air Temperature Calibration

Once pixel types within the lake water boundary are defined, the number of water/ice pixels can be counted. The variations in the water/ice coverage can be obtained by analyzing the time series of satellite data. However, there are still some uncertain pixels despite the reduction of most of the cloud pixels. In addition, thin ice and mixed pixels on the lakeshore are unstable factors to distinguish between water/ice pixels. Therefore, air temperature removes outliers in the time series of water/ice coverage [13]. The basic principle is that the ice coverage on day t ($I_{(t)}$) in the lake should not decrease when the mean air temperature on day t ($T_{(t)}$) decreases, and $I_{(t)}$ should not increase when $T_{(t)}$ increases. However, the mean air temperature of a particular day cannot be chosen because it is not representative, rather it fluctuates frequently and greatly. Therefore, the mean air temperature within several days (2–50 days) before day t was calculated, and the highest correlation coefficient (−0.83) was found between the mean air temperature within 30 days before day t ($T_{(t,30m)}$)

and $I_{(t)}$. Hence, $T_{(t,30m)}$ was used as a dynamic air temperature correction to compare with the $T_{(t)}$ on day t, and then $I_{(t)}$ can be corrected according to the following equation:

$$\begin{cases} I_{(t)} = I_{(t-1)}, & \text{if } T_{(t)} < T_{(t,30m)} \text{ AND } I_{(t)} < I_{(t-1)} \\ I_{(t)} = I_{(t-1)}, & \text{if } T_{(t)} > T_{(t,30m)} \text{ AND } I_{(t)} > I_{(t-1)} \end{cases} \quad (7)$$

After correcting $I_{(t)}$, the water coverage can be determined by $W_{(t)} = 1 - I_{(t)} - C_{(t)}$, where $C_{(t)}$ is the cloud coverage on day t. To preserve the original pixel type as much as possible, Equation (7) was corrected only for the $I_{(t)}$ from November to March.

3.1.5. Extraction of Lake Ice Phenology

The $I_{(t)}$ during the ice-covered period is usually close to 1. The cloud pixels and mixed pixels also cause large fluctuations in $I_{(t)}$, which further affect the accuracy of the automatic extraction of ice phenology. Therefore, unlike a previous study [34] that used $I_{(t)}$ as the criterion, this paper used the $W_{(t)}$ to judge ice phenological dates. As shown in Equation (8), the freeze-up start (FUS) and freeze-up end (FUE) are defined as the date when $W_{(t)}$ first fell below 0.8 and 0.2 [35], respectively. Correspondingly, the break-up start (BUS) and break-up end (BUE) are defined as the date when $W_{(t)}$ was again above 0.2 and 0.8, respectively. Four periods can be calculated based on the four ice phenology dates. The freezing-up duration (FUD) is the duration process of reducing the $W_{(t)}$ in the lake (FUE–FUS). Corresponding to FUD, we chose BUE-BUS to calculate the break-up duration (BUD), while the ice cover duration (ICD) equaled BUE-FUS. The complete freezing duration (CFD) [36] is the period the lake surface is completely frozen; it equals BUS-FUE. In addition, to avoid errors in the automatic extraction results from less-than-ideal quality pixels during the ice-free period, Equation (8) discriminates these dates as $T_{(t)}$ drops to 0 °C each year.

$$\begin{cases} \text{FUS} = \text{Min}(t), & \text{if } W_{(t)} < 0.8 \\ \text{FUE} = \text{Min}(t), & \text{if } W_{(t)} < 0.2 \\ \text{BUS} = \text{Max}(t), & \text{if } W_{(t)} > 0.2 \\ \text{BUE} = \text{Max}(t), & \text{if } W_{(t)} > 0.8 \end{cases} \quad (8)$$

3.2. Statistical Analysis

The linear regression was used to analyze the trends in the eight ice phenology characteristics of Wuliangsu Lake. Their trends were established for each characteristic. In addition, the mean absolute error (*MAE*), root mean square error (*RMSE*), and correlation coefficient (*R*) were calculated for comparisons of ice phenology dates for Wuliangsu Lake derived from filed observations and MODIS products.

4. Results

4.1. Determining the Optimal Threshold

The threshold was set at 0.09–0.17 with a 0.01 interval (Section 3.1.2). The water/ice pixels in the MODIS products were extracted cyclically according to the dynamic threshold and single-band threshold methods. The classification results were transformed into water/ice coverage and compared with those of selected Landsat and Sentinel-2 imageries in the freezing/thawing period. The results show that the error was smallest when the threshold a was 0.13 (Figure 5), at which the MAE = 5.84% and R = 0.98.

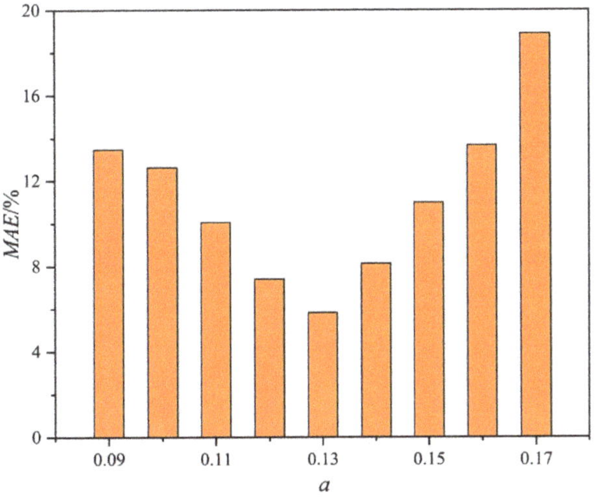

Figure 5. Mean absolute error (*MAE*) of the comparison between the water area extracted by MODIS product through the dynamic threshold *a* and that extracted by Landsat/Sentinel-2 product.

4.2. Cloud Removal and Filling

Figure 6 depicts the filling of the cloud-covered pixels using the threshold classification results according to Section 3.1.3. Before filling the cloud-covered pixels, the mean cloud coverage ($\overline{C}_{(t)}$) of Terra and Aqua satellite products was 43.41% in 2017. After combining the two satellite products (S1 in Figure 3), $\overline{C}_{(t)}$ reduced to 28.68%. Using the temporal continuity method in three days (S2 in Figure 3), $\overline{C}_{(t)}$ reduced again by 83.47%. Finally, $\overline{C}_{(t)}$ reduced to 4.39%, according to the temporal continuity method in five days (S3 in Figure 3). This three-step cloud removal process reduced the $\overline{C}_{(t)}$ by a total of 91.16%, which largely minimized the impact of clouds on water/ice classification results. It is noteworthy that the $C_{(t)}$ was still greater than 0.2 for several consecutive days after the complete cloud removal process (underlined box of Figure 6). The reason is that the calculation based on temporal continuity was needed in filling cloud-covered pixels for S2 and S3. Still, a higher $C_{(t)}$ for many consecutive days after S1 processing indirectly affected the cloud removal effect of S2 and S3. Therefore, if Aqua and Terra satellites are cloud-covered in the same area for many consecutive days, it would be difficult to remove the cloud and fill the area with natural surface categories.

Table 1 summarizes the cloud coverage of Wuliangsu Lake using MODIS data from October to next April between 2013 and 2022. We observed that 2016 had the highest cloud coverage during the study period after implementing S1 cloud removal. It is partly because the original $\overline{C}_{(t)}$ before the cloud removal process was higher in the study period (41.30% and 52.01%). In addition, some data were missing for the Terra satellite in the 2016 h26v04 region which affected the calculation of the complete time series. Although the S1 cloud removal result indirectly affected those of S2, S2 was still the most effective step among the cloud removal processes, achieving 75.82% cloud removals. After the complete cloud removal process, the lowest $\overline{C}_{(t)}$ was 2.95% in 2020, 8% lower in most years, thereby achieving a better filling cloud-covered pixels' effect (Table 1). At the same time, this process filled in the actual pixels in the dataset with quality limitations, which vastly increased the accuracy of the extracted ice phenology data.

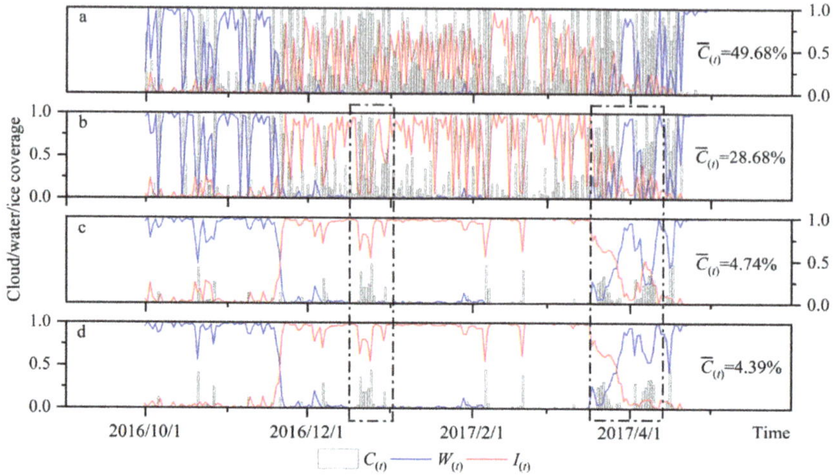

Figure 6. Cloud removal results for Wuliangsu Lake in 2017. (**a**) Original Aqua MODIS; (**b**) after S1 calculation; (**c**) after S2 calculation; (**d**) after S3 calculation ($C_{(t)}$, $W_{(t)}$, and $I_{(t)}$ denote cloud, water, and ice coverage, respectively; $\overline{C}_{(t)}$ denotes the mean cloud coverage. The underlined box marks high $C_{(t)}$).

Table 1. Comparison of cloud cover with and without cloud removal for Wuliangsu Lake from 2013 to 2022.

	Mean	Maximum	Minimum
Original Terra	38.67%	42.30% (2015)	31.54% (2014)
Original Aqua	47.71%	52.20% (2013)	38.72% (2020)
S1	28.83%	33.06% (2016)	22.99% (2014)
S2	6.97%	9.52% (2018)	3.19% (2020)
S3	6.42%	8.93% (2018)	2.95% (2020)

4.3. Air Temperature Correction

The cloud/water/ice coverage variations were obtained after filling the cloud-covered pixels. Although it is possible to clearly distinguish the ice phenology characteristics of Wuliangsu Lake in the last 10 years based on visual interpretation, unfilled cloud pixels and mixed pixels for the ice-covered period may lead to erroneous dates for the automatic extraction using Equation (8). Hence, an air temperature correction becomes necessary.

Taking 2017 as an example, Figure 7 compares cloud/water/ice coverage with and without temperature correction. Equation (8) automatically discriminated when $T_{(t)} < 0$ °C for the first time on 28 October 2016. As $T_{(t)}$ stabilized below 0 °C, $W_{(t)} < 0.8$ for the first time on 20 November 2016 and was, hence, extracted as the FUS date. When $T_{(t)}$ decreased rapidly and $W_{(t)}$ fell below 0.2 for the first time on 22 November 2016, the FUE date was extracted accordingly. Since $W_{(12/4)} = 21.26\%$ on 4 December 2016, it was automatically extracted as the BUS date (Equation (7)). Then, 30 March 2017 was the BUE date when $W_{(t)}$ was later higher than 0.8.

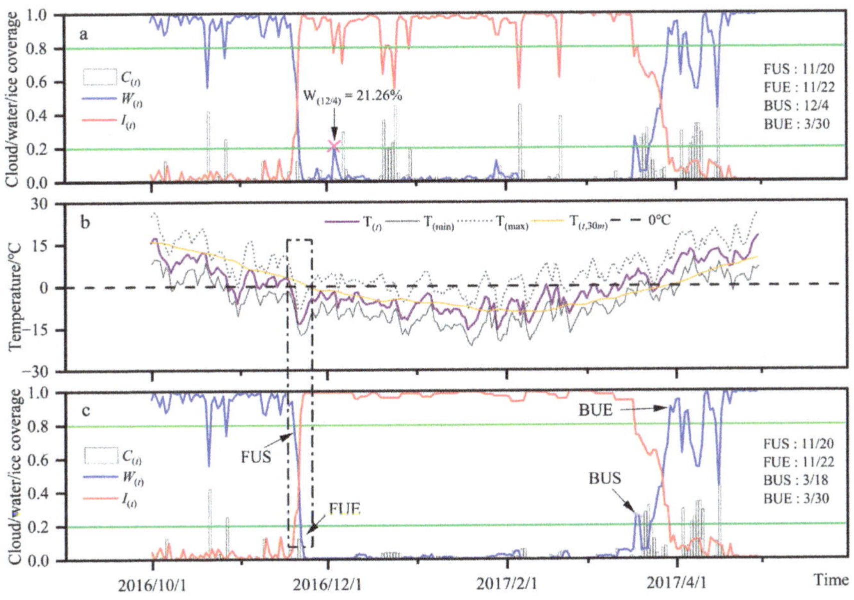

Figure 7. Daily cloud/water/ice coverage and air temperatures in Wuliangsu Lake in 2017. (**a**) Without temperature correction. (**b**) Daily mean air temperature ($T_{(t)}$), daily minimum air temperature ($T_{(\min)}$), daily maximum air temperature ($T_{(\max)}$), and mean air temperature of the previous 30 days ($T_{(t,30m)}$) after calculating at the meteorological stations near Wuliangsu Lake. (**c**) With air temperature correction. (The two green lines indicate the two thresholds for extracting ice phenology dates, and the four arrows from left to right indicate freeze-up start (FUS), freeze-up end (FUE), break-up start (BUS), and break-up end (BUE), respectively).

Among them, $W_{(t)}$ was high on 2016/12/4 due to the cloud shadow in the part of Wuliangsu Lake in the MOD09GQ and MYD09GQ datasets (Figure 7a). It resulted in a lower reflectance than normal in that region, automatically classified as water pixels through the threshold. It further caused a wrong ice phenology date using Equation (8). After air temperature correction for $I_{(t)}$ from November to March (Figure 7c), the $\overline{C}_{(t)}$ decreased from 4.39% to 2.96% in 2017. During the ice-covered period, the $C_{(t)}$ was approximately 0, and the sum of water-to-ice coverage was approximately 1. Thus, 18 March 2017 can be extracted accurately as the BUS dates. It proved the capability of air temperature correction, and selecting a stable $T_{(t,30m)}$ as the dynamic air temperature correction can effectively reduce the influence of cloud shadow and mixed pixels. This case would further improve the accuracy of Equation (8) for the automatic extraction of ice phenology dates.

4.4. Lake Wetland Ice Phenology

4.4.1. Ice Phenology

The ice phenology characteristics of Wuliangsu Lake during the study period are summarized in Table 2. On average, the Wuliangsu Lake started to freeze on 17 November, with the earliest FUS date being 9 November (2013, 2015) and the latest on 24 November (2014). The mean FUE date was 25 November, and the earliest FUE date was 16 November (2013), 15 days away from the latest FUE date. Wuliangsu Lake is shallow, generally having a short FUD (mean FUD = 7.1 days). The shortest FUD was two days, which occurred in 2017 (Figure 7). In addition, Wuliangsu Lake usually started melting on 25 March, with the earliest BUS appearing on 10 March (2013, 2021) and the latest on 25 March (2015). The CFD range and mean were 105–118 and 111.8 days, respectively. The average BUE date was 30 March, with the earliest and latest on 19 March (2021) and 30 March (2015, 2017), respectively. They resulted in a mean ICD of 126.5 days, while the longest ICD was 139 days in 2016.

Table 2. Ice phenology of Wuliangsu Lake between 2013 and 2022. (Date is expressed in mm/dd).

	FUS	FUE	BUS	BUE	FUD	BUD	CFD	ICD
2013	11/9	11/16	3/10	3/21	7	11	114	132
2014	11/22	11/28	3/17	3/24	6	7	108	121
2015	11/24	12/1	3/25	3/30	7	5	113	125
2016	11/9	11/24	3/22	3/29	15	7	117	139
2017	11/20	11/22	3/18	3/30	2	12	116	130
2018	11/17	11/27	3/14	3/23	10	9	106	125
2019	11/18	11/21	3/20	3/24	3	4	118	125
2020	11/19	11/24	3/18	3/22	5	4	113	122
2021	11/22	11/25	3/10	3/19	3	9	105	117
2022	11/14	11/27	3/16	3/24	13	8	108	129
Average	11/17	11/25	3/17	3/25	7.1	7.6	111.8	126.5
Max	11/24	12/1	3/25	3/30	15	12	118	139
Min	11/9	11/16	3/10	3/19	2	4	105	117

4.4.2. Verification of Ice Phenology Data

Field observations were conducted over four years (2016–2018, 2022) in Wuliangsu Lake, and the FUS, BUS, and BUE dates were recorded. Figure 8 depicts the comparisons between remote sensing results and field observation data. The $R = 0.99$, $MAE = 2.10$ days, and the root mean square error ($RMSE$) = 2.85 days. The agreement in Figure 8 argued for the feasibility of this study's algorithm. Compared with the results of Zhang et al. [12], the current study recorded a small error in the ice phenology of Wuliangsu Lake. This achievement is the main benefit of enhancing the accuracy of extracting lake boundaries and filling cloud-covered pixels, thus obtaining more accurate lake surface cover types.

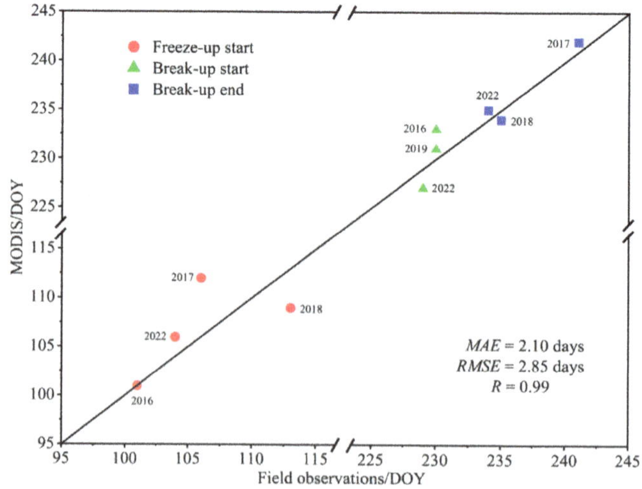

Figure 8. Comparisons of ice phenology dates for Wuliangsu Lake derived from filed observation and MODIS products (DOY means the day of the year relative to 1 August, R is the correlation coefficient, MAE is the mean absolute error, and $RMSE$ is the root mean square error).

5. Discussion

5.1. Error Analysis

5.1.1. Satellite Products

Step 4 in Figure 3 can obtain classification thresholds that are more suitable for the conditions of the water bodies in Wuliangsu Lake, but the error still existed after the optimal threshold $a = 0.13$ was selected (Figure 5). The reason is related to the spatio-temporal resolution of the satellite products. For example, in Figure 9, the lake water and ice areas extracted from Landsat-8/Sentinel-2 products on 2 March 2021 were about 129.56 and 125.58 km^2 based on the NDWI and NDSI formulas, respectively. It means the water area on 2 March 2021 was about 3.98 km^2. However, the water area extracted from the MODIS product was about 6.06 km^2 based on the threshold method, while the error in water coverage calculation was about 1.61%. This difference is mainly due to the complex shoreline shape of Wuliangsu Lake, and the inevitably adulterated mixed pixels with the 250 m spatial resolution in the MOD09GQ and MYD09GQ datasets (Figure 9c), despite the lake water boundary obtained using the NDWI formula. The same threshold classification and cloud removal process was conducted for the red band of the MOD09GA and MYD09GA datasets (Sections 3.1.2 and 3.1.3). Due to its 500 m spatial resolution, fewer pure pixels at the reed/ice boundary and more mixed water/ice pixels at the water/ice boundary resulted in more pixel misclassifications than the MOD09GQ and MYD09GQ datasets (Figure 9d). In addition, the opening/closing period of Wuliangsu Lake ice cover is usually less than 10 days, while the temporal resolution of Landsat-7 and Landsat-8 imageries is 16 days. Therefore, the open/close date cannot be captured annually, leading to a lack of optical imageries with higher spatial resolution, unconducive to comparing MODIS products. Although this paper combines Sentinel-2 imageries with a revisit period of 5 days, it still needs to consider the influence of clouds on the three optical satellite products. Thus, the quantity and quality of satellite images can cause errors in the statistical results after threshold classification.

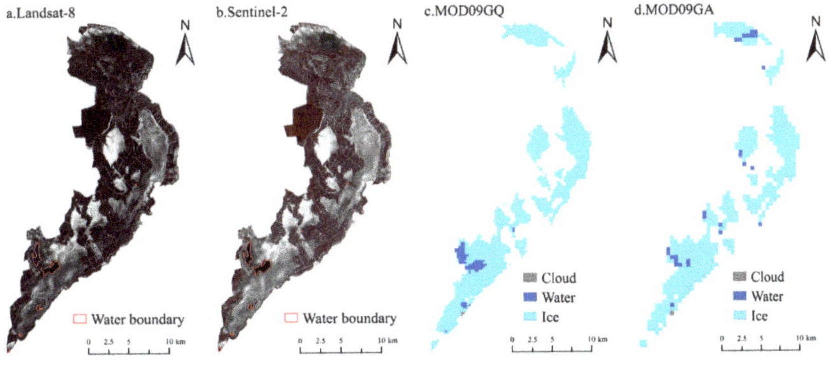

Figure 9. Comparisons of the water area extracted by Landsat-8, Sentinel-2, and MODIS products (selected for 2 March 2021).

5.1.2. Observation Principle

The residual difference between field observations and MODIS data can be attributed to the observation principle. The angle of view limits field observations, relying more on the individual subjective discrimination of lake surface cover types. In contrast, satellite remote sensing can more comprehensively and objectively assess the whole lake. However, it is limited by clouds, thin ice, and mixed pixels. Meanwhile, filling the cloud-covered pixels by spatial and temporal characteristics of the MODIS products maximizes the error of two days on the extracted ice phenology dates. This phenomenon occurs because pixel filling requires two days of previous and subsequent classification results (Equation (6)). In addition, the thresholds of $W_{(t)}$ for judging the ice phenology dates were 0.2 and 0.8, instead of 0 and 1. Thus, this results in an advance or delay in the extracted dates, such as later FUS and BUS dates, and earlier FUE and BUE dates, which further affect the calculation of CFD and ICD. More importantly, during the lake surface's freezing/thawing process, a "night freezing-day thawing" always occurs, affecting remote sensing and field results. For example, the absolute error of the FUS in 2017 was six days because the freezing–thawing cycle occurred at the beginning of the freezing period (Figure 7). The temporal resolution of MODIS products could not capture the variations on the lake surface within a day, and the field observers could not fully assess the overall surface condition of the lake, thus causing the maximum error during the comparison.

5.2. Trends of Ice Phenology

Figure 10 illustrates the linear trends of the ice phenology characteristics of Wuliangsu Lake. The FUS and the FUE dates show a delay trend (Figure 10a), with respective mean rates of 0.27 and 0.24 d/y. The BUS date is advanced by 0.26 d/y (Figure 10b), and the BUE date evinces a more obvious advancing trend with a 0.44 d/y rate. The CFD has a significant shortening trend (Figure 10c), with a mean rate of 0.50 d/y, which corresponds to the delay in the FUE date and the advance of the BUS date. The most dramatic variation in ice phenology is the ICD (Figure 10c), decreasing at −0.71 d/y. The FUD and BUD trends are unclear (Figure 10d), with −0.03 and −0.18 d/y mean rates, respectively. However, the FUD fluctuation is high, which is associated with climate change and the peculiarity of Wuliangsu Lake. For example, the FUS date was determined on 20 November when the $T_{(t)}$ was negative for two consecutive days in 2017, and the FUD was only two days (Figure 7c). The reason for this situation is that, on the one hand, the shallow water body of Wuliangsu Lake has a small heat capacity. Therefore, the rapid air temperature fluctuation will significantly impact the water body (Figure 7b). On the other hand, although the daily mean air temperature has just reduced to 0 °C, the minimum air temperature has dropped to a level relatively lower than the freezing point because of the large temperature difference between day and night (mean = 12.4 °C). It benefits the "night freezing-day thawing" cycle on the lake surface, and a stable ice cover forms only when the water column is well-mixed.

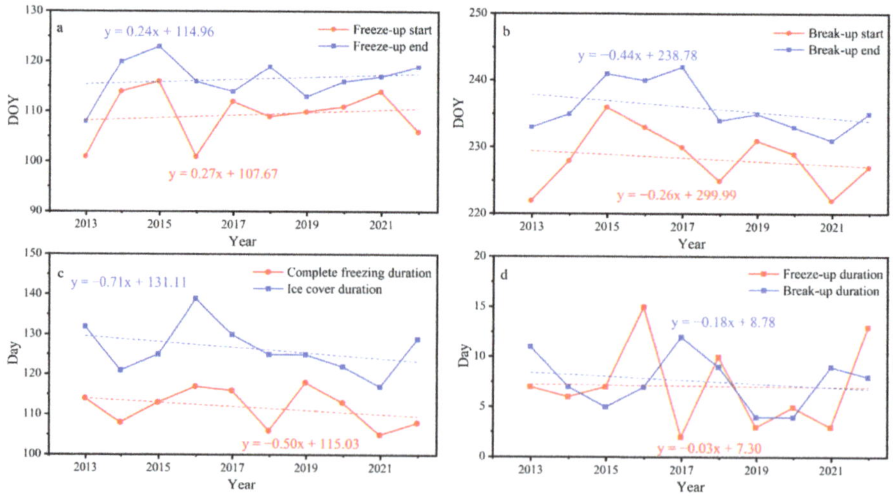

Figure 10. The ice phenology characteristics and their trends for Wuliangsu Lake from 2013 to 2022. (**a**) Freeze-up dates and their trends. (**b**) Break-up dates and their trends. (**c**) Complete freezing duration and ice cover duration and their trends. (**d**) Freeze-up and break-up durations and their trends. (Date is based on the annual cumulative day method, for example, 14 November 2021 is the 106th day of 2022).

6. Conclusions

Ice phenology in the shallow Wuliangsu Lake, with a significant portion of wetland in the northwest China arid area, was investigated using optical satellite data with high spatial and temporal resolution. A new method was developed and proved to effectively remove the influence of aquatic plants and clouds from the investigation domain.

Landsat-7, Landtsat-8, and Sentinel-2 imageries were first employed to extract the lake water boundary according to the NDWI formula and visual judgement. Moreover, based on the dynamic threshold and single-band threshold methods, an optimal threshold of $a = 0.13$ was determined for distinguishing water/ice pixels within the lake water boundary using the MODIS reflectivity products. Then, the time series of water/ice coverage was obtained. The series compared well with the results of Landsat-7, Landsat-8, and Sentinel-2 imageries during the opening/closing period of ice cover, with the MAE of 5.87%. In addition, the cloud-covered pixels in optical satellite imageries were removed using the state_1km_1 parameter of MODIS products, which filled 85.14% of the cloud-covered pixels and reduced the mean cloud coverage to less than 8% in most years. Finally, the influence of the remaining cloud and mixed pixels was further reduced by the air temperature correction, thus improving the accuracy of the ice phenology date.

According to these processes, we extracted the ice phenology characteristics of Wuliangsu Lake between 2013 and 2022, which correlated well with the field observations, with $R = 0.99$ and $MAE = 2.10$ days for the FUS, BUS, and BUE dates. We observed that Wuliangsu Lake usually starts to freeze from mid to late November every year, lasting until late March. The overall trend of FUS and FUE dates in the Wuliangsu Lake was delayed, with a mean rate of −0.27 and −0.24 d/y, respectively, and the BUS and BUE dates were advanced, with 0.26 and 0.44 d/y mean rates, respectively. Among the ice phenology characteristics, the most significant trend was the ICD, meaning 126.5 days and decreases by a mean rate of −0.71 d/y. Overall, the FUD and BUD trends are not clear; they exhibit a mean duration of 7.1 and 7.6 days, respectively, which are smaller than those found in most lakes on the Qinghai-Tibet Plateau [8]. This difference could be attributed to the peculiar shallowness of Wuliangsu Lake, which has a small heat capacity of the water body itself and requires less heat for freezing and melting, so the duration is shorter.

The present method has some advantages over previous studies. First, it can effectively remove the influence of aquatic plants in the lake wetland on the water/ice pixels' discrimination and accurately identify the lake water boundary rather than the outer lake boundary. Secondly, it can reduce the influence of clouds on optical satellite imageries, satisfying the need for the automatic extraction of lake wetland ice phenology. However, the effect of mixed pixels, thin ice, and partially unremoved cloud pixels on pixel type discrimination could not be completely avoided. Although, the reflectance MODIS product with the highest spatial resolution reflectance dataset of the MODIS product was chosen to distinguish between water and ice. This issue is expected to be further analyzed as technology improves and field observations increase. Nevertheless, based on the technical basis of this study, it is possible to perform calculations on longer time scales in the future with acceptable accuracy, and then explore the relationship between wetland lake ice phenology and climate change.

Author Contributions: Formal calculation, analysis, writing—original draft, P.H.; analysis, supervision, conceptualization, P.L.; writing—review and conceptualization, B.C.; writing—review and editing, L.Z., Q.W. and Z.L. All authors have read and agreed to the published version of the manuscript.

Funding: This research was supported by the National Key Research and Development Program of China (Grant No. 2019YFE0197600), the LiaoNing Revitalization Talents Program (Grant No. XLYC2007033), the National Natural Science Foundation of China (Grant Nos. 41922045, 41876213 and 52211530038), and the Academy of Finland (350576).

Data Availability Statement: MODIS products can be downloaded from https://ladsweb.modaps.eosdis.nasa.gov/ (accessed on 6 May 2022); Landsat and Sentinel-2 products can be downloaded from https://earthexplorer.usgs.gov/ (accessed on 6 May 2022); The meteorological data from the NMIC of the CMA can be downloaded from http://data.cma.cn (accessed on 6 May 2022).

Acknowledgments: We thank the editor and anonymous reviewers for their comments, which considerably improved this work.

Conflicts of Interest: The authors declare no conflict of interest.

References

1. Campbell, I.C.; Poole, C.; Giesen, W.; ValboJorgensen, J. Species diversity and ecology of Tonle Sap Great Lake, Cambodia. *Aquat. Sci.* **2006**, *68*, 355–373. [CrossRef]
2. Kim, J.G. Response of sediment chemistry and accumulation rates to recent environmental changes in the Clear Lake watershed, California, USA. *Wetlands* **2003**, *23*, 95–103. [CrossRef]
3. Leppäranta, M.; Wen, L.J. Ice Phenology in Eurasian Lakes over Spatial Location and Altitude. *Water* **2022**, *14*, 1037. [CrossRef]
4. Bernhardt, J.; Engelhardt, C.; Kirillin, G.; Matschullat, J. Lake ice phenology in Berlin-Brandenburg from 1947–2007: Observations and model hindcasts. *Clim. Change* **2012**, *112*, 791–817. [CrossRef]
5. Rigosi, A.; Carey, C.C.; Ibelings, B.W.; Brookes, J.D. The interaction between climate warming and eutrophication to promote cyanobacteria is dependent on trophic state and varies among taxa. *Limnol. Oceanogr.* **2014**, *59*, 99–114. [CrossRef]
6. Zdorovennova, G.; Palshin, N.; Golosov, S.; Efremova, T.; Belashev, B.; Bogdanov, S.; Terzhevik, A. Dissolved Oxygen in a Shallow Ice-Covered Lake in Winter: Effect of Changes in Light, Thermal and Ice Regimes. *Water* **2021**, *13*, 2435. [CrossRef]
7. Scheffers, B.R.; Meester, L.D.; Bridge, T.C.L.; Hoffmann, A.A.; Pandolfi, J.M.; Corlett, R.T.; Butchart, S.H.M.; Pearce-Kelly, P.; Kovacs, K.M.; Dudgeon, D.; et al. The broad footprint of climate change from genes to biomes to people. *Science* **2016**, *354*, aaf7671. [CrossRef] [PubMed]
8. Kropáček, J.; Maussion, F.; Chen, F.; Hoerz, S.; Hochschild, V. Analysis of ice phenology of lakes on the Tibetan Plateau from MODIS data. *Cryosphere* **2013**, *7*, 287–301. [CrossRef]
9. Qiu, Y.; Xie, P.; Leppäranta, M.; Wang, X.; Lemmetyinen, J.; Lin, H.; Shi, L. MODIS-based Daily Lake Ice Extent and Coverage dataset for Tibetan Plateau. *Big Earth Data* **2019**, *3*, 170–185. [CrossRef]
10. Walsh, S.E.; Vavrus, S.J.; Foley, J.A.; Fisher, V.A.; Wynne, R.H.; Lenters, J.D. Global patterns of lake ice phenology and climate: Model simulations and observations. *J. Geophys. Res.-Atmos.* **1998**, *3*, 28825–28837. [CrossRef]
11. Brown, L.C.; Duguay, C.R. The response and role of ice cover in lake-climate interactions. *Prog. Phys. Geogr.* **2010**, *34*, 671–704. [CrossRef]
12. Han, W.; Huang, C.; Duan, H.; Gu, J.; Hou, J. Lake Phenology of Freeze-Thaw Cycles Using Random Forest: A Case Study of Qinghai Lake. *Remote Sens.* **2020**, *12*, 4098. [CrossRef]
13. Zhang, S.; Pavelsky, T.M. Remote Sensing of Lake Ice Phenology across a Range of Lakes Sizes, ME, USA. *Remote Sens.* **2019**, *11*, 1718. [CrossRef]
14. Geldsetzer, T.; Sanden, J.V.D.; Brisco, B. Monitoring lake ice during spring melt using RADARSAT-2 SAR. *Can. J. Remote Sens.* **2010**, *36*, S391–S400. [CrossRef]
15. Meier, W.N.; Dai, M. High-resolution sea-ice motions from AMSR-E imagery. *Ann. Glaciol.* **2006**, *44*, 352–356. [CrossRef]
16. Mack, S.; Padman, L.; Klinck, J. Extracting tidal variability of sea ice concentration from AMSR-E passive microwave single-swath data: A case study of the Ross Sea. *Geophys. Res. Lett.* **2013**, *40*, 547–552. [CrossRef]
17. Tanaka, Y.; Tateyama, K.; Kameda, T.; Hutchings, J.K. Estimation of melt pond fraction over high-concentration Arctic sea ice using AMSR-E passive microwave data. *J. Geophys. Res. Oceans* **2016**, *121*, 7056–7072. [CrossRef]
18. Tuttle, S.E.; Roof, S.R.; Retelle, M.J.; Werner, A.; Gunn, G.E.; Bunting, E.L. Evaluation of Satellite-Derived Estimates of Lake Ice Cover Timing on Linnévatnet, Kapp Linné, Svalbard Using In-Situ Data. *Remote Sens.* **2022**, *14*, 1311. [CrossRef]
19. Šmejkalová, T.; Edwards, M.; Dash, J. Arctic lakes show strong decadal trend in earlier spring ice-out. *Sci Rep. UK* **2016**, *6*, 38449. [CrossRef]
20. Cai, Y.; Ke, C.Q.; Li, X.; Zhang, G.; Duan, Z.; Lee, H. Variations of Lake Ice Phenology on the Tibetan Plateau From 2001 to 2017 Based on MODIS Data. *J. Geophys. Res. Atmos.* **2019**, *124*, 825–843. [CrossRef]
21. Williamson, A.G.; Banwell, A.F.; Willis, I.C.; Arnold, N.S. Dual-satellite (Sentinel-2 and Landsat 8) remote sensing of supraglacial lakes in Greenland. *Cryosphere* **2018**, *12*, 3045–3065. [CrossRef]
22. Zhang, X.; Wang, K.; Kirillin, G. An Automatic Method to Detect Lake Ice Phenology Using MODIS Daily Temperature Imagery. *Remote Sens.* **2021**, *13*, 2711. [CrossRef]
23. Wu, Y.H.; Duguay, C.R.; Xu, L.L. Assessment of machine learning classifiers for global lake ice cover mapping from MODIS TOA reflectance data. *Remote Sens. Environ.* **2021**, *253*, 112206. [CrossRef]
24. Messager, M.L.; Lehner, B.; Grill, G.; Nedeva, I.; Schmitt, O. Estimating the volume and age of water stored in global lakes using a geo-statistical approach. *Nat. Commun.* **2016**, *7*, 13603. [CrossRef]
25. Sharma, S.; Blagrave, K.; Filazzola, A.; Imrit, M.A.; Hendricks Franssen, H.J. Forecasting the Permanent Loss of Lake Ice in the Northern Hemisphere Within the 21st Century. *Geophys. Res. Lett.* **2021**, *48*, e2020GL091108. [CrossRef]

26. Tai, X.N.; Wang, N.L.; Wu, Y.W.; Zhang, Y.J. Lake ice phenology variations and influencing factors of Selin Co from 2000 to 2020. *J. Lake Sci.* **2022**, *34*, 334–348. [CrossRef]
27. Lu, P.; Cao, X.W.; Li, G.Y.; Huang, W.F.; Leppäranta, M.; Arvola, L.; Li, Z.J. Mass and Heat Balance of a Lake Ice Cover in the Central Asian Arid Climate Zone. *Water* **2020**, *12*, 2888. [CrossRef]
28. Huang, W.F.; Zhao, W.; Zhang, C.; Leppranta, M.; Li, Z.J.; Li, R.; Lin, Z.J. Radiative penetration dominates the thermal regime and energetics of a shallow ice-covered lake in an arid climate. *Cryosphere* **2021**, *16*, 1793–1806. [CrossRef]
29. Duan, X.; Wang, X.; Mu, Y.; Ouyang, Z. Seasonal and diurnal variations in methane emissions from Wuliangsu Lake in arid regions of China. *Atmos. Environ.* **2005**, *39*, 4479–4487. [CrossRef]
30. Cheng, B.; Xie, F.; Lu, P.; Huo, P.Z.; Leppäranta, M. The role of lake heat flux in the growth and melting of ice. *Adv. Polar Sci.* **2021**, *32*, 364–373. [CrossRef]
31. McFeeters, S.K. The use of the Normalized Difference Water Index (NDWI) in the delineation of open water features. *Int. J. Remote Sens.* **1996**, *17*, 1425–1432. [CrossRef]
32. Hall, D.K.; Riggs, G.A.; Salomonson, V.V. *Algorithm Theoretical Basis Document (ATBD) for the MODIS Snow and Sea Ice-Mapping Algorithms*; NASA: Greenbelt, MD, USA, 2001. Available online: https://modis-snow-ice.gsfc.nasa.gov/?c=atbd&t=atbd (accessed on 15 January 2022).
33. Gafurov, A.; Bárdossy, A. Cloud removal methodology from MODIS snow cover product. *Hydrol. Earth Syst. Sci.* **2009**, *13*, 1361–1373. [CrossRef]
34. Yang, Q.; Song, K.; Wen, Z.; Hao, X.; Fang, C. Recent trends of ice phenology for eight large lakes using MODIS products in Northeast China. *Int. J. Remote Sens.* **2019**, *40*, 5388–5410. [CrossRef]
35. Yang, Q.; Song, K.; Ha, X.; Wen, Z.; Tan, Y.; Li, W. Investigation of spatial and temporal variability of river ice phenology and thickness across Songhua River Basin, northeast China. *Cryosphere* **2020**, *14*, 3581–3593. [CrossRef]
36. Duguay, C.R.; Bernier, M.; Gauthier, Y.; Kouraev, A. *Remote Sensing of Lake and River Ice*; Wiley: Hoboken, NJ, USA, 2014; ISBN 9781118368909.

water

MDPI

Article

Ice Velocity in Upstream of Heilongjiang Based on UAV Low-Altitude Remote Sensing and the SIFT Algorithm

Enliang Wang [1,2], Shengbo Hu [1], Hongwei Han [1,2,*], Yuang Li [1], Zhifeng Ren [1] and Shilin Du [1]

1 School of Water Conservancy and Civil Engineering, Northeast Agricultural University, Harbin 150030, China; hljwel@126.com (E.W.); hushengboandy@163.com (S.H.); liyuang9804@163.com (Y.L.); rzf921006@163.com (Z.R.); dslneau16@163.com (S.D.)

2 Heilongjiang Provincial Key Laboratory of Water Resources and Water Conservancy Engineering in Cold Region, Harbin 150030, China

* Correspondence: hanhongwei@neau.edu.cn; Tel.: +86-131-5980-8919

Abstract: In river management, it is important to obtain ice velocity quickly and accurately during ice flood periods. However, traditional ice velocity monitoring methods require buoys, which are costly and inefficient to distribute. It was found that UAV remote sensing images combined with machine vision technology yielded obvious practical advantages in ice velocity monitoring. Current research has mainly monitored sea ice velocity through GPS or satellite remote sensing technology, with few reports available on river ice velocity monitoring. Moreover, traditional river ice velocity monitoring methods are subjective. To solve the problems of existing time-consuming and inaccurate ice velocity monitoring methods, a new ice velocity extraction method based on UAV remote sensing technology is proposed in this article. In this study, the Mohe River section in Heilongjiang Province was chosen as the research area. High-resolution orthoimages were obtained with a UAV during the ice flood period, and feature points in drift ice images were then extracted with the scale-invariant feature transform (SIFT) algorithm. Moreover, the extracted feature points were matched with the brute force (BF) algorithm. According to optimization results obtained with the random sample consensus (RANSAC) algorithm, the motion trajectories of these feature points were tracked, and an ice displacement rate field was finally established. The results indicated that the average ice velocities in the research area reached 2.00 and 0.74 m/s, and the maximum ice velocities on the right side of the river center were 2.65 and 1.04 m/s at 16:00 on 25 April 2021 and 8:00 on 26 April 2021, respectively. The ice velocity decreased from the river center toward the river banks. The proposed ice velocity monitoring technique and reported data in this study could provide an effective reference for the prediction of ice flood disasters.

Keywords: unmanned aerial vehicles; SIFT algorithm; brutal force matching; RANSAC algorithm; ice velocity

Citation: Wang, E.; Hu, S.; Han, H.; Li, Y.; Ren, Z.; Du, S. Ice Velocity in Upstream of Heilongjiang Based on UAV Low-Altitude Remote Sensing and the SIFT Algorithm. *Water* **2022**, *14*, 1957. https://doi.org/10.3390/w14121957

Academic Editor: Guido Paliaga

Received: 4 May 2022
Accepted: 16 June 2022
Published: 18 June 2022

1. Introduction

Rivers north of 30° N in China generally develop ice hazards at the stages of winter river closure and spring river opening [1]. With increasing ice concentration and ice velocity in a given river section, the impact on hydraulic structures in the river increases. Upon river thawing in spring, specific river sections can become blocked by large amounts of ice, such as narrow river courses, shoals, sharp and continuous bends, and the front edges of unfrozen ice sheets. Ice accumulates to form ice jams or ice dams, which often severely block water flow sections and notably raise upstream water levels, resulting in ice jam flooding. Under long-term severe ice conditions, the impact of ice drift in spring can cause varying degrees of damage to hydraulic structures such as piers and bank revetments, threaten the lives and property of people, and further increase the difficulty of ice flood prevention [2]. Therefore, feasible and accurate ice velocity monitoring technology constitutes a major problem urgently requiring a solution in the field of water disasters. Ice velocity monitoring is not only the main priority in the fields of major natural disaster monitoring and mitigation, but is also one of the major problems related to water engineering safety, the national economy, livelihoods of people, and social development [3].

At present, some achievements had been made regarding the velocity measurements of sea ice and glaciers. Sea ice velocity monitoring methods are mainly divided into two categories: field observation methods and remote sensing monitoring methods [4]. Of these categories, field observation methods mainly include marker monitoring methods and global positioning system (GPS)-based monitoring methods [5,6], whereas remote sensing monitoring methods are largely categorized into optical and microwave remote sensing monitoring methods [7,8]. The methods applied in river ice velocity monitoring are mostly field observation methods [9], for example, visual tracking methods [10]. These methods rely on a simple principle but require highly experienced observers. Moreover, field observation methods exhibit notable subjectivity and high time and labor costs, and these methods are unsuitable for large-scale and long-term tracking. In the GPS monitoring method [11], large ice blocks with a good shape are selected, GPS sensors are installed atop these ice blocks, and the instantaneous speed of the ice blocks equipped with GPS sensors is monitored in real time with a high accuracy through the positioning mode involving a satellite, GPS sensors, and base station. For example, Song Chunshan et al. [12] measured the ice drift velocity in the upper reaches of the Heilongjiang River with the GPS monitoring method in 2019 and studied the drift ice blocking phenomenon in each reach during the ice flood period based on the velocity variation. However, the GPS tracking method can only obtain the velocity of a single ice block but cannot obtain the ice velocity field in large-scale areas, and the observation process (sensor placement and recovery) results in a high time cost and difficulty. Deng Xiao et al. [13] set up a camera on the shore, manually selected two images at a certain time interval in the obtained monitoring footage, and determined the moving distance of drift ice to calculate the ice drift velocity. However, due to the influence of side view, images typically suffer the problem of image geometric distortion, which affects the monitoring accuracy. In addition, it takes a long time to process data manually.

With the continuous improvement of unmanned aerial vehicles (UAVs) and sensor technology, small UAVs have been widely deployed in low-altitude remote sensing measurements [14–17]. Compared to the visual and GPS-based tracking methods, the multisensor UAV platform exhibits the characteristics of high efficiency, low cost, flexible operation, and strong suitability in complex river environments. This technique not only overcomes the difficulty of image data acquisition but also avoids the image distortion phenomenon attributed to side shooting from the shore, which improves the accuracy of the original data and provides a new technical means for ice velocity monitoring within a large range.

To solve the above problems, this paper proposes the combination of the SIFT and RANSAC algorithms to track feature points and researches the velocity vector of drift ice on this basis. To verify the feasibility of the proposed method, the Mohe section located in the upper reaches of Heilongjiang Province is selected as the observation area. An UAV is employed to obtain high-resolution video data during the ice flood period. Images are processed in OpenCV software that are automatically extracted at 1-s intervals in video. Subsequently, the ice plane velocity field in the observation area is extracted to perform ice dynamic analysis, thus providing a suitable technique and reference data for ice flood disaster early warnings and prediction.

2. Materials and Methods

2.1. Overview of the Research Area

The Heilongjiang River is located in the northernmost part of China, with a wide drainage area and abundant water resources. The drainage basin is located in the cold temperate zone. The ice thickness can reach 2 m in winter, and the ice surface is covered with snow exhibiting an average thickness of 30 cm [18]. Due to the unique geographical location of Heilongjiang, the upper reaches of the Heilongjiang River flow from low latitudes to high latitudes [19] through mountainous areas with narrow and tortuous river courses. There is abundant rainfall and runoff but insufficient observation equipment and measurement means. During the ice flood period, ice mainly accumulates in the Mohe section to form ice dams [20]. Therefore, the research area is the Mohe section in Heilongjiang Province, and the takeoff and landing point of the UAV is an open area southeast of the Mohe hydrological station (as shown in Figure 1).

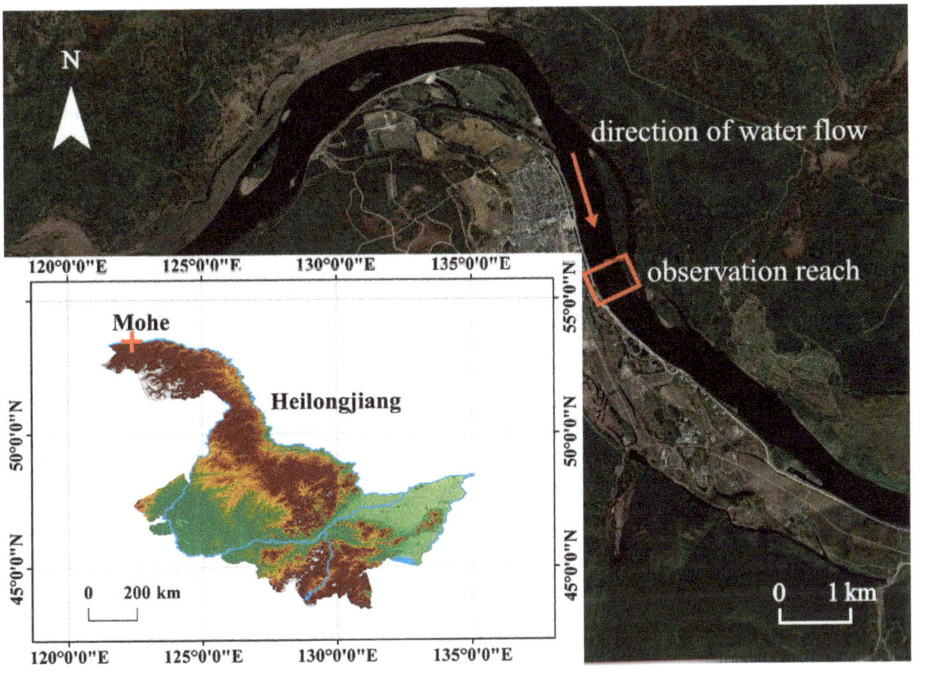

Figure 1. Schematic diagram of observation geographical location.

2.2. Research Method

At 16:00 on 25 April 2021 and 8:00 on 26 April 2021, a DJI quad-rotor small UAV (Mavic 2 Pro) was employed for fixed-point shooting. The UAV is equipped with a Hasselblad L1d-20c (CMOS) aerial camera with an effective resolution of 20 million pixels and can record 4K high-dynamic range (HDR) footage. The weather conditions captured in each aerial photograph were sunny, the wind speed was lower than 10.7 m/s, the flight altitude was set to 400 m, the elevation difference between the take-off point and water surface was 5 m, the lens was oriented vertically downward, and the UAV hovered during shooting.

In this study, the tracking process of ice feature points included four main steps. Step 1. Feature point selection. First, a large number of ice feature points were extracted from the phase I image (a certain frame of image in the observation video) with the SIFT operator, and ice feature points were extracted from the phase II image (the frame of image after 1 s in the observation video) in the same way.

Step 2. Feature point matching. Based on the Euclidean distance, a BF matcher was adopted to track the same feature points in the above two images.

Step 3. Match result filtering. The obtained matching feature point pairs were filtered with the RANSAC algorithm to remove any incorrect matching feature point pairs.

Step 4. Ice velocity determination. The relationship between the UAV height and the actual length represented by a unit pixel was determined through calibration, the distance of the filtered feature point pairs was transformed into the moving distance of drift ice, and, finally, the ice drift velocity was determined.

2.2.1. Feature Point Selection

The SIFT algorithm is a feature point detection algorithm based on the scale space first proposed by Lowe [21] in 1999 and formally put forward after summary and improvement in 2004 [22]. A difference of Gaussian (DOG) scale space is constructed. The scale space describes the original image at different levels, and the different levels are represented by various scale parameters. Koenderink [23] and Lindeberg [24] demonstrated that the Gaussian kernel is the only linear kernel for scale space construction.

The Gaussian function of scale transformation is defined as:

$$G(x,y,\sigma) = \frac{1}{2\pi\sigma^2} e^{-(x^2+y^2)/2\sigma^2} \tag{1}$$

Image scale space function $L(x, y, \sigma)$ is expressed as:

$$L(x,y,\sigma) = G(x,y,\sigma) \otimes I(x,y) \tag{2}$$

where $I(x, y)$ is the input image, $G(x, y, \sigma)$ is a variable-scale Gaussian operator, $\otimes$ is the convolution operation, and σ is the mean square deviation of the Gaussian normal distribution, which is denoted as the spatial scale factor. The larger the σ value, the more blurred the image becomes. Conversely, the smaller the σ value, the clearer the image becomes. To ensure that the extreme points in the scale space attain a suitable uniqueness and stability, Lowe [22] proposed the DOG concept, which is the difference between two adjacent scale images.

The calculation equation of the differential Gaussian image is as follows:

$$D(x,y,\sigma) = (G(x,y,k\sigma) - G(x,y,\sigma)) \otimes I(x,y) = L(x,y,k\sigma) - L(x,y,\sigma) \tag{3}$$

where k is a constant factor of the difference between adjacent scale images, and the k value is related to the number of pictures (S) in each layer of the scale space, i.e., $k = 2^{1/S}$.

After construction of the DOG pyramid, it is necessary to select key points from the pyramid scale space, namely, local feature extreme points in the image space. Each sampling point is compared to 8 adjacent points in the current scale layer and 18 points in the upper and lower adjacent scale layers to determine whether the considered sampling point is an extreme point in the current scale space and two-dimensional space [25].

Because the extreme points detected with the above method are distributed in the discrete space, the extreme points in the known discrete space are interpolated to extreme points in the continuous space through subpixel interpolation. To obtain the exact position of an extreme point, the DOG function is expanded according to the Taylor formula as follows:

$$D(X) = D + \frac{\partial D^T}{\partial X} X + \frac{1}{2} X^T \frac{\partial^2 D}{\partial X^2} X \tag{4}$$

where D is the $D(x, y, \sigma)$ value at the feature point position and $X = (x, y, \sigma)^T$ is the offset relative to this feature point. The extreme value of $D(x)$ is determined, i.e., $\frac{\partial D(X)}{\partial X} = 0$, and the position of the extreme point is:

$$\hat{X} = -\frac{\partial^2 D^{-1}}{\partial X^2} \frac{\partial D}{\partial X} \tag{5}$$

The extreme value at the corresponding extreme point is:

$$D\left(\hat{X}\right) = D + \frac{1}{2} \frac{\partial D^T}{\partial X} \hat{X} \tag{6}$$

If $\left|D\left(\hat{X}\right)\right|$<0.03, this feature point is removed as an unstable feature point. Because the DOG operator can produce strong edge responses, it is necessary to apply the Hessian matrix at the feature points to eliminate points with an unstable edge response [26]. The Hessian matrix is defined as:

$$H = \begin{bmatrix} D_{xx} & D_{xy} \\ D_{xy} & D_{yy} \end{bmatrix} \quad (7)$$

where D_{xx}, D_{xy}, and D_{yy} denote derivations of x or y.

Among the accurately located stable feature points, the parameter direction is specified for each feature point based on the gradient direction distribution characteristics of its neighborhood pixels, and the feature vector is then generated relative to this direction so that the SIFT-determined feature points remain invariant to image rotation.

Gradient modulus:

$$m(x,y) = \sqrt{(L(x+1,y) - L(x-1,y))^2 + (L(x,y+1) - L(x,y-1))^2} \quad (8)$$

Gradient direction:

$$\theta(x,y) - \arctan\left[\frac{L(x,y+1) - L(x,y-1)}{L(x+1,y) - L(x-1,y)}\right] \quad (9)$$

Through the above process, the feature points contain three basic components of information: position, scale, and direction. Moreover, the feature points are invariant to translation, scaling, and rotation. To ensure that the feature points remain unchanged under various conditions, such as solar radiation level and viewing angle, a descriptor is established for each feature point. The feature point is adopted as the center, the coordinate axis of the image is rotated to match the direction along which the feature point is located, the neighborhood near the feature point is divided into 4 × 4 subwindows, the gradient information in each subwindow is obtained, an 8-dimensional direction histogram is generated, a 128-dimensional feature descriptor is finally established, and the descriptor subvector elements are normalized to improve the resistance to light changes [27].

2.2.2. Feature Point Matching

The BF matcher is a simple matching algorithm. Suppose the set of all feature points in image P_1 is $\{a_1, a_2, \ldots a_n\}$, and the set of all feature points in image P_2 is expressed as $\{b_1, b_2, \ldots b_n\}$. When the BF matcher is applied to match the feature points in images P_1 and P_2, the Euclidean distance from the feature vectors of each feature point in image P_1 to each feature point in image P_2 is calculated. The result of the algorithm is an $n \times m$-dimensional matrix $\begin{bmatrix} d_{11} & \cdots & d_{1m} \\ \vdots & & \vdots \\ d_{n1} & \cdots & d_{nm} \end{bmatrix}$, where d_{ij} is the Euclidean distance between the feature vectors of feature points a_i and b_j [28].

2.2.3. Match Result Filtering

After all feature points are matched, incorrect matching points inevitably occur, which can affect the accuracy of ice velocity calculation. In this paper, the RANSAC algorithm is implemented to further optimize the matching results, and the matching feature points are screened through continuous iteration.

The RANSAC algorithm is an iterative algorithm proposed by Fischler and Bolles [29] to estimate the parameters of mathematical models. This algorithm has been widely applied in line fitting and plane fitting. Correct data contained in the considered data are recorded as inner points, and abnormal data (or noise) are recorded as outer points [30]. The probability of obtaining the correct model through the RANSAC algorithm is:

$$Z = 1 - \left(1 - P^K\right)^M \tag{10}$$

where P is the probability of the inner points, K is the minimum number of data points required to solve the model, and M is the number of iterations. The above equation yields the following:

$$M = \frac{\log(1 - Z)}{\log(1 - P^K)} \tag{11}$$

Finally, optimal matching feature point pairs are obtained, i.e., the feature point pairs with the same name between the input phase I ice image and matching phase II ice image.

2.2.4. Ice Velocity Calculation

After all feature point pairs with the same feature descriptor are obtained, the horizontal displacement of ice can be extracted, and the feature point pairs with the same feature descriptor can then be transformed into the ice velocity according to the pixel scale and image interval, as follows:

$$S = \sqrt{(X_2 - X_1)^2 + (Y_2 - Y_1)^2} \tag{12}$$

$$V = \frac{S \cdot R}{T \cdot 100} \tag{13}$$

where X_1, X_2, Y_1, and Y_2 are the coordinates of the feature point pairs with the same feature descriptor, S is the moving distance of the feature point in the image (px), R is the pixel scale (cm/px), T is the time between the two images (s), and V is the ice velocity (m/s).

To realize quantitative analysis of the ice velocity, first, the UAV is deployed to shoot A1-sized paper panels at different heights to determine the shooting height of UAV remote sensing images and the actual area represented by a unit pixel. After the actual area is squared, a fitting equation between the unit pixel length and actual length in remote sensing images at the different heights is obtained, as shown in Figure 2.

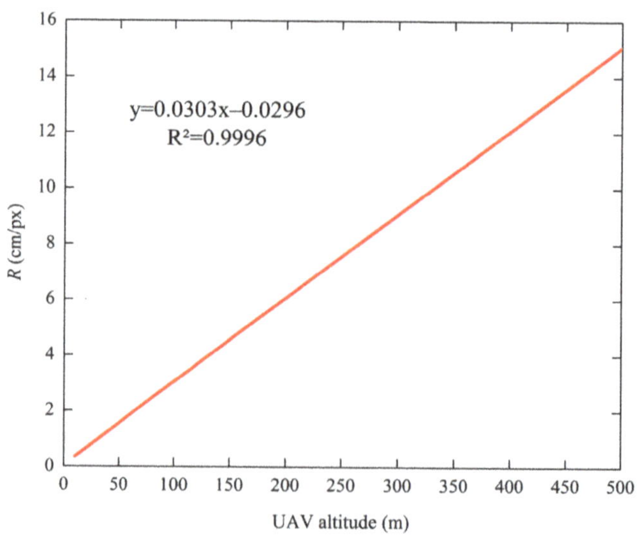

Figure 2. Fitting curve between the unit pixel length and actual length in remote sensing images at the different heights.

3. Results

Through feature point selection, matching, and filtration of the two obtained orthophoto images, 1429 pairs of effective feature points (as shown in Figure 3) were extracted from the river section at 16:00 on 25 April 2021, and 2726 pairs of effective feature points (as shown in Figure 4) were extracted from the river section at 8:00 on 26 April 2021. According to the principle of feature point selection, there were sparse or even absent feature extreme points on the river surface, but the distribution of the feature points on the ice surface was relatively dense and stable, which satisfied the basic requirements of ice velocity analysis. At 8:00 on 26 April 2021, due to the large amount of broken ice in the river section, the selected effective feature point density was higher than that at 16:00 on 25 April 2021.

Figure 3. Distribution of feature point pairs at 16:00 on 25 April 2021.

Figure 4. Distribution of feature point pairs at 8:00 on 26 April 2021.

The ice velocity in the Mohe section in Heilongjiang at 16:00 on 25 April and at 8:00 on 26 April 2021 was extracted, as shown in Figures 5 and 6, respectively. The distribution characteristics of the ice velocity are similar. Affected by the river bend, areas with the highest ice velocity were concentrated on the right side of the river center, and the ice velocity tended to decrease toward the banks on both sides. Moreover, there occurred a sudden change in the ice flow velocity near the river banks on both sides, which was likely caused by collision between drift ice and the bank slopes. Figure 7a,b show a statistical histogram of the ice velocity at 16:00 on 25 April and 8:00 on 26 April 2021. The figure shows that at 16:00 on 25 April 2021, the ice velocity was mostly in the range 2.40~2.50 m/s, accounting for 30.02% of all velocity values. The maximum, minimum, and average ice velocities were2.65 m/s, 1.11 m/s, and 2.00 m/s, respectively. At 8:00 on 26 April 2021, the ice velocity was largely in the range 0.90~1.00 m/s, accounting for 50.33% of all velocity values. The maximum, minimum, and average ice velocities were 1.04 m/s, 0.38 m/s, and 0.74 m/s, respectively. Compared to 16:00 on 25 April 2021, the standard deviation of the ice velocity at 8:00 on 26 April 2021 was 0.18, and the ice velocity distribution was more uniform.

Figure 5. Plane velocity field at 16:00 on 25 April 2021 (define the transverse pixel coordinate as x, and the longitudinal pixel coordinate as y, the coordinate directions are as shown in this figure).

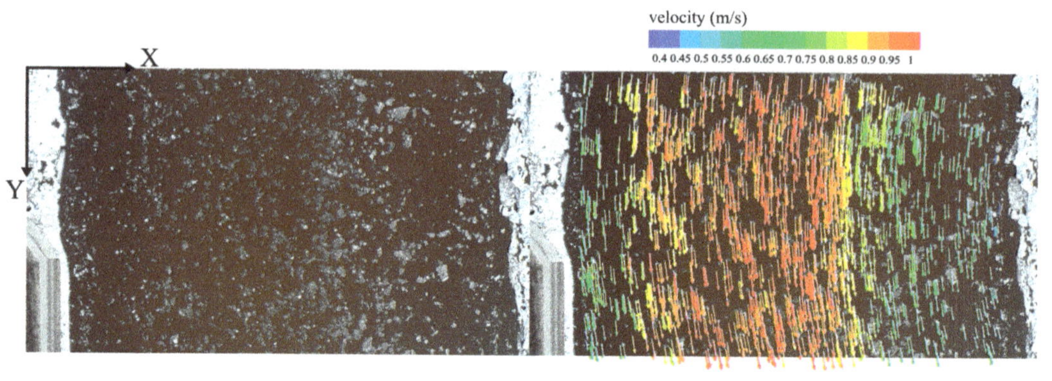

Figure 6. Plane velocity field at 8:00 on 26 April 2021 (define the transverse pixel coordinate as x, and the longitudinal pixel coordinate as y, the coordinate directions are as shown in this figure).

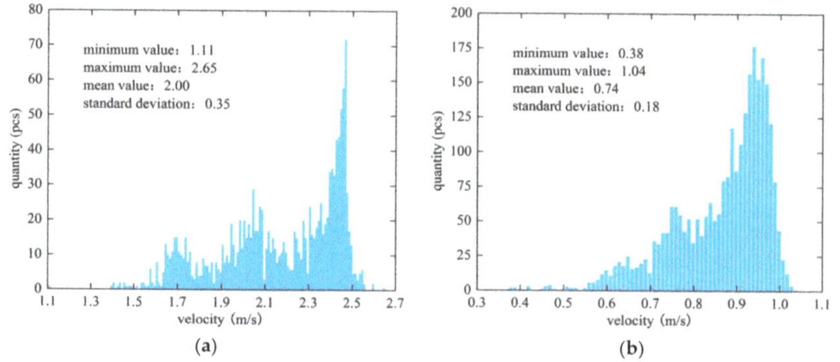

Figure 7. Statistical histogram of ice plane velocity (**a**) 16:00, 25 April 2021; (**b**) 8:00, 26 April 2021.

4. Discussion

4.1. Ice Velocity Analysis

At 18:35 on 24 April 2021, the upstream ice dam burst, and the water level in the study area encompassing the Mohe section changed sharply. From 18:35 on 24 April 2021 to 5:00 on 25 April 2021, ice jams were formed many times downstream of the Mohe section research area, resulting in rapid water level rise in the research area. Until 5:00 on 25 April 2021, the Mohe section in Heilongjiang remained fully open, the water level in the research area decreased rapidly, and the ice velocity was high. At 16:00 on 25 April, the average velocity of ice was 2.00 m/s. At 8:00 on 26 April 2021, the water level in the research area dropped to the normal level, the ice flood period in Heilongjiang had basically ended (as shown in Figure 8), the ice velocity was low, and the average velocity of ice reached 0.74 m/s, which was in line with the actual situation.

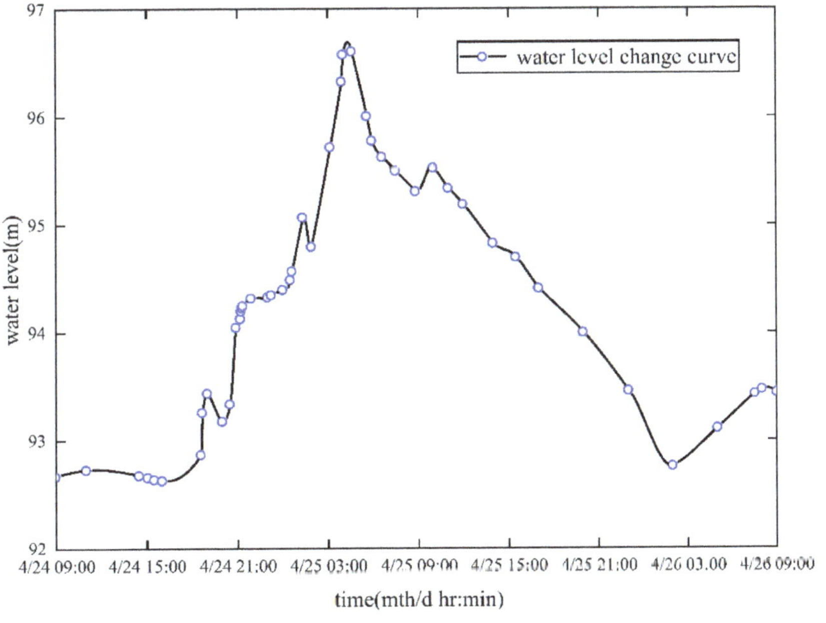

Figure 8. Water level change chart.

To study the distribution characteristics of the ice velocity during the ice flood period in the Mohe section of Heilongjiang more comprehensively, the transverse and longitudinal components of the ice velocity during the different periods were calculated by tracking the displacement components of the effective characteristic points in the corresponding coordinate system. For this calculation, V_x is the transverse component of ice velocity, V_y is the longitudinal component of ice velocity, and the positive and negative symbols indicate the direction of the velocity component. According to the schematic diagram of the observation geographical location, the observation area is located at a river bend. As indicated by the transverse velocity component shown in Figure 9a,c, the ice velocity upstream of the river was biased to the right side of the river, which can easily damage the right bank of the river, form ice accumulation, seriously block the water crossing section, and threaten the lives and property of people. For example, the ice accumulation and damage to the bank slope are shown in Figure 10, which are in line with the actual site conditions. As indicated by the longitudinal velocity component depicted in Figure 9b,d, the ice velocity was oriented along the river channel, and there occurred no ice reflux phenomenon affected by vortexing.

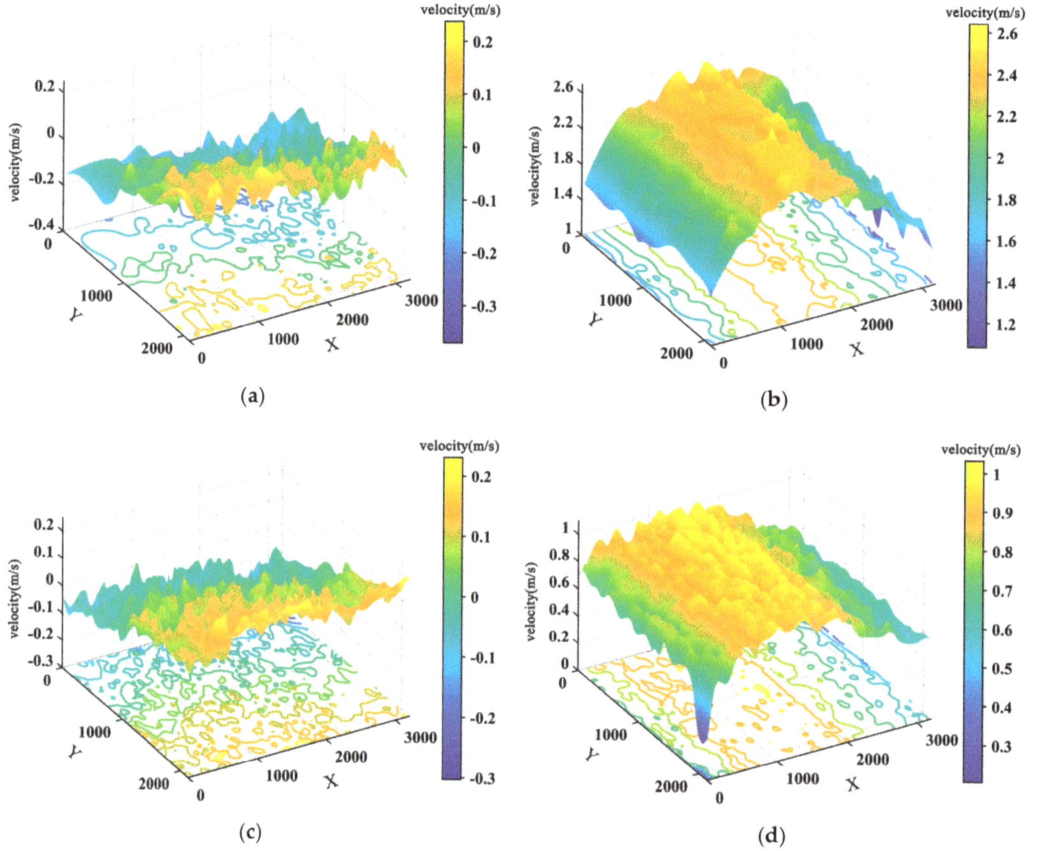

Figure 9. Distribution law of ice velocity components: (**a**) the transverse components of the ice velocity at 16:00 on 25 April 2021; (**b**) the longitudinal components of the ice velocity at 16:00 on 25 April 2021; (**c**) the transverse components of the ice velocity at 8:00 on 26 April 2021; (**d**) the longitudinal components of the ice velocity at 8:00 on 26 April 2021.

Figure 10. The photo of the right bank (**a**) before ice flood period, (**b**) during ice flood period.

To highlight the significance of this research, the related research results are compared. Deng Xiao [13] carried out a series of studies on the ice velocity in the Heilongjiang River. Under the assumption that the ice flows horizontally, the ice velocity was confirmed by manual processing. It needs to be noted that the average velocity of ice was 1.62 m/s and the image data were recorded at 10:28 on 2 May 2016. The ice movement at different times are shown in Figure 11. Although the ice velocity can be determined, this method can only manually acquire the velocity for a single ice block, and this method typically suffers the problem of image geometric distortion, which affects the monitoring accuracy and is time consuming. In addition, Chave [31] installed a real time data acquisition system in Lake St. Pierre in the St. Lawrence River to measure the ice velocity. The system can work around the clock. Unfortunately, the instruments often work underwater and cannot be moved. Therefore, the cost of installation and maintenance is not economical. However, the proposed method, UAV combined with SIFT and RANSAC algorithms, can fill these gaps. The velocity of ice blocks can be determined simultaneously, and a large number of in situ monitoring instruments are not required. In addition, the proposed method is suitable for ice running at high speed.

Figure 11. Manual extraction of ice position: (**a**) original position of target ice at 10:28 on 2 May 2016, (**b**) position of target ice after 10 s [13].

4.2. Ice Velocity Verification

Upon completion of feature point matching in ice images obtained with the UAV, although the matching results are further screened with the RANSAC algorithm, ice may collide with other ice blocks or bank slopes, thus changing the local extreme points, which can lead to incorrect matching or the absence of feature point pairs in images. To quantitatively evaluate the accuracy of ice velocity determination in the research area, ice was selected in the river section, feature points with the same name were extracted through visual interpretation (as shown in Figure 12), the ice velocity was manually calculated, and the obtained calculation value was compared to the ice velocity within 1 m before and after the section based on the SIFT algorithm, as shown in Figure 13a,b, respectively. The errors were less than 0.02 m/s. It was determined that the ice velocity determined based on the SIFT algorithm was consistent with the actual situation.

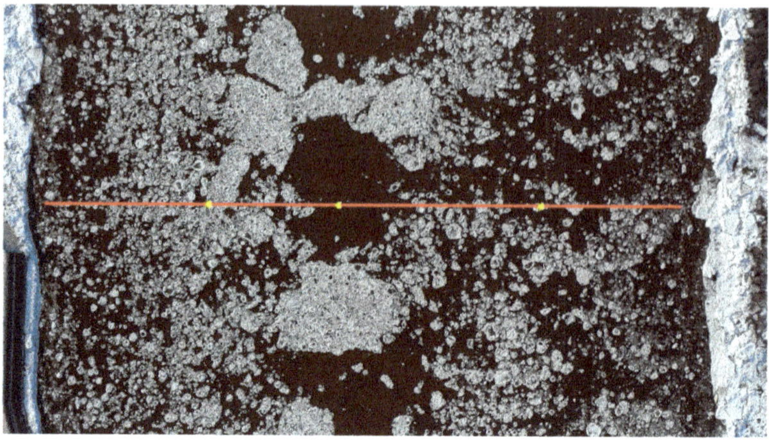

Figure 12. Characteristic points on transverse section of the river channel.

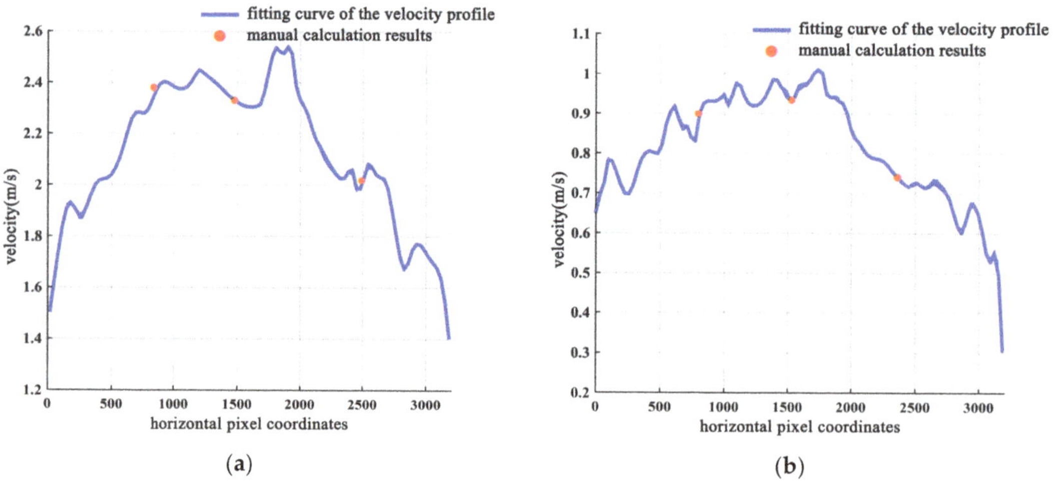

Figure 13. Comparison of plane velocity field based on SIFT algorithm and manual calculation: (**a**) 16:00, 25 April 2021; (**b**) 8:00, 26 April 2021.

5. Conclusions

To monitor and realize in-depth analysis of ice change information, this paper tracks ice surface feature points based on UAV remote sensing images and the SIFT algorithm, establishes the corresponding ice velocity field, and finally carries out in-depth analysis and interpretation of ice movement based on the established ice velocity field. The following conclusions are obtained:

1. Feature points were selected as local extreme points. Only ice surfaces, edges, and corners can produce dense and stable feature points. The SIFT algorithm provided the advantages of high precision and notable robustness. The SIFT and RANSAC algorithms were jointly employed to track the feature points, which realized ice velocity monitoring.
2. At 16:00 on 25 April 2021, the ice velocity was mostly in the range 2.40~2.50 m/s, accounting for 30.02% of all velocity values. The maximum, minimum, and average ice velocities were2.65 m/s, 1.11 m/s, and 2.00 m/s, respectively. At 8:00 on 26 April 2021, the ice velocity was mainly in the range 0.90~1.00 m/s, accounting for 50.33%

of all velocity values. The maximum, minimum, and average ice velocities were 1.04 m/s, 0.38 m/s, and 0.74 m/s, respectively. The areas with the highest ice velocity in the studied reach were concentrated on the right side of the river center, and the ice velocity tended to decrease toward the banks on both sides. In addition, based on the ice velocity field, ice accumulation formation on the bank slope was highly related to the ice velocity.

3. Under the influence of illumination changes and ice collisions, the local extreme points varied. The matching results were further screened with the RANSAC algorithm. The error between the ice velocity obtained with the SIFT algorithm and the manually calculated ice velocity was less than 0.02 m/s, which was in line with the actual situation.
4. Compared to traditional ice velocity monitoring methods, the proposed method attained a high precision, and greatly reduced time and labor costs. The research method in this paper provides a new technical means for large-scale ice change information monitoring. In the future, we can perform further research on ice jams and ice dams based on this technique.

Author Contributions: Conceptualization, E.W. and H.H.; methodology, E.W., S.H. and H.H.; formal analysis, S.H. and H.H.; investigation, E.W., S.H., Y.L., Z.R. and S.D.; data curation, S.H., Y.L., Z.R. and S.D.; writing—original draft preparation, E.W. and S.H.; writing—review and editing, H.H.; funding acquisition, E.W. All authors have read and agreed to the published version of the manuscript.

Funding: This research was supported by the National Natural Science Foundation of China (41876213), the National Key R&D Program of China (No.2018YFC0407301-03) and the Natural Science Foundation of Heilongjiang Province of China (No. LH2020E004).

Institutional Review Board Statement: Not applicable.

Informed Consent Statement: Not applicable.

Data Availability Statement: The data are available in the case that they are required.

Conflicts of Interest: The authors declare no conflict of interest.

References

1. Goldberg, M.D.; Li, S.; Lindsey, D.T.; Sjoberg, W.; Zhou, L.; Sun, D. Mapping, Monitoring, and Prediction of Floods Due to Ice Jam and Snowmelt with Operational Weather Satellites. *Remote Sens.* **2020**, *12*, 1865. [CrossRef]
2. Wang, J.; Shi, F.-y.; Chen, P.-p.; Wu, P.; Sui, J. Impact of bridge pier on the stability of ice jam. *J. Hydrodyn.* **2015**, *27*, 865–871. [CrossRef]
3. Wang, R.; Huang, D.; Zhang, X.; Wei, P. Combined pattern matching and feature tracking for Bohai Sea ice drift detection using Gaofen-4 imagery. *Int. J. Remote Sens.* **2020**, *41*, 7486–7508. [CrossRef]
4. Heid, T.; Kääb, A. Evaluation of existing image matching methods for deriving glacier surface displacements globally from optical satellite imagery. *Remote Sens. Environ.* **2012**, *118*, 339–355. [CrossRef]
5. Dorrer, E.; Hofmann, W.; Seufert, W. Geodetic results of the Ross Ice Shelf Survey expeditions, 1962–1963 and 1965–66. *J. Glaciol.* **1969**, *8*, 67–90. [CrossRef]
6. Manson, R.; Coleman, R.; Morgan, P.; King, M. Ice velocities of the Lambert Glacier from static GPS observations. *Earth Planets Space* **2000**, *52*, 1031–1036. [CrossRef]
7. Chaouch, N.; Temimi, M.; Romanov, P.; Cabrera, R.; McKillop, G.; Khanbilvardi, R. An automated algorithm for river ice monitoring over the Susquehanna River using the MODIS data. *Hydrol. Processes* **2014**, *28*, 62–73. [CrossRef]
8. Heygster, G.; Alexandrov, V.; Dybkjær, G.; Hoyningen-Huene, W.v.; Girard-Ardhuin, F.; Katsev, I.; Kokhanovsky, A.; Lavergne, T.; Malinka, A.; Melsheimer, C. Remote sensing of sea ice: Advances during the DAMOCLES project. *Cryosphere* **2012**, *6*, 1411–1434. [CrossRef]
9. Shen, H.T.; Liu, L. Shokotsu River ice jam formation. *Cold Reg. Sci. Technol.* **2003**, *37*, 35–49. [CrossRef]
10. Ozsoy-Cicek, B.; Kern, S.; Ackley, S.F.; Xie, H.; Tekeli, A.E. Intercomparisons of Antarctic sea ice types from visual ship, RADARSAT-1 SAR, Envisat ASAR, QuikSCAT, and AMSR-E satellite observations in the Bellingshausen Sea. *Deep Sea Res. Part II* **2011**, *58*, 1092–1111. [CrossRef]
11. Gleason, S. Towards sea ice remote sensing with space detected GPS signals: Demonstration of technical feasibility and initial consistency check using low resolution sea ice information. *Remote Sens.* **2010**, *2*, 2017–2039. [CrossRef]
12. Song, C.S.; Zhu, X.Y.; Han, H.W.; Lin, L.B.; Yao, Z. The influence of riverway characteristics on the generation and dissipation of ice dam in the upper reaches of Heilongjiang River. *J. Hydrol. Eng.* **2020**, *51*, 1256–1266. (In Chinese)

13. Deng, X.; Wang, D.R.; Feng, J.; Zhang, L.; Qin, J.M. Image correction algorithm optimization and system design for ice density monitoring. *J. Taiyuan Univ. Technol.* **2020**, *51*, 310–315. (In Chinese)
14. Bushaw, J.D.; Ringelman, K.M.; Johnson, M.K.; Rohrer, T.; Rohwer, F.C. Applications of an unmanned aerial vehicle and thermal-imaging camera to study ducks nesting over water. *J. Field Ornithol.* **2020**, *91*, 409–420. [CrossRef]
15. Hunter III, J.E.; Gannon, T.W.; Richardson, R.J.; Yelverton, F.H.; Leon, R.G. Integration of remote-weed mapping and an autonomous spraying unmanned aerial vehicle for site-specific weed management. *Pest Manag. Sci.* **2020**, *76*, 1386–1392. [CrossRef] [PubMed]
16. Shafer, M.W.; Vega, G.; Rothfus, K.; Flikkema, P. UAV wildlife radiotelemetry: System and methods of localization. *Methods Ecol. Evol.* **2019**, *10*, 1783–1795. [CrossRef]
17. Kalke, H.; Loewen, M. Support vector machine learning applied to digital images of river ice conditions. *Cold Reg. Sci. Technol.* **2018**, *155*, 225–236. [CrossRef]
18. Wang, H.; Sun, J. Variability of Northeast China river break-up date. *Adv. Atmos. Sci.* **2009**, *26*, 701–706. [CrossRef]
19. Fu, H.; Liu, Z.; Guo, X.; Cui, H. Double-frequency ground penetrating radar for measurement of ice thickness and water depth in rivers and canals: Development, verification and application. *Cold Reg. Sci. Technol.* **2018**, *154*, 85–94. [CrossRef]
20. Guo, X.; Wang, T.; Fu, H.; Guo, Y.; Li, J. Ice-jam forecasting during river breakup based on neural network theory. *J. Cold Reg. Eng.* **2018**, *32*, 04018010. [CrossRef]
21. Lowe, D.G. Object recognition from local scale-invariant features. In Proceedings of the Seventh IEEE International Conference on Computer Vision, Kerkyra, Greece, 20–27 September 1999; pp. 1150–1157. [CrossRef]
22. Lowe, D.G. Distinctive image features from scale-invariant keypoints. *Int. J. Comput. Vision* **2004**, *60*, 91–110. [CrossRef]
23. Koenderink, J.J. The structure of images. *Biol. Cybern.* **1984**, *50*, 363–370. [CrossRef] [PubMed]
24. Lindeberg, T. Detecting salient blob-like image structures and their scales with a scale-space primal sketch: A method for focus-of-attention. *Int. J. Comput. Vision* **1993**, *11*, 283–318. [CrossRef]
25. Divya, S.; Paul, S.; Pati, U.C. Structure tensor-based SIFT algorithm for SAR image registration. *IET Image Process.* **2019**, *14*, 929–938. [CrossRef]
26. Chang, L.; Duarte, M.M.; Sucar, L.E.; Morales, E.F. A Bayesian approach for object classification based on clusters of SIFT local features. *Expert Syst. Appl.* **2012**, *39*, 1679–1686. [CrossRef]
27. Zhao, J.; Zhang, X.; Gao, C.; Qiu, X.; Tian, Y.; Zhu, Y.; Cao, W. Rapid mosaicking of unmanned aerial vehicle (UAV) images for crop growth monitoring using the SIFT algorithm. *Remote Sens.* **2019**, *11*, 1226. [CrossRef]
28. Saxena, A.; Jat, M.K. Analysing performance of SLEUTH model calibration using brute force and genetic algorithm–based methods. *Geocarto Int.* **2020**, *35*, 256–279. [CrossRef]
29. Fischler, M.A.; Bolles, R.C. Random sample consensus: A paradigm for model fitting with applications to image analysis and automated cartography. *Commun. ACM* **1981**, *24*, 381–395. [CrossRef]
30. Sun, Q.; Zhang, Y.; Wang, J.; Gao, W. An improved FAST feature extraction based on RANSAC method of vision/SINS integrated navigation system in GNSS-denied environments. *Adv. Space Res.* **2017**, *60*, 2660–2671. [CrossRef]
31. Chave, R.A.J.; Lemon, D.D.; Fissel, D.B.; Dupuis, L.; Dumont, S. Real-time measurements of ice draft and velocity in the St. Lawrence River. In Proceedings of the Oceans' 04 MTS/IEEE Techno-Ocean'04 (IEEE Cat. No. 04CH37600), Kobe, Japan, 9–12 November 2004; pp. 1629–1633. [CrossRef]

water

MDPI

Article

The Influence of the Internal Properties of River Ice on Ground Penetrating Radar Propagation

Hongwei Han [1,2], Yu Li [1], Wanyun Li [1], Xingchao Liu [1], Enliang Wang [1,2] and Haiqiang Jiang [1,2,*]

[1] School of Water Conservancy and Civil Engineering, Northeast Agricultural University, Harbin 150030, China
[2] Heilongjiang Provincial Key Laboratory of Water Resources and Water Conservancy Engineering in Cold Region, Harbin 150030, China
* Correspondence: hqjiang@neau.edu.cn; Tel.: +86-188-4579-4662

Abstract: Ground penetrating radar (GPR) has proven to be a very effective method for examining ice thickness. However, two preconditions must be met for this approach to be useful; the round-trip travel time of electromagnetic (EM) waves and radar transmission speed in the ice must be known. These issues are problematic because many factors affect radar transmission speed in ice, including impurities, physical properties such as porosity and density, and temperature. Results show that if these factors are not taken into account and a signal velocity of 0.17 m/ns in pure ice is used to estimate thickness, overestimates will result. We carried out a series of GPR surveys using dual channel host 200 MHz shielded antennas at the Toudaoguai Hydrological Station on the Yellow River, China, and collected samples to analyze ice impurities and physical properties. The results show that the ice crystal types include frazil, granular, and column at the Toudaoguai Hydrologic Station section. Our analysis of ice gas bubble and sediment content showed that the gas bubble volume content is between 11.95 and 13.0% in the frazil ice and between 7.9% and 8.6% in granular and columnar ice. At the same time, the ice sediment content ranged between 0.11‰ and 0.57‰, and the average was 0.24‰ in granular and columnar ice, which was about one-tenth that of the suspended sediment concentration in water. Additionally, a combination of GPR data as well as ice impurities, porosity, density, and temperature enabled us to provide insights on the variability of radar transmission speed and the equivalent dielectric permittivity in river ice. Our extensive observations reveal that radar transmission speed falls between 0.141 m/ns and 0.164 m/ns and that the equivalent dielectric permittivity of river ice increases in concert with ice temperature.

Keywords: river ice; thickness; GPR; temperature; Yellow River

Citation: Han, H.; Li, Y.; Li, W.; Liu, X.; Wang, E.; Jiang, H. The Influence of the Internal Properties of River Ice on Ground Penetrating Radar Propagation. *Water* **2023**, *15*, 889. https://doi.org/10.3390/w15050889

Academic Editor: Frédéric Frappart

Received: 11 December 2022
Revised: 9 February 2023
Accepted: 23 February 2023
Published: 25 February 2023

1. Introduction

Freshwater ice cover is a fundamental characteristic of northern temperate aquatic systems and is associated with the least productive months of the year [1]. Seasonal freshwater ice will still be present when the accumulated number of freezing days is sufficiently large, while perennial ice cover can be found at high altitudes and at high polar latitudes. Brooks et al. [2] studied the areal extent and volume of Northern Hemisphere freshwater ice in rivers and lakes and showed that the total covered area at peak thickness, north of the January 0 °C isotherm (excluding the Greenland ice sheet), is 1.7×10^6 km^2. Additionally, the total volume of Northern Hemisphere freshwater ice is 1.6×10^3 km^3. The extent and duration of freshwater ice cover has a tremendous impact on human life, including influencing large-scale climate change [3]. At the same time, flood forecasting and mitigation are influenced when ice cover breaks up on a river [4,5]. Freshwater ice is closely related to our life; accurate measurement of freshwater ice thickness is important if we are to determine the production activities of ice cover. The traditional method to estimate ice thickness is the use of a drill-hole, but this approach is relatively inefficient when generating large amounts of data.

Ground penetrating radar (GPR) is a noninvasive geophysical technique that can be used to detect electrical discontinuities in the shallow subsurface. This noninvasive geophysical technique was first used in the early 20th century by German scientists to investigate the nature of buried features [6,7]. Concomitant with technology development, GPR systems became commercially available and digital data acquisition was first feasible in the 1980s [8]. This approach is now widely applied across a range of fields, including underground archaeology, soil science, and engineering. GPR also plays an important role in the detection of ice and snow; in glaciology, for example, several research studies [9–16] have employed this approach to measure the thickness of glaciers. A range of investigations have also used GPR to measure snow thickness [17–19], and this approach has also been applied to snowpack analysis to determine physical properties, measure liquid water content [20–22], estimate density [23–25], and determine snow water equivalent [26,27]. Applications of GPR to river [28–30], sea [31], and reservoir ice [32] have primarily focused on thickness measurements and the detection of floating or grounded ice conditions [33,34]. Because the dielectric permittivities of ice (ε = 3–4), freshwater (ε = 81), and sediment (ε = 5–40) contrast greatly with one another, GPR can be used to detect the ice–water and water–sediment interface via changes in electromagnetic (EM) signals when high-frequency waves penetrate through these boundaries. The GPR can also be used to estimate winter ebullition bubble volume in lake ice, which can make an accurate assessment of gas emissions from lakes [35]. The dielectric permittivity of ice is not a fixed value; it varies depending on physical properties, including density, gas content, and textural structure. Thus, if we want to accurately measure the ice thickness by GPR, there is a strong need to carry out a detailed study of the influence of the internal properties of river ice on ground penetrating radar propagation.

The dielectric permittivity of river ice is relatively complex under natural conditions, as it is influenced by its physical properties and inner structure. Thus, river ice thickness can be measured accurately via GPR under the premise of determining the dielectric permittivity. We use GPR in this study to investigate the effects of impurities, physical properties (porosity and density), and temperature on the dielectric permittivity of river ice.

2. Research Area and Methods

2.1. Overview of the Research Area

The Yellow River is the third-longest river in Asia and the sixth-longest in the world. Originating on the high Qinghai-Tibetan Plateau in Qinghai Province in western China, the Yellow River flows north, turns south, and then bends east for a total of 5464 km before draining into the Bohai Sea near the city of Dongying in Shandong Province, encompassing a basin area of ca. 752,000 km^2. The upper part of this drainage basin is located on the Qinghai-Tibetan Plateau, the middle part is located on the Loess Plateau, and the lower part is on the North China Plain [36]. The Toudaoguai cross-section lies in the terminal mainstream channel of the upper Yellow River, as shown in Figure 1; specifically, this section is located in the zone where the river transitions from a meandering to a straight channel, the latter of which has a length of about 16 km [37]. The winters are bitterly cold and long in this region, with air temperatures below zero from November to March [38]. As air temperature changes, ice formation, development, movement, accumulation, thaw, and disappearance all exert important impacts on river engineering and management. Within the Toudaoguai cross-section, freeze-up of the Yellow River often occurs in mid-December, while ice break-up takes place towards the end of March the following year [39]. During the break-up period, ice floods lead to frequent jams and dams within the region.

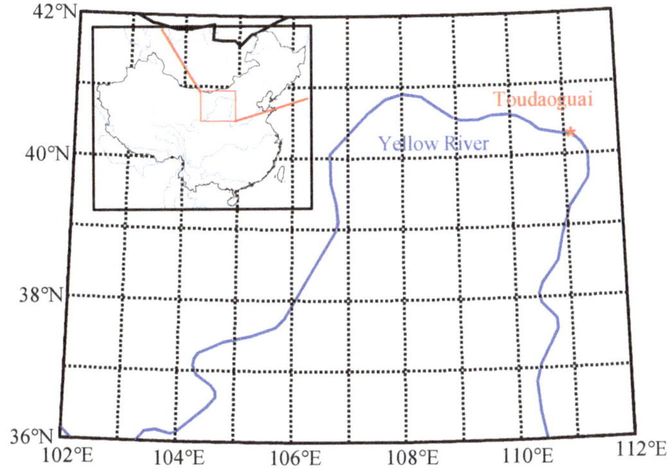

Figure 1. Location of our study area within the Toudaoguai cross-section of the Yellow River.

2.2. Research Method

We performed radar measurements using a RIS-K2 system (IDS Ltd., Pisa, Italy) and a survey with dual channel host 200 MHz shielded antennas. We operated GPR systems in continuous reflection mode and used an odometer wheel to measure and mark distances; all GPR profile data were collated with a laptop computer and processed using the software GresWin2.

GPR surveys were conducted along the Toudaoguai Hydrological Station river section on 19 January and 15, 20, 24, 25, and 27 February 2014, and four measurements were made every day in order to eliminate the discreteness of test results. The main river channel at this section spans about 340 m, while the shoal is about 140 m wide. We detected four survey lines (Figure 2) during the field investigation; two of these (L-1, L-2) were perpendicular to the course of river, while another two (L-3, L-4) ran parallel to the river course.

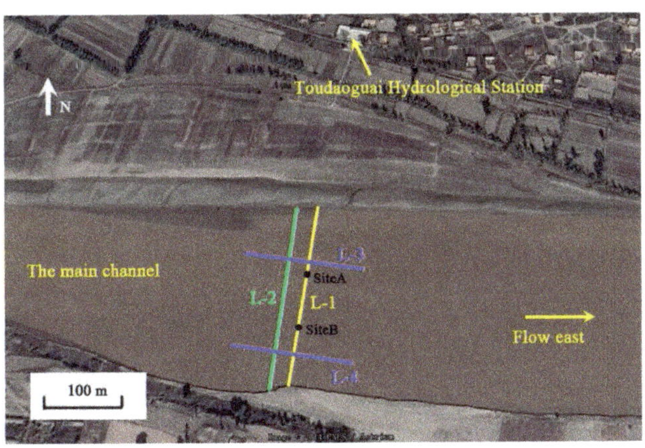

Figure 2. Aerial view of the Toudaoguai Hydrological Station section. Four survey lines (L-1, L-2, L-3, L-4) were used in the main channel during the field investigation. Site A and Site B are the locations of ice core sampling points respectively. Landscape image source: Google Earth, 25 October 2013.

As shown in Figure 3, four 2.0 m long temperature sensor chains were used to determine the ice and water temperature along L-1 line from 8 January 2014. The temperature probe spacing was 0.01 m, and the measuring accuracy was 0.1 °C. Because the ice thickness at the hydrological observation point was less than undisturbed ice cover, we controlled the L-1 line to be more than 1 m away from hydrological observation points in actual operation in order to ensure the accuracy of inversion radar ice thickness measurement. We verified GPR reflections by drilling ice thickness (Figure 3c), and the precision of manual measurement of ice thickness was 0.001 m. In order to study the influence of the internal structure of the river ice on the EM velocity, ice specimens were sampled by a Kovacs drill (Roseburg, OR, USA) with a corer of 0.09 m diameter at Site A and Site B on 15 February 2014.

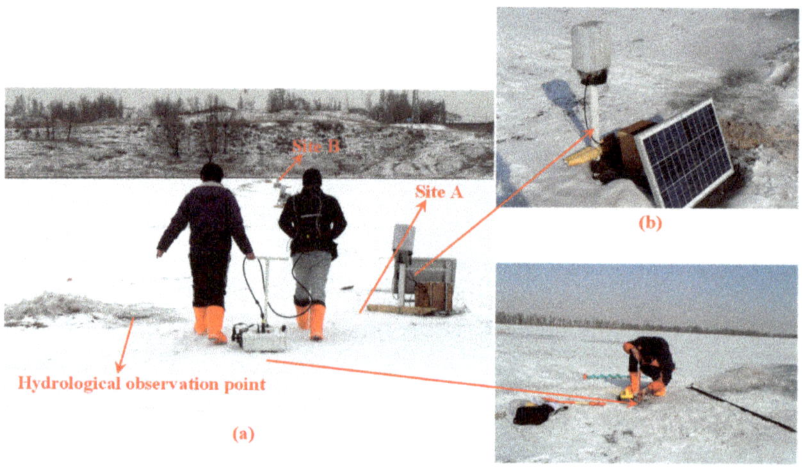

Figure 3. Pictures of the test site. (**a**) GPR survey. (**b**) Temperature measuring device. (**c**) Measurement of ice thickness using drilling method.

The standard time-voltage wiggle traces of GPR signals are shown in Figure 4; comparing the time domains of these two signals shows the same waveform in ice cover. The first and largest echo (A_1) is due to the air–ice interface, while the second largest echo (A_2) that appears later is due to reflections of the ice–water interface. The time interval between A_1 and A_2 represents the two-way travel time through the ice cover.

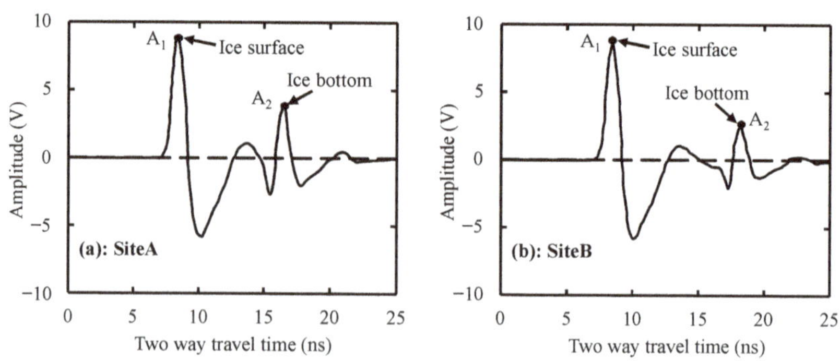

Figure 4. The standard time-voltage wiggle traces of the GPR signal at Site A (**a**) and Site B (**b**) that measured on 15 February 2014.

Based on EM wave propagation theory, EM velocity in a single medium can be expressed as follows:

$$V = \frac{c}{\sqrt{\varepsilon}} \tag{1}$$

where c is the EM velocity in a vacuum, 0.3 m/ns, while ε denotes the dielectric permittivity of the medium, 1 for air and 3.17 for pure ice.

Nevertheless, analyses of the physical properties of ice revealed the existence of vertical stratification. Ice cover can therefore be assumed to comprise a multilayer dielectric system, with homogeneous properties in each layer. Thus, the electrical conductivity of ice can be neglected.

The RIS-K2 system coupled with dual-channel hosted 200 MHz shielded antennas is a monostatic radar system. Thus, round-trip travel times of EM waves from transmission to reception within ice cover can be obtained as follows:

$$T = t_1 + t_2 + t_3 + \ldots + t_{\mathrm{n}} = \frac{2}{c}(h_1\sqrt{\varepsilon_1} + h_2\sqrt{\varepsilon_2} + h_3\sqrt{\varepsilon_3} + \ldots + h_{\mathrm{n}}\sqrt{\varepsilon_{\mathrm{n}}}) = \frac{2}{c}\sum_{\mathrm{i}=1}^{\mathrm{n}} h_i\sqrt{\varepsilon_i} \tag{2}$$

and

$$T = \frac{2H\sqrt{\varepsilon_{dz}}}{c} \tag{3}$$

Solving Equations (2) and (3), the equivalent dielectric permittivity (ε_{dz}) of ice cover is as follows:

$$\varepsilon_{dz} = \left(\sum_{\mathrm{i}=1}^{\mathrm{n}} \frac{h_i\sqrt{\varepsilon_i}}{H}\right)^2 \tag{4}$$

In these expressions, T refers to two way travel time (ns), c is the EM velocity in a vacuum, i is the interface number, t_i is the ns from interface $(i-1)$ to interface (i), ε_i is the dielectric permittivity of ice from interface $(i-1)$ to interface (i), H is the total depth (m), h_i is the depth (m) of ice from interface $(i-1)$ to interface (i), and ε_{dz} is the equivalent dielectric permittivity of ice cover.

Kovacs et al. [40] developed an empirical relationship connecting ice density with the dielectric permittivity if river ice is assumed to be a composite material comprised of ice and air, as follows:

$$\varepsilon_{\mathrm{ice}} = (1 + 0.845\rho)^2 \tag{5}$$

where ρ is measured ice density and $\varepsilon_{\mathrm{ice}}$ is the ice dielectric permittivity.

Shokr [41] modified the two-phase dielectric mixing model for ice, suggesting that when the volume fraction of inclusions is less than 0.1, the dielectric permittivity of granular ice (ε_m) can be calculated as follows:

$$\varepsilon_m = \varepsilon_h\left[1 + 3v_{\mathrm{gas}}\frac{(\varepsilon_{\mathrm{gas}} - \varepsilon_h)}{(\varepsilon_{\mathrm{gas}} + 2\varepsilon_h)}\right] \tag{6}$$

Similarly, the dielectric permittivity of columnar ice can be calculated as follows:

$$\varepsilon_{mx} = \varepsilon_h\left[1 + 2v_{\mathrm{gas}}\frac{(\varepsilon_{\mathrm{gas}} - \varepsilon_h)}{(\varepsilon_{\mathrm{gas}} + \varepsilon_h)}\right] \tag{7}$$

$$\varepsilon_{mz} = \varepsilon_h + v_{\mathrm{gas}}(\varepsilon_{\mathrm{gas}} - \varepsilon_h) \tag{8}$$

In these expressions, ε_{mx} and ε_{mz} refer to the dielectric permittivities of columnar ice in the x direction (perpendicular to the ice column) and z direction (along the ice column), while $\varepsilon_{\mathrm{gas}}$ is the dielectric permittivity of gas bubbles, ε_h is the d dielectric permittivity of pure ice, and ν_{gas} is the volume content of gas bubbles. Thus, gas bubble volume content relative to porosity can be calculated as follows [42]:

$$\nu_{gas} = \frac{\rho_0 - \rho}{\rho_0} \times 100\% \quad (9)$$

In this expression, ρ_0 is the density of pure ice (916.8 kg/m^3), while ρ is the measured density of ice cover. The results of these calculations show that gas bubble volume content is less than 0.1; thus, Equations (6)–(8) can be used to calculate the dielectric permittivity of river ice.

3. Results

3.1. Ice Formation and Ice Physical Properties

The formation of river ice is a complex process that depends on a combination of hydrodynamic and atmospheric conditions. In the Toudaoguai Hydrological Station section, observed hydrological data show that ice cover forms from the left side of the river to the right; as air temperature decreases, a thin cover of ice forms in gently flowing areas near to the left bank of the river. As the left part of the river iced over, frazil particles and slush in turbulent water floated to the surface to create frazil pans on the right side of the river, before congealing into ice cover in mid-December 2013.

We collected a series of 0.09 m diameter ice cores for textural and structural analyses on 15 February 2014; at this time, the ice thickness was 0.600 m at Site A and 0.780 m at Site B. Vertical thin sections viewed under polarized and normal light revealed the crystalline structure and defects (i.e., gas bubbles and sand particles) in river ice. We collected three types of ice—frazil, granular, and columnar—from the Toudaoguai Hydrological Station section (Figure 5). The ice forming at the beginning of winter at Site A comprised granular ice in a layer about 0.05 m thick. Columnar and frazil ice were alternately distributed beneath this layer, while at Site B, the ice comprised frazil ice, attributed to slush accumulation (Figure 6). According to the variation of the thickness of frazil slush and field observation, it was found that the frazil slush suspended in water and transported under the ice cover throughout the winter; during the initial freeze-up stage, frazil slush in the flow accumulated along the underside of the ice cover due to jams, forming a maximum slush thickness of 1.70 m by 31 December 2013. Following the initiation of ice cover, energy exchange over the top of this layer and frazil slush led to the formation of frazil ice.

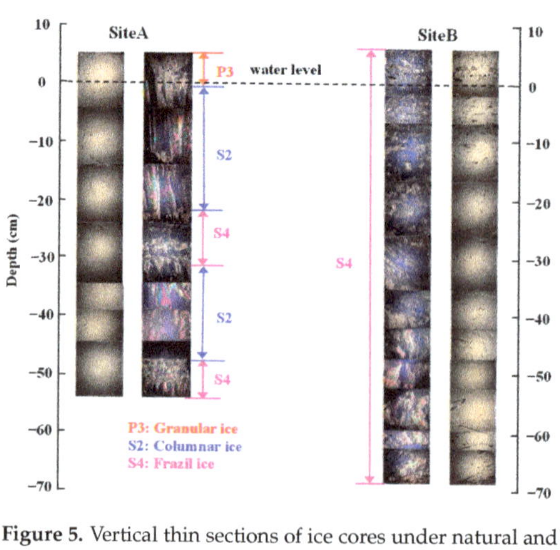

Figure 5. Vertical thin sections of ice cores under natural and polarized light collected on 15 February 2014 at Site A and Site B. The origin on the *y*-axis corresponds with the water level.

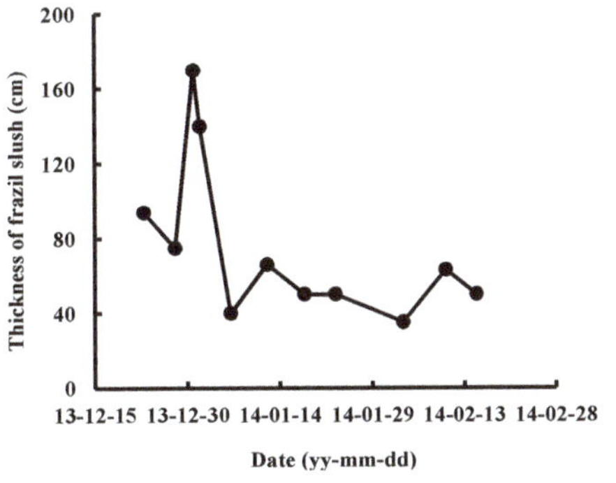

Figure 6. Thickness of frazil slush under the ice cover at Site B.

We measured ice density using the mass-volume method. A gas/ice horizontal thin section image was photographed under normal light, and the ice gas content was calculated using MATLAB software. Vertical density profiles and ice core gas contents are shown in Figure 7; it is noteworthy that frazil ice from Site B has a higher gas content than the pure thermal growth ice that contains intermittent frazil ice from Site A. Results show that the gas content of the frazil ice layer at core Site A is about 1.3% larger than adjacent columnar ice layers. Ice density is also negatively correlated with gas content; the density at Site B was less than 900 kg/m^3 but increased with depth as gas content decreased.

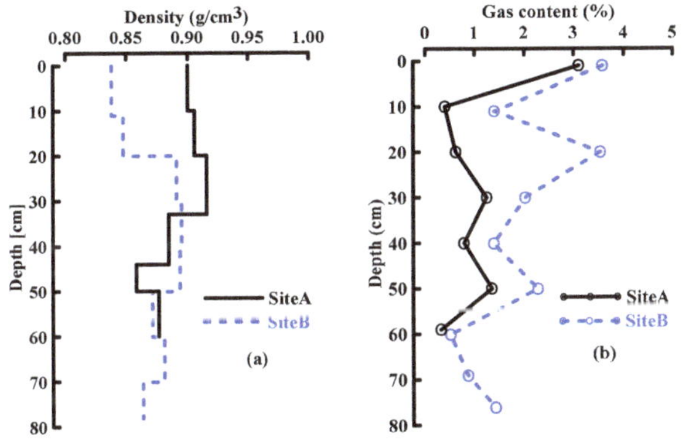

Figure 7. Vertical profiles of density (**a**) and gas content (**b**) for ice cores collected on 15 February 2014.

3.2. Interpretation of GPR Image

Radar investigations comprise measurements of wave propagation time; thus, once EM wave propagation speed is known, it is possible to accurately derive ice cover reflector depth. The EM propagation speed is then determined by the electrical properties of a material; in pure ice, for example, EM speed ranges between 0.1667 m/ns and 0.1700 m/ns [19]. In this study, depth on the right vertical axis of Figure 8 was estimated using a signal velocity of 0.17 m/ns for freshwater ice.

As shown in Figure 8, radar images provide clear cross-sections of ice cover. As expected, we observed significant reflections from the air–ice and ice–water interfaces; the distribution of ice thickness was uneven in the L-1 and L-2 surveys because cover was thinner on the left side (Figure 8a,b) and a 25 m wide section of grounded ice was present near to the left bank of the shallow water area. In this area, radar signal also revealed a strong and continuous reflection from the ice–sediment interface. Ice thickness was also not homogeneous on the right section because this region comprised frazil ice attributed to slush accumulation. In addition, survey L-4, parallel to the river course, showed that frazil ice thickness was not homogeneous (Figure 8d), while the radar result from survey line L-3 showed a more uniform distribution of ice thickness (Figure 8c).

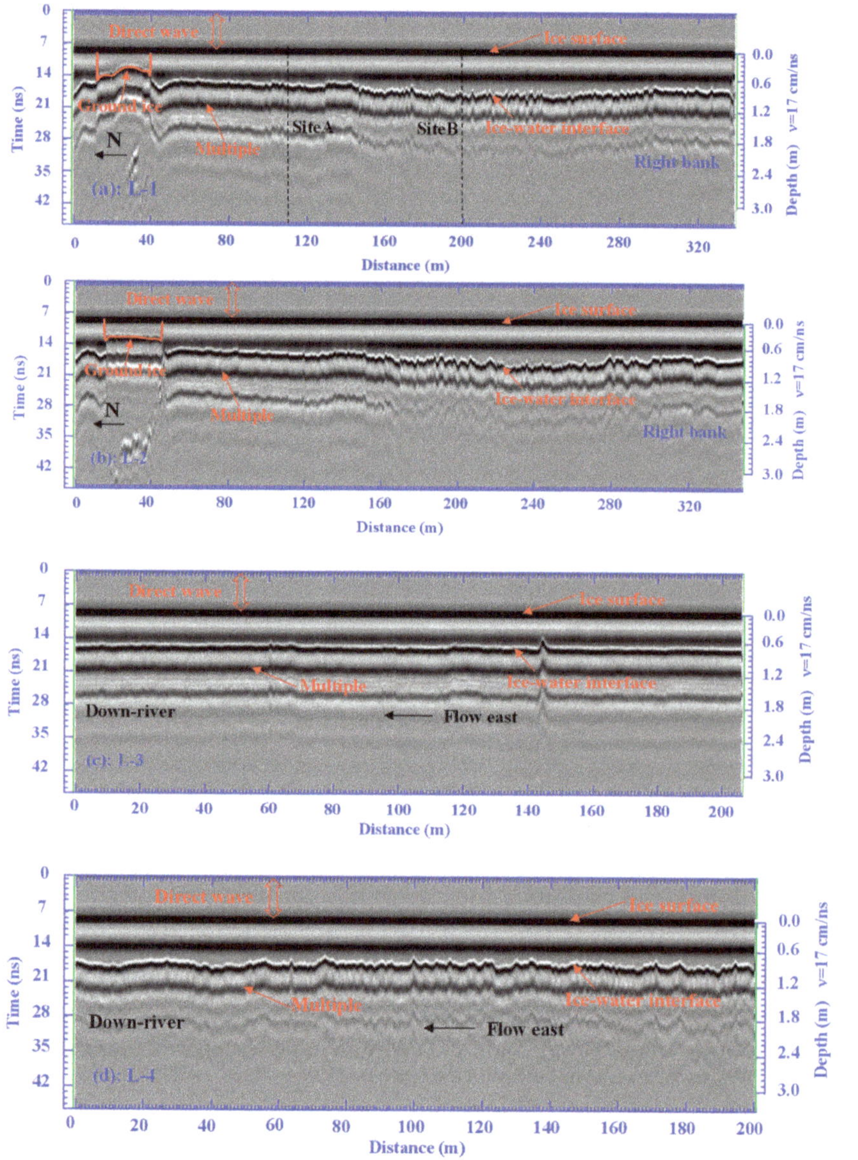

Figure 8. The 200 MHz GPR image profiles measured on 15 February 2014, with direct air and surface-coupled waves retained. The left side is north in images (**a**,**b**), while the left side is east in (**c**,**d**); the location of ice core drilling points in (**a**) is marked with a dotted line. Depth on the right vertical axis was estimated by converting two-way travel time to distance for observed reflection using a signal velocity of 0.17 m/ns in freshwater ice.

3.3. Dielectric Permittivities of River Ice

Because gas bubbles are randomly distributed spheres in frazil ice, similar to the granular case, the formula for the latter can be used to calculate the dielectric permittivity for the former. Thus, applying Equations (5)–(9) as well as the physical properties of ice at Site A and Site B on 15 February 2014, we calculated the dielectric permittivities for river ice. Significant differences were not found between the results of the two calculations (Figure 9) for frazil ice at Site B, while the difference was larger for the columnar ice at Site A.

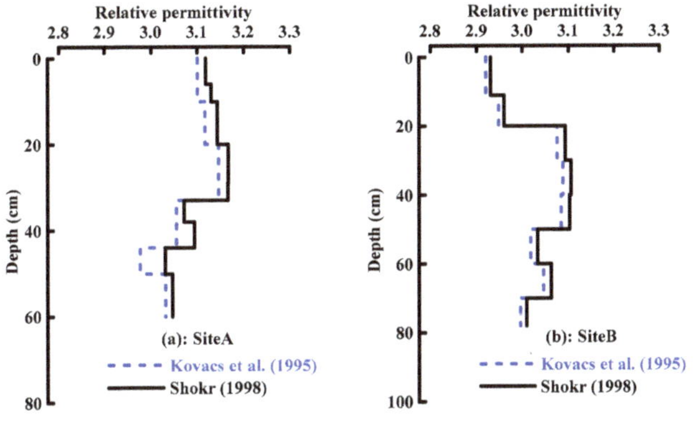

Figure 9. Dielectric permittivities of river ice at Site A (**a**) and Site B (**b**) calculated using the equations presented by Kovacs et al. [40] and Shokr [41].

The equivalent dielectric permittivities of river ice calculated using the radar method were derived using Equation (4) based on the round-trip travel times and drilled ice thicknesses. As shown in Table 1, the equivalent dielectric permittivities of river ice calculated using the radar method were higher than results derived from empirical formulae based on ice density [40]. At the same time, the gas bubble volume content was between 11.95% and 13.0% at Site A and between 7.9% and 8.6% at Site B. Measurements of in situ ice taken on 15 February 2014 show that the average temperature was −2.45 °C at Site A and −3.24 °C at Site B. It is therefore likely that the river ice was composed of crystals, gas bubbles, and impurities and did not contain liquid water at this time. In particular, the equivalent dielectric permittivity of this river ice was so high that it was likely caused by impurities; Evans [43] discussed how the presence of impurities in snow increased values of the dielectric permittivity, and the same is likely true for ice. Previous results show that the annual average suspended sediment concentration in Yellow River water at the Toudaoguai Hydrological Station was about 2.5‰ (2.5 kg/m^3) between 1987 and 2010 [44]. The suspended sediment in river water mainly comprises silts and sands, which participate in ice freezing and thawing processes. We therefore analyzed the ice sediment content (Figure 10); the results show that sediment concentration ranged between 0.11‰ and 0.57‰ and that the average at Site A was 0.24‰. Although the sediment concentration of ice was about one-tenth that of the suspended sediment concentration in water, particles in the former (i.e., silts and sands) had significantly higher dielectric permittivities [45,46]. It is therefore likely that sediment content influences the equivalent dielectric permittivities of river ice when the radar method is used for calculations. Unfortunately, however, our existing experimental data are insufficient to build a model that encompasses ice sediment concentration and enables the estimation of the river ice dielectric permittivity.

Table 1. The dielectric permittivity of river ice at Site A and Site B on 15 February 2014, as calculated using three different methods.

Regime	Site A		Site B	
	ε_{dz} [1]	ε_{avg} [2]	ε_{dz} [1]	ε_{avg} [2]
Kovacs et al. [40]	3.1	3.1 (±0.06)	3.0	3.0 (±0.06)
Shokr [41]	3.1	3.1 (±0.05)	3.0	3.0 (±0.06)
Radar method	—	3.5 (±0.06)	—	3.3 (±0.03)

Note(s): [1] ε_{dz} is the equivalent dielectric permittivity; [2] ε_{avg} is the mean dielectric permittivity, with standard deviation in parentheses.

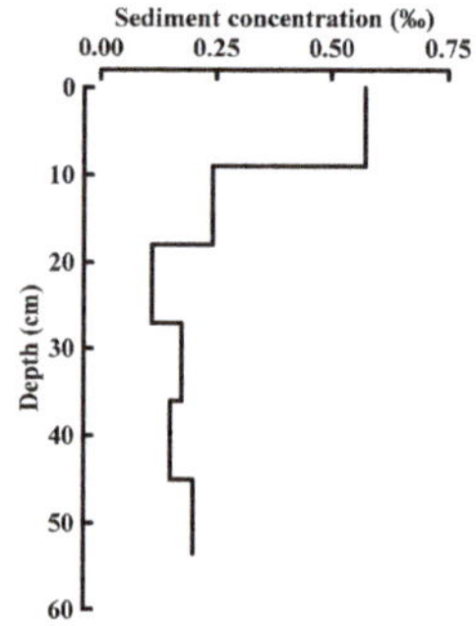

Figure 10. Vertical profiles of ice sediment concentration from Site A.

4. Discussion

In addition to impurities, it is well-known that temperature can also impact the dielectric permittivity of ice. This is because the internal structure changes as ice temperature increases and approaches the melting point. Thus, in order to analyze the relationship between the ice dielectric permittivity and temperature, we first evaluated the latter for both air and ice. Results show (Figure 11) that air temperature rose gradually between 10 January 2014 and 1 February 2014, especially the daytime temperature after 21 January 2014, which became positive. This increase in air temperature also resulted in a rise in ice temperature and melting; as a result, ice surface melting was initiated at both Site A and Site B. Bottom melting at the ice–water interface also led to the thinning of ice at Site A during this period. Another period of increasing temperatures was seen between 9 February 2014 and 25 February 2014; ice temperature rose close to 0 °C near to the end of this period of increase (Figure 12).

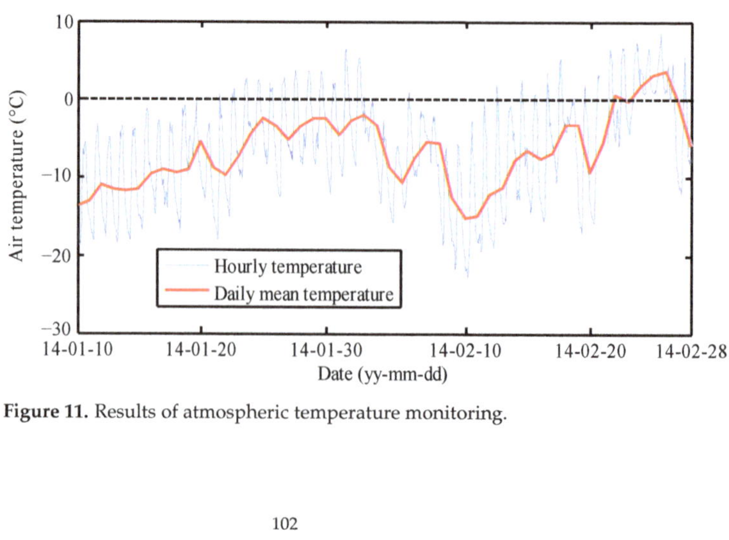

Figure 11. Results of atmospheric temperature monitoring.

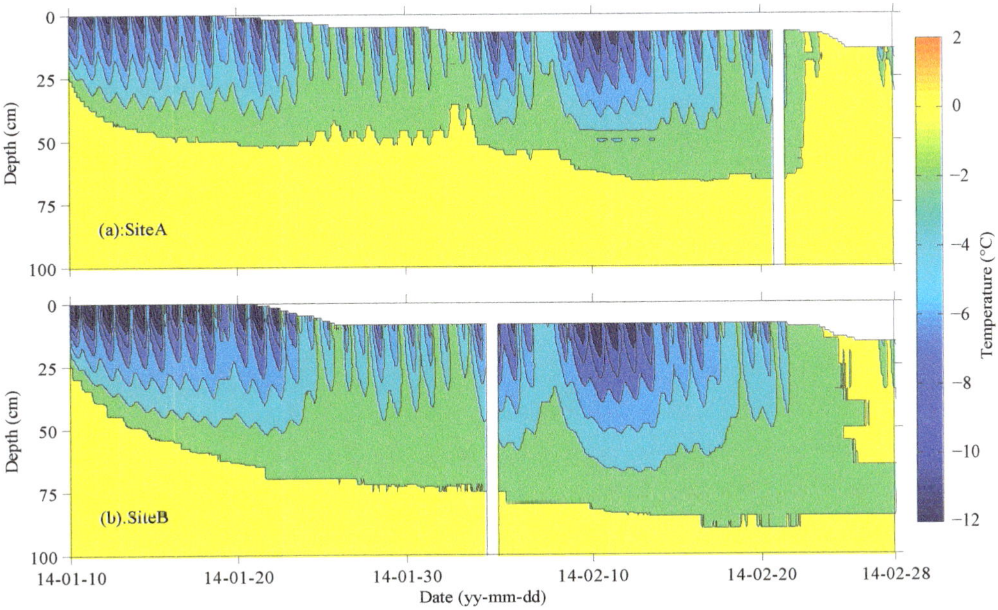

Figure 12. Internal ice and water temperature evolution at Site A (**a**) and Site B (**b**).

We computed radar transmission speeds in river ice using travel times and ice thicknesses measured at a number of locations. Results show (Figure 13) that these speeds in river ice were between 0.141 m/ns and 0.164 m/ns, lower than those observed in 'pure' ice. This result is consistent with previously published data; for example, Moorman and Michel [46] noted that ice melting on Bulb and Kame lakes had transmission speeds of 0.133 m/ns and 0.140 m/ns, respectively. These workers were also able to clarify that the slower velocity measured in river ice compared to 'pure' ice is the result of unfrozen water content. Indeed, even outside the melting period, radar transmission speeds measured in natural ice are normally lower than the 0.17 m/ns that characterizes 'pure' ice. Hausmann and Behm [47] measured transmission speeds of cave ice between 10.157 m/ns and 0.169 m/ns, while Stoyanovich et al. [48] reported ice EM propagation velocity to be 0.154 m/ns. These data also suggest that radar transmission speeds in natural ice are highly variable; consequently, the dielectric permittivity of natural ice is not always equal to 3.17. Our results also suggest that the dielectric permittivity of ice increases in concert with ice temperature; data show that average ice temperature was −2.85 °C at Site A and −3.32 °C at Site B on 19 January 2014, during the coldest period of the whole freezing cycle. At the same time, the radar transmission speed of a pulse through river ice was 0.163 m/ns, while the equivalent dielectric permittivity was around 3.4 at Site A and Site B based on drill-hole thickness data. Data show that as spring approached, ice temperature gradually increased and was generally above −2.00 °C on 24 February 2014. Finally, after two days of rapid increases, ice temperature was close to 0 °C on 27 February 2014. Equivalent dielectric permittivities also increased rapidly over this period (Figure 13), especially when the ice temperature was close to 0 °C. This is because the dielectric permittivity of water is obviously higher than ice, while the inevitable presence of unfrozen water within ice causes an increase in the dielectric permittivity as the temperature approaches 0 °C. Previous studies that have used GPR to assess winter roads have shown that radar transmission speeds are strongly dependent on ice thickness; thus, the inherent nature of depth analysis means that the presence of unfrozen water in ice will be the main factor influencing calculations [49].

You et al. [50] also reported the average velocities decreased slightly with a rise in the unfrozen water content in the ice.

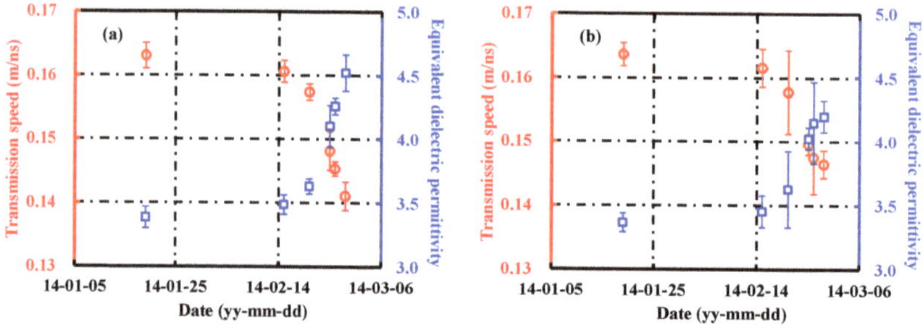

Figure 13. Radar transmission speed and equivalent dielectric permittivity calculated by measured ice thickness and round-trip travel times: (**a**) Site A; (**b**) Site B.

During our whole observation period, the minimum equivalent dielectric permittivity of the Yellow River ice was 3.4 ± 0.1 when the average ice temperature was between −2.85 °C and −3.32 °C on 19 January 2014, which was higher than laboratory measurements of the artificial ice reported in previous studies. Bohleber et al. [51,52] reported that the mean value of the relative permittivity of the artificial ice grown from ultra-pure water is 3.18 ± 0.01 at −20 °C. We believe that the difference of the dielectric permittivity between the river ice and pure ice is due to the different physical properties of ice. Regardless of ice impurities or the effect of ice temperature, radar transmission speeds in river ice can ultimately be attributed to the influence of the internal structure on the dielectric permittivity. The influence of ice temperature on radar transmission speed is affected by varying the percentage of liquid water in the ice; indeed, establishing a relationship between the liquid water percentage and the ice dielectric permittivity is the most scientific method to evaluate radar transmission speeds in river ice. However, it is difficult to measure the percentage of liquid water directly, which limits our ability to establish a relationship between the dielectric permittivity and the liquid water percentage. Thus, our approach to address the relationship between ice temperature and the dielectric permittivity of river ice provides a method for assessing high-temperature ice, which can be used to evaluate radar transmission speeds. Accurate measurement of ice thickness can nevertheless be achieved via the use of GPR at the end of the winter.

5. Concluding Remarks

We conducted a series of GPR measurements in 2014 on river ice thicknesses at the Toudaoguai Hydrological Station section of the upper Yellow River, China. This field investigation utilized four survey lines, of which two were perpendicular to the river course, while another two ran parallel. Our results show that ice cover formed by the accumulation of frazil ice was not homogeneous in the right part of river, while ice cover that formed via thermodynamic processes was more uniformly distributed on the left side of the river.

In order to use GPR to measure ice thickness, it is necessary to know the round-trip travel time of EM waves as well as radar transmission speed. Because it is possible to obtain accurate round-trip travel times in ice as part of the measurement process, radar transmission speed, determined by the dielectric permittivity, is the main factor that influences accuracy. Thus, in order to ensure measurement precision, in situ calibration is necessary to determine radar transmission speed or the equivalent dielectric permittivity of ice. We therefore investigated the effects of variable physical properties on the dielectric permittivity of river ice; the results reveal that equivalent dielectric permittivities of river ice calculated using the radar method are greater than those based on empirical formulae for ice density [40] and volume fraction [41], directly attributable to the concentration of sediment in ice. The effect of ice temperature on the dielectric permittivity of ice is affected by changes in the percentage of liquid water in ice; our results show that the equivalent dielectric permittivity of ice increases in concert with ice temperature.

Thus, while GPR can be used to rapidly measure the total thickness of ice, it cannot be used to determine changes in spatial thickness caused by uneven ice surface fluctuations. We therefore recommend that future studies employ real-time kinematic techniques to measure ice surface elevation changes and combine them with total ice thicknesses obtained using GPR measurements to evaluate fluctuations in the surfaces and bases of ice.

Author Contributions: Conceptualization, H.H. and E.W.; methodology, H.H., E.W., X.L. and H.J.; formal analysis, H.H. and W.L.; investigation, H.H.; data curation, H.H., Y.L., W.L. and X.L.; writing—original draft preparation, H.H. and Y.L.; writing—review and editing, H.J.; funding acquisition, E.W. All authors have read and agreed to the published version of the manuscript.

Funding: This research was supported by Natural Science Foundation of Heilongjiang Province of China (No. LH2020E004), the National Natural Science Foundation of China (41876213), and the Project to Support the Development of Young Talent by Northeast Agricultural University.

Data Availability Statement: The data are available upon request.

Conflicts of Interest: The authors declare no conflict of interest.

References

1. Frenette, J.J.; Thibeault, P.; Lapierre, J.F.; Hamilton, P.B. Presence of algae in freshwater ice cover of fluvial LAC Saint-Pierre (St. Lawrence River, Canada). *J. Phycol.* **2008**, *44*, 284–291. [CrossRef] [PubMed]
2. Brooks, R.N.; Prowse, T.D.; O'Connell, I.J. Quantifying Northern Hemisphere freshwater ice. *Geophys. Res. Lett.* **2013**, *40*, 1128–1131. [CrossRef]
3. Prowse, T.; Alfredsen, K.; Beltaos, S.; Bonsal, B.; Duguay, C.; Korhola, A.; McNamara, J.; Vincent, W.F.; Vuglinsky, V.; Weyhenmeyer, G.A. Arctic freshwater ice and its climatic role. *AMBIO* **2011**, *40*, 46–52. [CrossRef]
4. Ferrick, M.G.; Mulherin, N.D. Framework for control of dynamic ice breakup by river regulation: Regulated Rivers. *Res. Manag.* **1989**, *3*, 79–92. [CrossRef]
5. Mahabir, C.; Hicks, F.; Fayek, A.R. Neuro-fuzzy river ice breakup forecasting system. *Cold Reg. Sci. Technol.* **2006**, *46*, 100–112. [CrossRef]
6. Reynolds, J.M. *An Introduction to Applied and Environmental Geophysics*; Wiley: New York, NY, USA, 1997; p. 796.
7. Neal, A. Ground-penetrating radar and its use in sedimentology: Principles, problems and progress. *Earth-Sci. Rev.* **2004**, *66*, 261–330. [CrossRef]
8. Annan, A.P.; Davis, J.L. Design and development of a digital ground penetrating radar system. In *Ground Penetrating Radar*; Canada Communication Group: Ottawa, ON, Canada, 1992; pp. 15–23.
9. Binder, D.; Brückl, E.; Roch, K.H.; Behm, M.; Schöner, W.; Hynek, B. Determination of total ice volume and ice-thickness distribution of two glaciers in the Hohe Tauern region, Eastern Alps, from GPR data. *Ann. Glaciol.* **2009**, *50*, 71–79. [CrossRef]
10. Cui, X.B.; Sun, B.; Tian, G.; Tang, X.Y.; Zhang, X.P.; Jiang, Y.Y.; Guo, J.X.; Li, X. Ice radar investigation at Dome A, East Antarctica: Ice thickness and subglacial topography. *Chin. Sci. Bull.* **2010**, *55*, 425–431. [CrossRef]
11. Saintenoy, A.; Friedt, J.M.; Booth, A.D.; Tolle, F.; Bernard, E.; Laffly, D.; Marlin, C.; Griselin, M. Deriving ice thickness, glacier volume and bedrock morphology of Austre Lovénbreen (Svalbard) using GPR. *Near Surf. Geophys.* **2013**, *11*, 253–261. [CrossRef]
12. Bohleber, P.; Sold, L.; Hardy, D.R.; Schwikowski, M.; Klenk, P.; Fischer, A.; Sirguey, P.; Cullen, N.J.; Potocki, M.; Hoffmann, H.; et al. Ground-penetrating radar reveals ice thickness and undisturbed englacial layers at Kilimanjaro's Northern Ice Field. *Cryosphere* **2017**, *11*, 469–482. [CrossRef]
13. Langhammer, L.; Rabenstein, L.; Bauder, A.; Maurer, H. Ground-penetrating radar antenna orientation effects on temperate mountain glaciers. *Geophysics* **2017**, *82*, H15–H24. [CrossRef]
14. Mccarthy, M.; Pritchard, H.; Willis, I.; King, E. Ground-penetrating radar measurements of debris thickness on Lirung Glacier, Nepal. *J. Glaciol.* **2017**, *63*, 543–555. [CrossRef]
15. Li, Y.; Li, Z.; Wang, N. Ice thickness sounded by ground penetrating radar on the Meikuang Glacier in the Eastern Kunlun Mountains. *J. Glaciol. Geocryol.* **2018**, *40*, 38–46. (in Chinese). [CrossRef]
16. Mishra, A.; Negi, B.D.S.; Banerjee, A.; Nainwal, H.C.; Shankar, R. Estimation of ice thickness of the Satopanth Glacier, Central Himalaya using ground penetrating radar. *Curr. Sci. India* **2018**, *114*, 785–791. [CrossRef]
17. Kanagaratnam, P.; Markus, T.; Lytle, V.; Heavey, B.; Jansen, P.; Prescott, G.; Gogineni, S.P. Ultrawideband radar measurements of thickness of snow over sea ice. *IEEE Trans. Geosci. Remote Sens.* **2007**, *45*, 2715–2724. [CrossRef]
18. Galley, R.J.; Trachtenberg, M.; Langlois, A.; Barber, D.G.; Shafai, L. Observations of geophysical and dielectric properties and ground penetrating radar signatures for discrimination of snow, sea ice and freshwater ice thickness. *Cold Reg. Sci. Technol.* **2009**, *57*, 29–38. [CrossRef]

19. Liu, H.; Takahashi, K.; Sato, M. Measurement of dielectric permittivity and thickness of snow and ice on a brackish lagoon using GPR. *IEEE J-STARS* **2014**, *7*, 820–827. [CrossRef]
20. Lundberg, A.; Thunehed, H. Snow wetness influence on impulse radar snow surveys: Theoretical and laboratory study. *Nord. Hydrol.* **2000**, *31*, 89–106. [CrossRef]
21. Bradford, J.H.; Harper, J.T.; Brown, J. Complex dielectric permittivity measurements from ground-penetrating radar data to estimate snow liquid water content in the pendular regime. *Water Resour. Res.* **2009**, *45*, W08403. [CrossRef]
22. Sundström, N.; Gustafsson, D.; Kruglyak, A. Lundberg, A. Field evaluation of a new method for estimation of liquid water content and snow water equivalent of wet snowpacks with GPR. *Hydrol. Res.* **2013**, *44*, 600–613. [CrossRef]
23. Previati, M.; Godio, A.; Ferraris, S. Validation of spatial variability of snowpack thickness and density obtained with GPR and TDR methods. *J. Appl. Geophys.* **2011**, *75*, 284–293. [CrossRef]
24. Forte, E.; Dossi, M.; Colucci, R.R.; Pipan, M. A new fast methodology to estimate the density of frozen materials by means of common offset GPR data. *J. Appl. Geophys.* **2013**, *99*, 135–145. [CrossRef]
25. Godio, A.; Frigo, B.; Chiaia, B.; Maggioni, P.; Freppaz, M.; Ceaglio, E.; Dellavedova, P. Integration of upward GPR and water content reflectometry to monitor snow properties. *Near Surf. Geophys.* **2018**, *16*, 1–10. [CrossRef]
26. Holbrook, W.S.; Miller, S.N.; Provart, M.A. Estimating snow water equivalent over long mountain transects using snowmobile-mounted ground-penetrating radar. *Geophysics* **2016**, *81*, WA183–WA193. [CrossRef]
27. Webb, R.W. Using ground penetrating radar to assess the variability of snow water equivalent and melt in a mixed canopy forest, Northern Colorado. *Front. Earth Sci.* **2017**, *11*, 482–495. [CrossRef]
28. Li, Z.; Li, C.; Yang, Y.; Zhang, B.; Deng, Y.; Li, G. Physical scheme and parametrization by using air temperature and ice thickness to improve accuracy of GPR propagation velocity in Yellow River ice. *J. Hydraul. Eng.* **2022**, *53*, 902–913. [CrossRef]
29. Best, H.; Mcnamara, J.P.; Liberty, L. Association of ice and river channel morphology determined using ground-penetrating radar in the Kuparuk River, Alaska. *Arct. Antarct. Alp. Res.* **2005**, *37*, 157–162. [CrossRef]
30. Fu, H.; Liu, Z.; Guo, X.; Cui, H. Double-frequency ground penetrating radar for measurement of ice thickness and water depth in rivers and canals: Development, verification and application. *Cold Reg. Sci. Technol.* **2018**, *154*, 85–94. [CrossRef]
31. Li, T.; Zhao, J.; Jiao, Y.; Hou, J.; Mu, L. Regional characteristics of sea ice thickness in Canadian shelf and Arctic Archipelago measured by Ground Penetrating Radar. *Acta Oceanol. Sin.* **2015**, *34*, 110–116. [CrossRef]
32. Li, Z.J.; Jia, Q.; Zhang, B.S.; Leppäranta, M.; Huang, W.F. Influences of gas bubble and ice density on ice thickness measurement by GPR. *Appl. Geophys.* **2010**, *7*, 105–113. [CrossRef]
33. Stevens, C.W.; Moorman, B.J.; Solomon, S.M.; Hugenholtz, C.H. Mapping subsurface conditions within the near-shore zone of an Arctic delta using ground penetrating radar. *Cold Reg. Sci. Technol.* **2009**, *56*, 30–38. [CrossRef]
34. Jones, B.M.; Gusmeroli, A.; Arp, C.D.; Strozzi, T.; Grosse, G.; Gaglioti, B.V.; Whitman, M.S. Classification of freshwater ice conditions on the Alaskan Arctic Coastal Plain using ground penetrating radar and TerraSAR-X satellite data. *Int. J. Remote Sens.* **2013**, *34*, 8267–8279. [CrossRef]
35. Fantello, N.; Parsekian, A.D.; Anthony, K.M.W. Estimating winter ebullition bubble volume in lake ice using ground-penetrating radar. *Geophysics* **2017**, *83*, H13–H25. [CrossRef]
36. Xu, J.; Cheng, D. Relation between the erosion and sedimentation zones in the Yellow River, China. *Geomorphology* **2002**, *48*, 365–382. [CrossRef]
37. Ran, L.; Wang, S.; Fan, X. Channel change at Toudaoguai Station and its responses to the operation of upstream reservoirs in the upper Yellow River. *J. Geogr. Sci.* **2010**, *20*, 231–247. [CrossRef]
38. Wang, T.; Yang, K.; Guo, Y. Application of artificial neural networks to forecasting ice conditions of the Yellow River in the Inner Mongolia Reach. *J. Hydrol. Eng.* **2008**, *13*, 811–816. [CrossRef]
39. Ji, H.L.; Zhang, A.D.; Gao, R.Z.; Zhang, B.S.; Xu, J. Application of the break-up date prediction model in the Inner Mongolia Reach of the Yellow River. *Adv. Sci. Technol. Water Resour.* **2012**, *32*, 42–45. (in Chinese). [CrossRef]
40. Kovacs, A.; Gow, A.J.; Morey, R.M. The in-situ dielectric constant of polar firn revisited. *Cold Reg. Sci. Technol.* **1995**, *23*, 245–256. [CrossRef]
41. Shokr, M.E. Field observations and model calculations of dielectric properties of Arctic sea ice in the microwave C-band. *IEEE Trans. Geosci. Remote Sens.* **1998**, *36*, 463–478. [CrossRef]
42. Gherboudj, I.; Bernier, M.; Hicks, F.; Leconte, R. Physical characterization of air inclusions in river ice. *Cold Reg. Sci. Technol.* **2007**, *49*, 179–194. [CrossRef]
43. Evans, S. Dielectric properties of ice and snow-A review. *J. Glaciol.* **1965**, *5*, 773–792. [CrossRef]
44. Ran, D.; Yao, W.; Shen, Z.; Ni, Y.; Dong, F. Analysis on the contribution rate of driving factors for the annual water and sediment variations at the Toudaoguai hydrological station in the Yellow River. *Adv. Water Sci.* **2015**, *26*, 769–778. (in Chinese). [CrossRef]
45. Davis, J.L.; Annan, A.P. Ground-penetrating radar for high-resolution mapping of soil and rock stratigraphy. *Geophys. Prospect.* **1989**, *37*, 531–551. [CrossRef]
46. Moorman, B.J.; Michel, F.A. Bathymetric mapping and sub-bottom profiling through lake ice with ground-penetrating radar. *J. Paleolimnol.* **1997**, *18*, 61–73. [CrossRef]
47. Hausmann, H.; Behm, M. Imaging the structure of cave ice by ground-penetrating radar. *Cryosphere* **2011**, *5*, 329–340. [CrossRef]

48. Stoyanovich, G.M.; Pupatenko, V.V.; Sukhobok, Y.A. Detection of ground ice using ground penetrating radar method. *Sci. Cold Arid Reg.* **2015**, *7*, 541–546. [CrossRef]
49. Annan, A.P.; Diamanti, N.; Redman, J.D.; Jackson, S.R. Ground-penetrating radar for assessing winter roads. *Geophysics* **2016**, *81*, WA101–WA109. [CrossRef]
50. You, Y.; Yang, M.; Yu, Q.; Pan, X.; Guo, L.; Wang, X.; Wu, Q. Factors influencing the reliability of grounded and floating ice distinguishing based on ground penetrating radar reflection amplitude. *Cold Reg. Sci. Technol.* **2018**, *154*, 1–8. [CrossRef]
51. Bohleber, P.; Wagner, N.; Eisen, O. Permittivity of ice at radio frequencies: Part I. Coaxial transmission line cell. *Cold Reg. Sci. Technol.* **2012**, *82*, 56–67. [CrossRef]
52. Bohleber, P.; Wagner, N.; Eisen, O. Permittivity of ice at radio frequencies: Part II. Artificial and natural polycrystalline ice. *Cold Reg. Sci. Technol.* **2012**, *83*, 13–19. [CrossRef]

water

MDPI

Article

Spatial Variations of Fabric and Microstructure of Blue Ice Cores at the Shear Margin of Dalk Glacier, Antarctica

Siyu Lu [1,2], Nan Zhang [1,*], Danhe Wang [3], Guitao Shi [3], Tianming Ma [4], Hongmei Ma [2], Chunlei An [2] and Yuansheng Li [1,2,*]

1 Polar Research Center, Institute for Polar Science and Engineering, College of Construction Engineering, Jilin University, Changchun 130021, China
2 Polar Research Institute of China, Shanghai 200136, China
3 Key Laboratory of Geographic Information Science, School of Geographic Sciences, East China Normal University, Shanghai 200241, China
4 School of Earth and Space Sciences, University of Science and Technology of China, Hefei 230026, China
* Correspondence: Correspondence: znan@jlu.edu.cn (N.Z.); liyuansheng@pric.org.cn (Y.L.)

Abstract: The study of the fabric and microstructure of ice at the shear margin of the Antarctic ice sheet is of great significance for understanding the ice flow and its contributions to sea level rise. In this study, twenty-three one-meter long ice cores were drilled from blue ice areas at the shear margin of the Dalk Glacier, Antarctica. The ice fabric and microstructure of these ice cores are analyzed using a G50 fabric analyzer. This study shows that the shallow ice cores in this region present a cluster fabric as a consequence of shear stress. The grain size decreases following the direction of the ice flow towards the exposed bedrock at the end of the glacier, due to the blocking and squeezing by the bedrock. The formation mechanism of the shallow ice layers is that the ice from the original accumulation area flows here, lifted by the bedrock and shaped by the summer ablation and denudation. The basal ice at the shear margin of the Dalk Glacier is strongly rubbed by the bedrock and demonstrates a cluster fabric. The analysis of stable water isotopes shows a weak negative correlation between shallow ice fabric and stable water isotopes with depth. Bedrock topography and shear stress have a greater influence on grain microstructure among different ice cores over long distances at shear margins.

Keywords: ice core; fabric; microstructure; glacial shear margin; Dalk Glacier

Citation: Lu, S.; Zhang, N.; Wang, D.; Shi, G.; Ma, T.; Ma, H.; An, C.; Li, Y. Spatial Variations of Fabric and Microstructure of Blue Ice Cores at the Shear Margin of Dalk Glacier, Antarctica. *Water* **2023**, *15*, 728. https://doi.org/10.3390/w15040728

Academic Editors: Zhijun Li, Tomasz Kolerski, Li Zhou, Xiaohong Shi, Zhengyong Zhang and Fang Li

Received: 23 January 2023
Revised: 8 February 2023
Accepted: 9 February 2023
Published: 12 February 2023

1. Introduction

The Antarctic and Greenland ice sheets are the most important contributors to sea level rise this century as a result of global warming [1–3]. The Antarctic ice sheet is formed by the accumulation of snow, which is transformed into firn and ice by a process called densification. After this process, ice is anisotropic under the initial natural conditions. During the process of ice sheet flow, large-scale anisotropy is formed in the ice, which is called fabric (also described in the literature as 'Crystallographic Preferred Orientation', 'lattice preferred orientation', or 'texture') [4–7]. Ice sheet flow comprises internal deformation and external deformation, which is mainly the basal slip caused by gravity and tidal forces [5,8,9]. Antarctic ice flow velocity is slow at the dome and fast at the shear margin. The shear margin also accelerates the melting of the ice shelf [10]. The fabric and microstructure of natural ice change under the influence of deformation during its movement from the dome to the margin of the ice sheet. This process comprises the formation and evolution of fabric, changes in grain size, and the dynamic recrystallization of microstructures [5,11–14].

Currently, researches on the ice fabric of Antarctic ice cores mainly focus on deep ice cores, such as the Vostok deep ice core, Dome C deep ice core, EDML deep ice core, and Dome Fuji deep ice core [15–19]. An ice sheet moves more slowly at domes and ice divides, where ice cores are less affected by ice flow. The flow of ice streams strongly affects the stability of the polar ice sheet [5,20–22]. The basal and lateral shear margins are the key regions that affect the dynamics of glacier flow [23,24]. The subglacial bedrock and the shear margin bedrock create drag forces at the interface with the glacier that impede the movement of the ice flow, which has a strong influence on an ice flow and mass transport. Especially in blue ice areas near the margin of the ice sheet, where the thickness of the ice sheet is rapidly getting thinner, the bedrock has a greater effect on the microstructure of the ice. In addition to the evolution of ice fabric, the bubbles trapped in ice also undergo significant changes such as elongation, shearing, and fragmentation during the glacial movement. The microscopic morphology and evolution of bubbles can be affected by the ice grains, and in turn, impact the development of ice grains.

In laboratory experiments, the deformation process of pure polycrystalline ice under shear forces simulates the fabric and microstructure of ice flow in the ice sheet margin region [25–29]. Under natural conditions, the ice fabric development in the ice sheet margin region is affected by temperature, impurity content, and structural heterogeneity, with depth and spatial variations [30–33]. To date, few studies have directly studied fabric from temperate and polar glacier margin regions [11,34–39]. In geophysical studies, ice radar-sounding techniques are used to detect large-scale ice fabric in glaciers [22,40–47]. For mountain glaciers, a study on fabric and microstructures of ice could provide a micro perspective for understanding the ice flow and predicting related disasters such as ice avalanches [48]. Solar radiative transfer [49], synthetic aperture radar, and optical satellite monitoring [50–52] are also sensitive to ice microstructures at different scales.

In this study, by using the G50 fabric analyzer [53], we analyzed the ice fabric and microstructure of twenty-tree one-meter-long ice cores drilled from the shear margin of the Dalk Glacier, Antarctica. In addition, the stable water isotopes were analyzed by a Picarro L-2130i Water Isotope Analyzer. The depth and spatial variations of the fabrics, microstructures, and stable water isotopes of ice cores in this shear margin were studied. Then, the formation mechanism and the influence of the subglacial bedrock on the ice were discussed. The results of this study can be used as a reference for comparisons between field study and laboratory experiments and improving the ice flow models at the shear margins of the ice sheet.

2. Materials and Methods

2.1. Sample Collection

The twenty-three 1 m-long shallow ice cores used in this study were drilled during 2019–2020 field season of 36th Chinese National Antarctic Research Expedition. The sampling site, as shown in Figure 1, was selected from the fast-flow area at the margin of the Dalk Glacier, East Antarctic. As seen in Figure 1, the ice flows in the direction from inland toward the coast. Drill sites IC, IW, and OIW were selected based on the ice flow direction. Ice cores numbered IC1 to IC7 were drilled in the direction of ice flow, from far to near the exposed subglacial bedrock. Ice cores labeled IC were drilled vertically downward from the glacier surface to a depth of 1 m. The IW and OIW ice cores were drilled from the basal regions of the ice flow at two different locations, in a horizontal direction, drilled into the ice 'wall'. The IW ice cores consisted of five separate cores in the front of an exposed bedrock. IW1 to IW3 are three cores from bottom to top at the same location, with adjacent cores spaced 20 cm apart, while IW4 to IW6 are three cores from bottom to top at another location, with adjacent cores spaced 20 cm apart. IW7 was a separate core from a third location. Nine OIW ice cores were taken from another three locations in a similar way, as shown in Figure 1. Detailed information about the basic information of the drill sites is shown in Table 1.

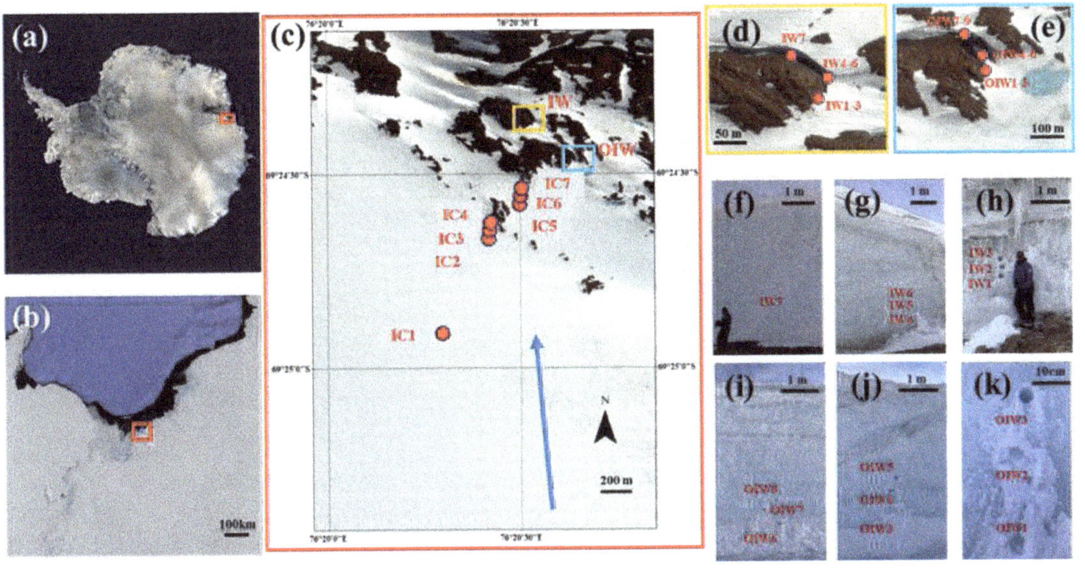

Figure 1. Maps of the drilling sites. (**a**) Map of Antarctica; (**b**) map of Dalk Glacier region; (**c**) map of the drill sites of the ice cores (the blue arrow represents the direction of ice flow); (**d**) drill sites of IW ice cores; (**e**) drill sites of OIW ice cores; (**f**) IW1-IW3; (**g**) IW4-IW6; (**h**) IW7; (**i**) OIW1-OIW3; (**j**) OIW4-OIW6; (**k**) OIW7-OIW9.

Table 1. Basic information of sampling sites.

Name	Longitude (E)	Latitude (S)	Altitude (m)	Drilling Direction
IC1	76°20′17.78″	69°24′54.68″	143	Vertical
IC2	76°20′25.05″	69°24′40.09″	139	Vertical
IC3	76°20′25.03″	69°24′39.85″	139	Vertical
IC4	76°20′24.79″	69°24′39.74″	143	Vertical
IC5	76°20′31.89″	69°24′33.59″	159	Vertical
IC6	76°20′32.46″	69°24′34.69″	161	Vertical
IC7	76°20′32.59″	69°24′35.83″	161	Vertical
IW1-IW3	76°20′28.23″	69°24′26.35″	27	Horizontal
IW4-IW6	76°20′29.43″	69°24′25.50″	20	Horizontal
IW7	76°20′24.69″	69°24′24.48″	19	Horizontal
OIW1-OIW3	76°20′46.72″	69°24′29.55″	155	Horizontal
OIW4-OIW6	76°20′46.87″	69°24′29.05″	151	Horizontal
OIW7-OIW9	76°20′45.39″	69°24′28.23″	152	Horizontal

All ice cores were drilled using a portable hand-held gasoline drill. After the ice cores were drilled, basic information was measured and recorded in the field. Afterward, the ice cores were put into insulated foam boxes and shipped back to the Ice Laboratory of the Polar Research Institute of China in a low-temperature container; throughout this process, the temperature of the ice was maintained at −20 °C. Unfortunately, we did not mark the relative direction between the cores and the ice flow when the ice cores were recovered in the field. It was no longer possible to determine the orientation information of the ice cores relative to the direction of ice flow. Furthermore, there were occasional breaks in the drilling process of each core, and the relative rotation angles of the upper and lower ice cores could not be identified since no mark was made at these breaks.

2.2. Thin Sections Preparation and Fabric Analysis

In the low-temperature laboratory at −20 °C, the ice cores were cut and subsampled using a band saw. The ice core samples were cut 1 cm thick and then made into thin sections by the microtome. The thin sections were glued to glass plates by dropping ultrapure water at four corners or all around them. Thin sections of IC ice cores were continuously made for the whole 1 m-long ice cores, and the direction was vertical to the surface (parallel to the drilling direction). For IW and OIW ice cores, only one sample was taken from each core for thin sections preparation since the rest of the cores were used for analysis of other proxies, such as gas analysis. Both horizontal and vertical thin sections were made for IW7 and OIW6.

The length of the thin sections ranges from 3 to 10 cm depending on the actual conditions of the ice samples. The thickness of thin sections was 200–400 μm. A total of 98 thin sections were made for all 23 ice cores for ice fabric and microstructure studies. Afterward, the thin sections were scanned and analyzed using a G50 ice fabric analyzer to obtain image data such as c-axes orientation maps and microstructure maps of the ice.

2.3. Stable Water Isotope Analysis

Stable water isotope analysis of the IC ice cores was performed by continuous sampling of the whole ice cores. IW and OIW ice cores took only 1 sample per core for isotope analysis. Ice samples for stable water isotope analysis were prepared during the process of making thin sections. Each of the samples were approximately 5–10 g. Then, the samples were melted in the ultra-clean laboratory. A total of 2 mL of the melted water samples were taken into clean sampling bottles and analyzed using Picarro L-2130i Water Isotope Analyzer.

We used 1 set of standard samples for quality control. Six secondary standard samples were prepared by mixing V-SMOW (absolute ratio $^{18}O/^{16}O$ = (2005.20 ± 0.43) × 10^{-6}, relative ratios $\delta^{18}O$ = 0‰ and δD = 0‰) and V-SLAP ($\delta^{18}O$ = −55.50‰ and δD = −428‰) in different ratios, and three of them with corresponding $\delta^{18}O$ were at −14.22‰, −19.88‰ and −27.53‰, respectively, and the corresponding δD was at −104.71‰, −148.62‰ and −208.63‰.

2.4. Image Data Processing Methods

The images obtained by the G50 fabric analyzer were used for grain and bubble microstructure analysis using MorphoLibJ [54], a deep learning plugin in Fiji Image J [55] software, and Trainable Weka Segmentation [56], a machine learning plugin in Fiji Image J software. To reduce noises, bubbles with area less than 0.1 mm^2 were removed. Incomplete small ice grains at the cutting edges of thin sections were removed when calculating grain sizes. The results of the grain and bubble analysis were also checked with the manually counted number of grains and bubbles.

3. Results

A total of 98 thin section samples were analyzed in this study. In this section, the results of six representative ice cores are presented in detailed figures. For IC1 and IC3 samples, the topmost and bottommost two samples of ice cores were selected, and IW5 and OIW5 ice cores were selected from horizontally drilled ice cores at the basal regions. In total, ten thin sections were chosen as typical samples, as shown from Figures 2–6. Data from all other samples were also analyzed and presented in other figures and curves in this paper.

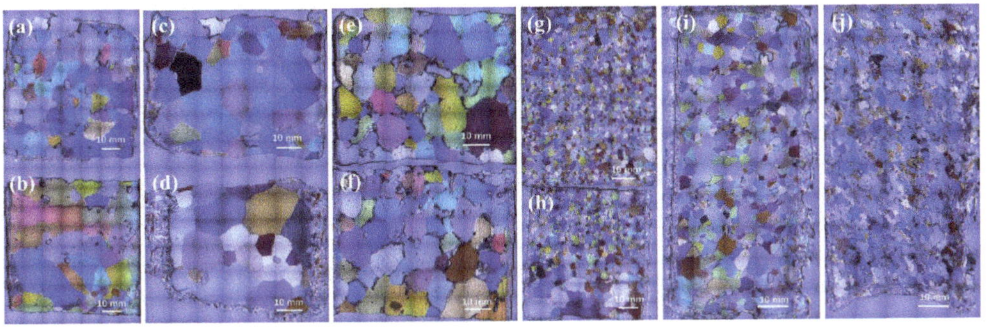

Figure 2. Microstructure maps of thin sections. (**a**) IC1-1; (**b**) IC1-2; (**c**) IC1-9; (**d**) IC1-10; (**e**) IC3-1; (**f**) IC3-2; (**g**) IC3-12; (**h**) IC3-13; (**i**) IW5; (**j**) OIW5.

3.1. Microstructures of Ice Thin Sections

The microstructure maps of the selected 10 typical thin sections are shown in Figure 2. As shown in the figure, the surrounding ice band is frozen pure water, which is used to fix the thin section. The areas that show different colors are ice grains. The transparent parts surrounded by ice grains are bubbles.

3.2. Ice Fabrics

The fabric data were analyzed using a G50 Fabric Investigator. The c-axis orientation of each grain was determined by manually selecting each ice grain, and the c-axis orientation histograms are shown in Figure 3a. The c-axis orientation data were plotted in polar coordinates to obtain the kernel contour maps, as shown in Figure 3b. Figure 3c shows the microstructure maps with colored orientations of each grain. Normally, when analyzing the fabric of deep ice cores, samples will be selected at an interval of dozens of meters. In this study, the IC1–IC7 ice cores were sampled continuously for fabric analysis. According to the kernel contour maps of the ice fabric, ice is mainly presented as a cluster fabric, and the clusters are usually 0° and 180°. However, the samples in different layers show some angular shift of cluster orientation.

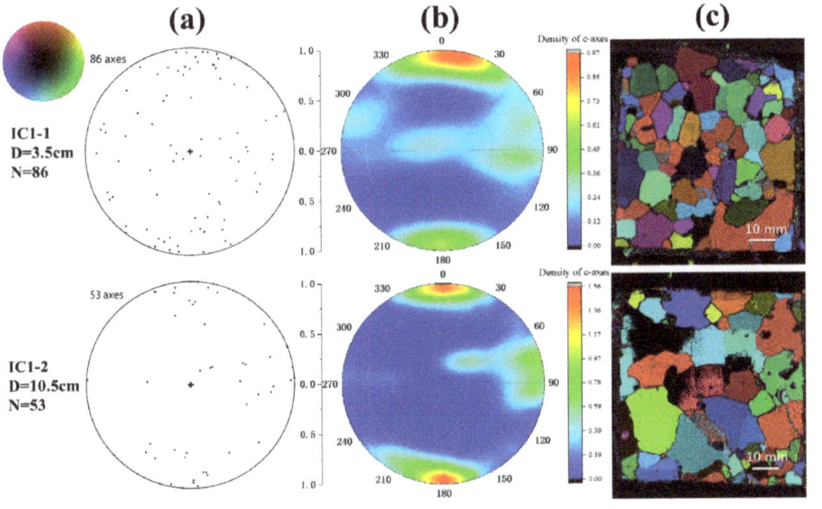

Figure 3. *Cont.*

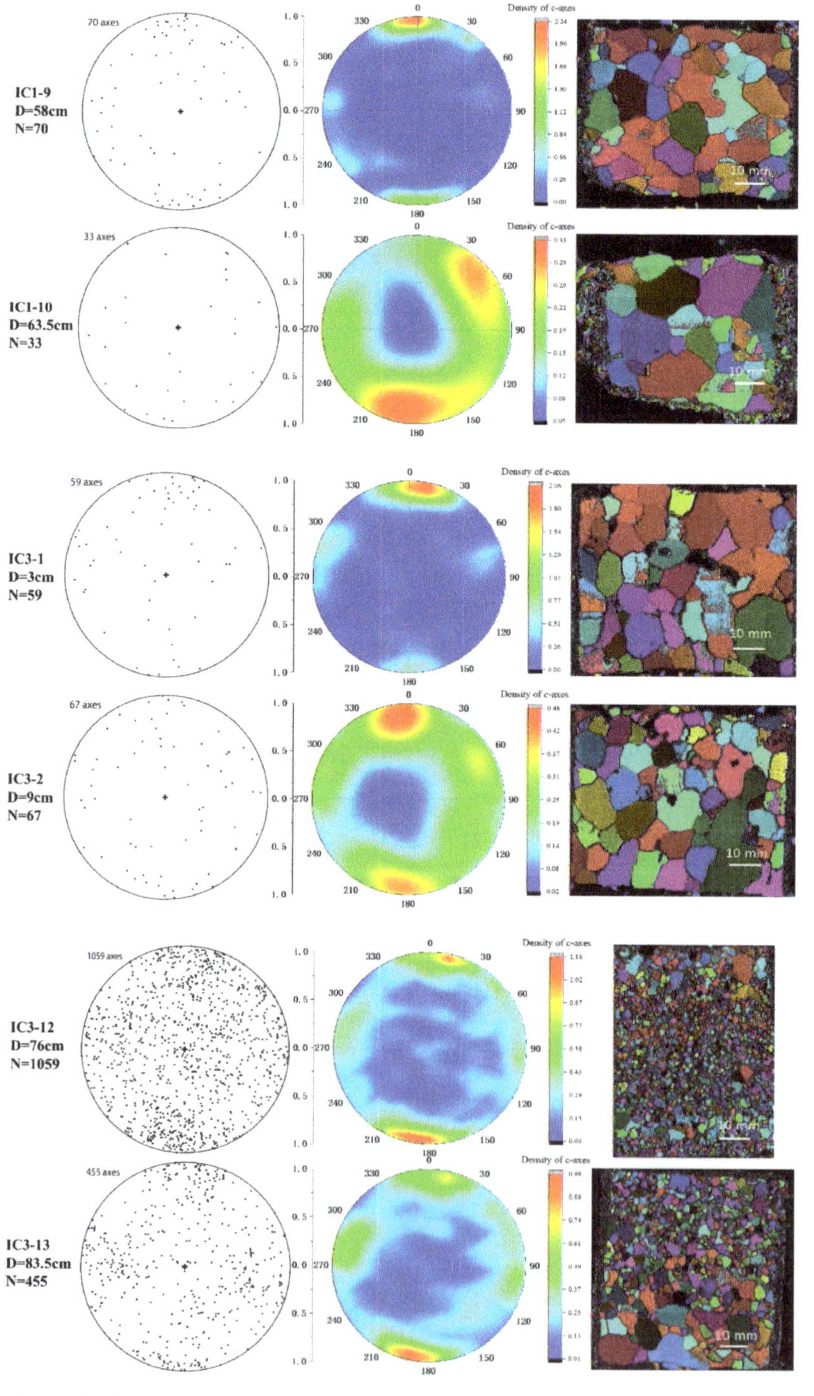

Figure 3. *Cont.*

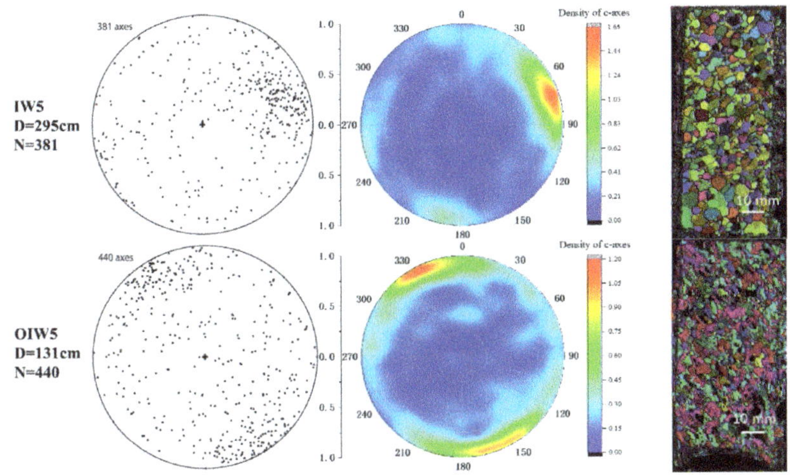

Figure 3. (**a**) The Schmidt diagrams (equatorial projections of *c*-axis orientations), each point represents the *c*-axis orientation of each grain; (**b**) the kernel contour maps of *c*-axis orientations of ice, the density of *c*-axes represents the strength of the fabric; (**c**) the microstructure maps with colored orientations of each grain, the legend of the map is shown in the color wheel at the top left of Figure 3. The left side shows the basic information for each thin section, including sample number, depth (D), and number of *c*-axis (N), one *c*-axis per grain.

3.3. Microstructure of Grains

Figure 4a shows the grain size frequency distributions. Grain circularity frequency distributions are presented in Figure 4b. Grain ellipse elongation frequency distributions are presented in Figure 4c. Circularity = $4\pi(A/P^2)$, where A = area, P = perimeter, and 1.0 is a perfect circle.

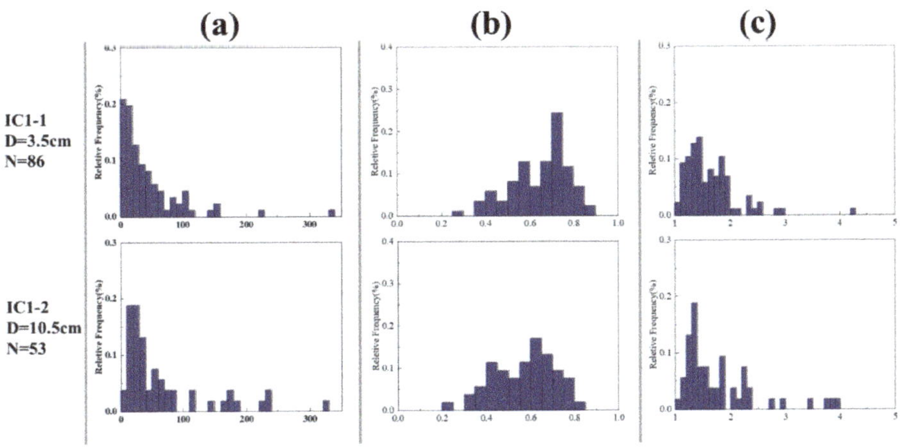

Figure 4. *Cont.*

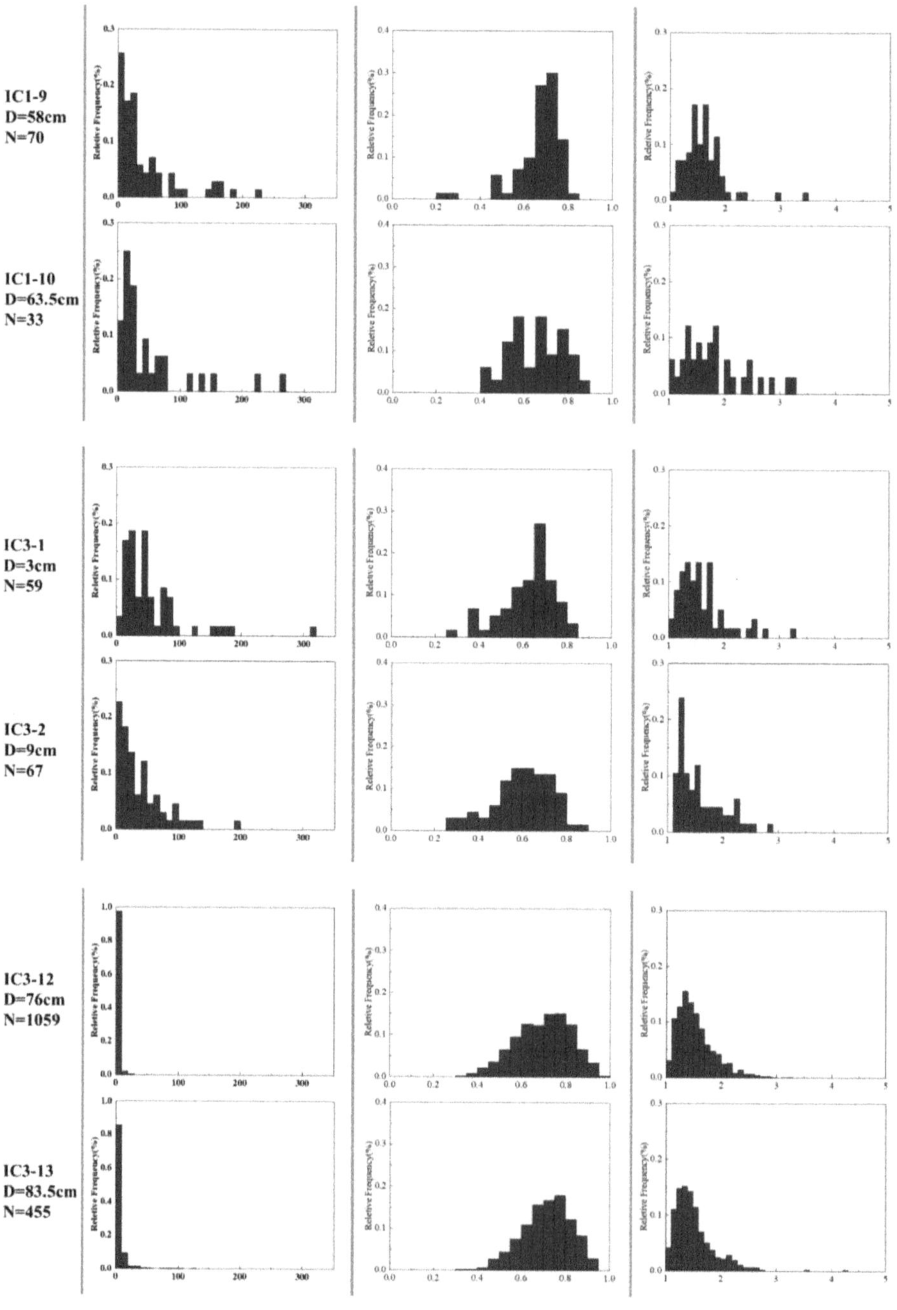

Figure 4. *Cont.*

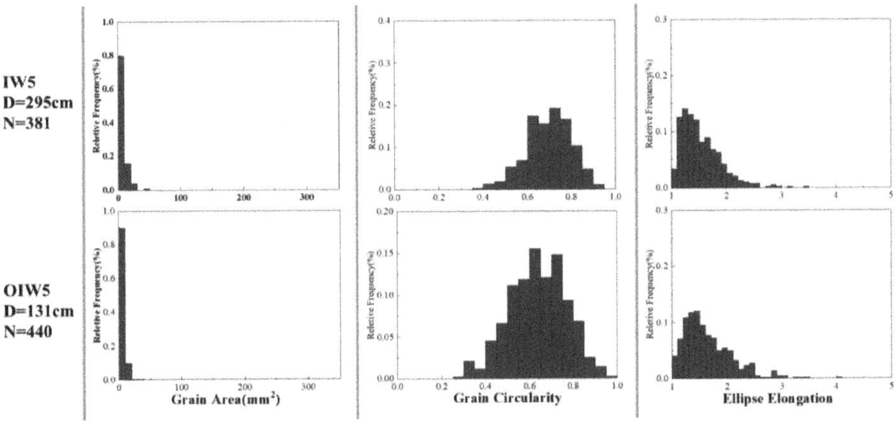

Figure 4. Grain frequency distribution diagrams. 'D' means depth, 'N' means number of grains. (**a**) Frequency distributions of grain size (mm^2); (**b**) frequency distributions of grain circularity; (**c**) frequency distributions of grain ellipse elongation.

As shown in Figure 4a, the grain size distributions are skewed, with the majority of grains plotting toward finer grain sizes and a tail extending toward larger grain sizes. As shown in Figure 4c, grain ellipse elongation ($d_{long\ axis}/d_{short\ axis}$) plots are skewed toward lower values, with some ratios extending toward values > 3. The averages of the grain axial ratio in these samples range from 1.5 to 1.8.

3.4. Microstructure of Bubbles

In general, the large-area scanning macroscope (LASM) method is used to analyze the properties of bubbles trapped in ice cores. In this study, the bubble characteristics were analyzed by analyzing microscopic images of ice thin sections scanned by a G50 fabric analyzer.

Bubble diameter frequency diagrams are skewed toward smaller bubble sizes. Most bubbles have diameters <1 mm, except the very shallow ice samples of IC ice cores (Figure 5a). Bubble circularity frequency diagrams are skewed to 1, with a small percentage of bubbles <0.5 (Figure 5b). The axial ratio ($d_{long\ axis}/d_{short\ axis}$) of bubbles distributes on a large scale, which means the bubbles have a more elongated shape characteristic.

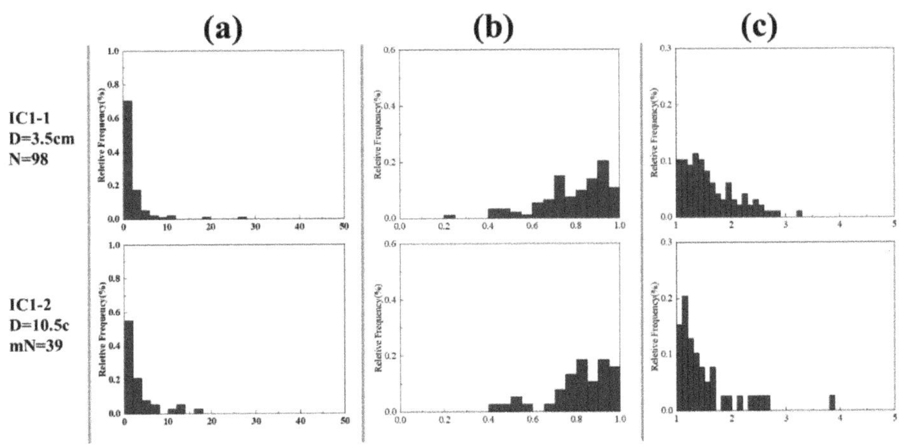

Figure 5. *Cont.*

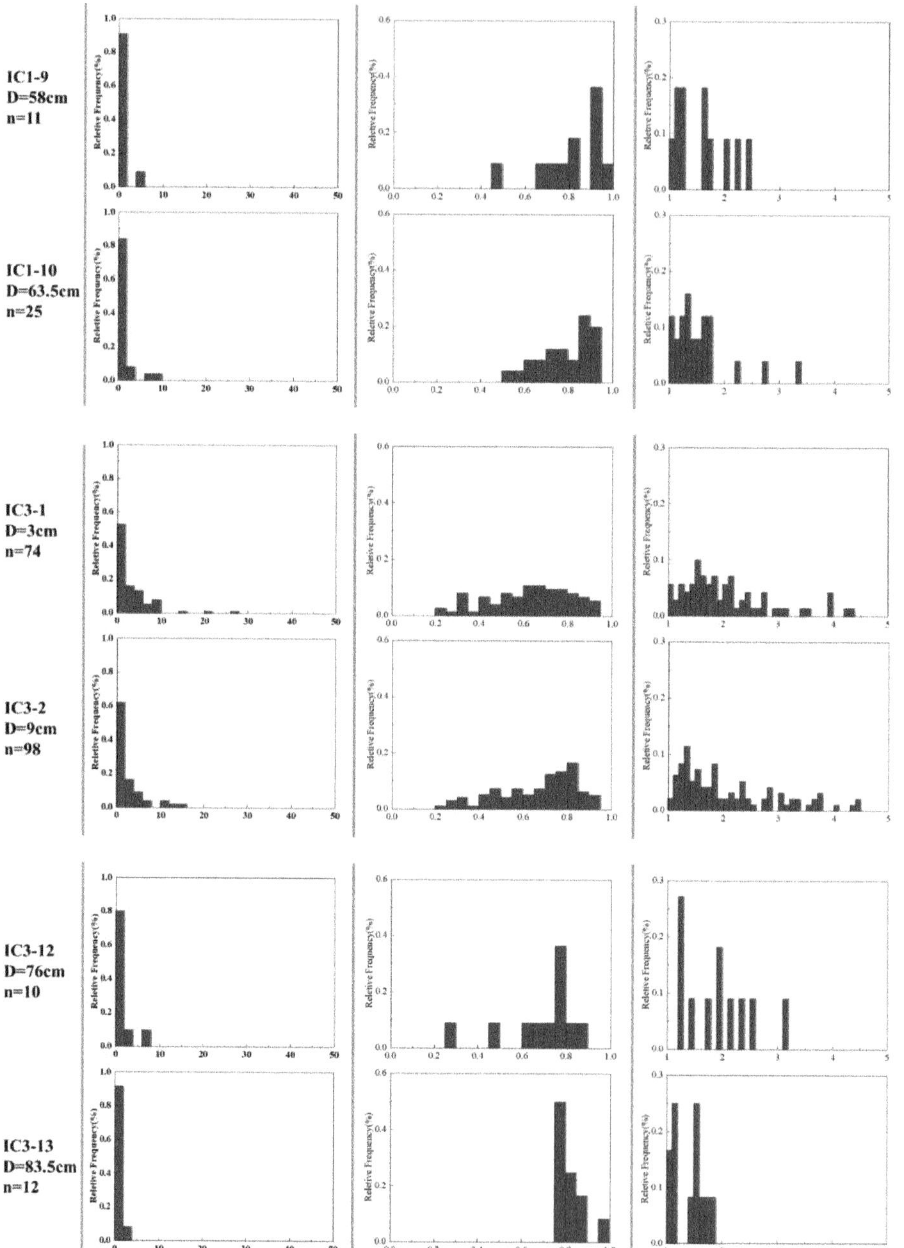

Figure 5. *Cont.*

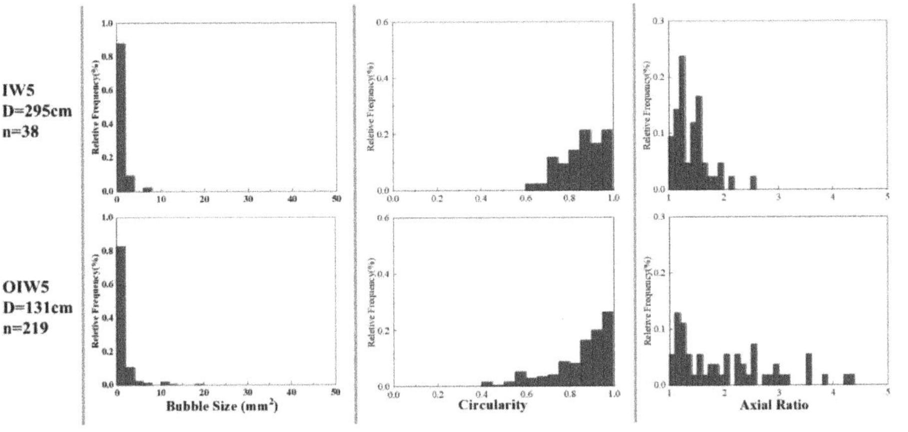

Figure 5. Bubble frequency distribution diagrams. 'D' means depth, 'n' means number of bubbles. (**a**) Frequency distributions of bubble size (mm^2); (**b**) frequency distributions of bubble circularity; (**c**) frequency distributions of bubble axial ratio.

3.5. Shape-Preferred Orientations (SPOs) of Ice Grains and Bubbles

Shape-preferred orientations (SPOs) are defined as the major axis orientation of each grain. In this study, the direction of the maximum Feret diameter is used to represent the shape preference orientations. The diameter passing through the center of a grain in any direction is called a Feret diameter. As shown in Figure 6, the angles of the maximum Feret diameters of ice grains and bubbles in the selected ten representative thin sections are plotted in the rose diagrams. Samples such as IW5 and OIW5 have a relatively strong grain SPO at angles around 20° anti-clockwise of the c-axis maximums. The SPOs of IC ice cores are at random angles compared with the direction of c-axis maximums.

3.6. Stable Water Isotopes of Ice

The variation trends of stable water isotopes ($\delta^{18}O$ and δD) of ice cores are generally consistent with atmospheric temperature changes. As a result, stable water isotopes are often used as a proxy to indicate paleoclimate change. Meanwhile, stable water isotopes of ice cores have been used to compare with the grain size of ice cores [39,57]. Based on the global atmospheric precipitation line GMWL, excess deuterium (d-excess) can be defined as d-excess = $\Delta d - 8 \times \delta^{18}O$ [58]. The variations of stable isotopes of IC cores with depth are plotted in Figure 7. The stable water isotopes fluctuate with depth.

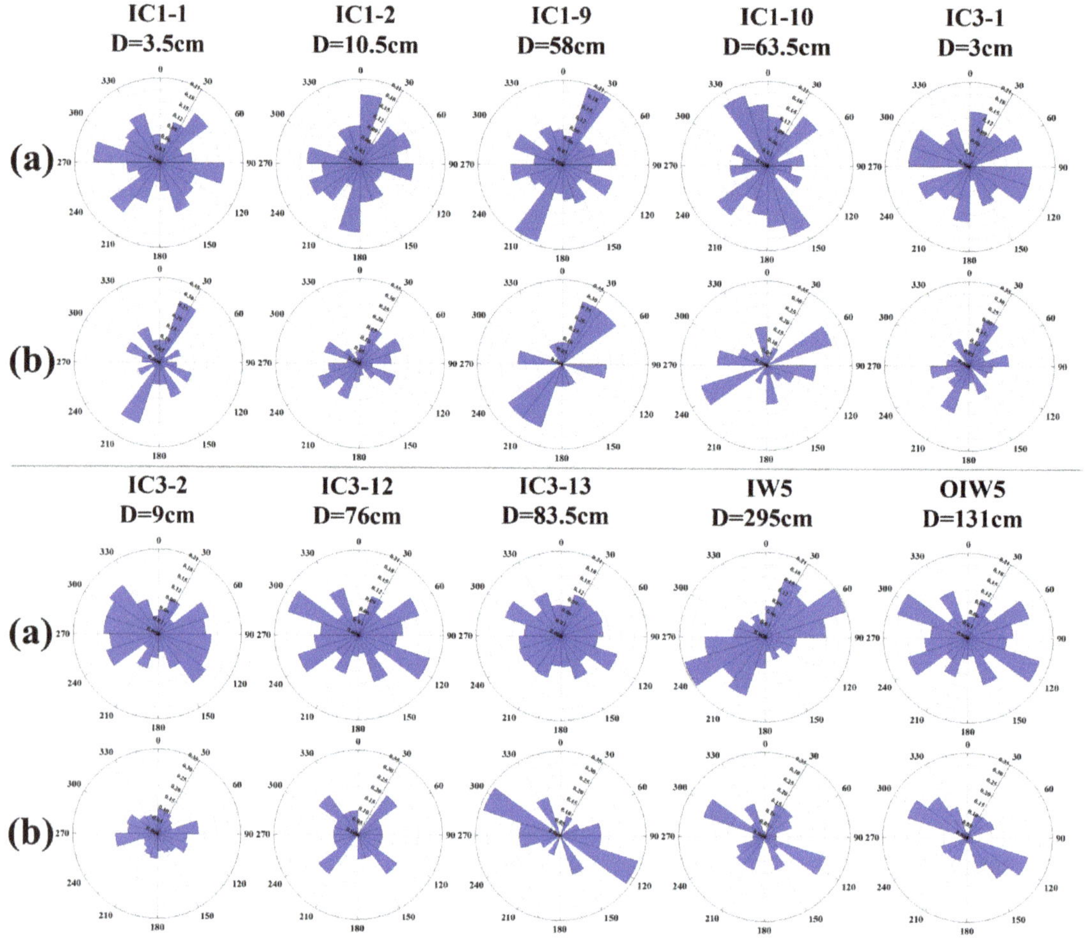

Figure 6. (**a**) Rose diagrams of shape-preferred orientations of ice grains; (**b**) rose diagrams of shape-preferred orientations of bubbles. 'D' means depth of the sample.

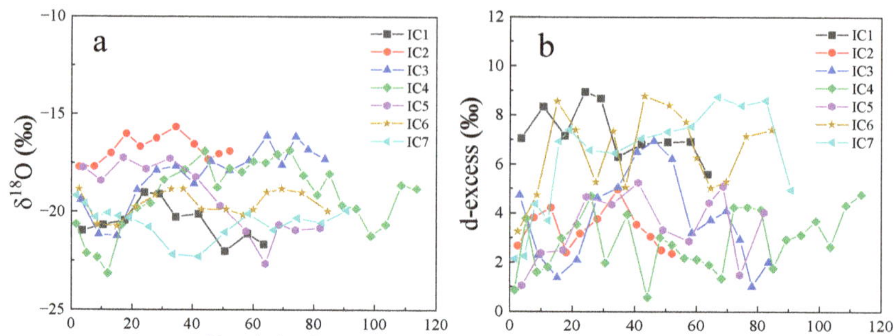

Figure 7. (**a**) Variations of $\delta^{18}O$ with depth of IC ice cores; (**b**) variations of d-excess with depth of IC ice cores.

4. Discussion

4.1. Evolution and Spatial Variations of Ice Fabrics

The deformation-induced fabric has a deformation kinematic symmetry [59]. Most of the IC thin sections show a cluster fabric with two *c*-axis maximum orientations. The strength of the fabric is not obviously correlated with the variation in depth. The angle of the *c*-axis maximum evolves with depth at an angle within 10° in most cases. Double cluster fabrics have been observed in shear-dominated regimes where discontinuous dynamic recrystallization is active, both in experimental deformation tests [26,29] and in natural conditions [34,37,39]. Like what is shown in studies of Whillans [34], shear margin ice has coarse grains, irregular grain boundaries, and double-maximum *c*-axes orientations, which indicate that grain boundary migration is the dominant process. Meanwhile, force and morphological analyses of the cluster fabric were also performed in the ice radar-sounding study [60]. The fabric of basal ice shows one or two *c*-axis maximums and has a significant difference between each ice core.

Although the fabric of ice cores is weak in many thin sections, there still exists an evolution trend of ice fabric even within a small scale of 1 m depth. Therefore, in fabric studies of deeper ice cores, especially in areas with fast ice flow velocity, there may be potential uncertainties using ice samples at an interval of tens or hundreds of meters. As a result, we should be more careful about the variations of ice fabric within a small depth scale when explaining the fabric evolution mechanism. It is suggested that when studying critical depth zones of deep ice cores, continuous thin section sampling of several meters should be conducted to analyze the evolution of fabric.

4.2. Depth and Spatial Variations of Grain Size of IC Ice Cores

Figure 8a shows the variations of IC grain sizes with depth. As shown in Figure 8a, the grain size of each ice core fluctuates with the increase in depth. Variations of averaged grain sizes with the relative distances among different IC ice cores are shown in Figure 8b. The averaged proxies of each IC ice core, including grain size, grain circularity, bubble size, bubble circularity, porosity, bubble axial ratio, $\delta^{18}O$, and d-excess, are shown in Figure 8d. From Figure 8d, it can be seen that the grain size of the seven IC ice cores shows a decreasing trend. It is further shown in Figure 8b that the decreasing rate of grain size from IC2 to IC4 ice cores is relatively fast, and the decreasing rate of grain size from IC5 to IC7 ice cores is relatively slow. The decreasing rates of grain size with distance are 0.71 mm^2/m from IC2 to IC4 and 0.16 mm^2/m from IC5 to IC7. These two sets of ice cores correspond to two different shear margin sites of exposed bedrock, and both decreasing rates of grain size are faster than that from IC1 to IC2 (0.012 mm^2/m).

Figure 8c shows a 3D-smooth projection of grain size to depth and ice flow (expressed by the number of IC ice cores). The smoothing method is adjacent-averaging with a smoothing parameter of 0.05 and a growth factor of 100. As shown in Figure 8c, the overall trend of grain size decreases with the increase in depth. The overall trend of grain size decreases as the ice cores get closer to the exposed bedrock. This result coincides with what we find in Figure 8b. In summary, the subglacial topography has a significant influence on the grain size in shallow glaciers at the shear margin.

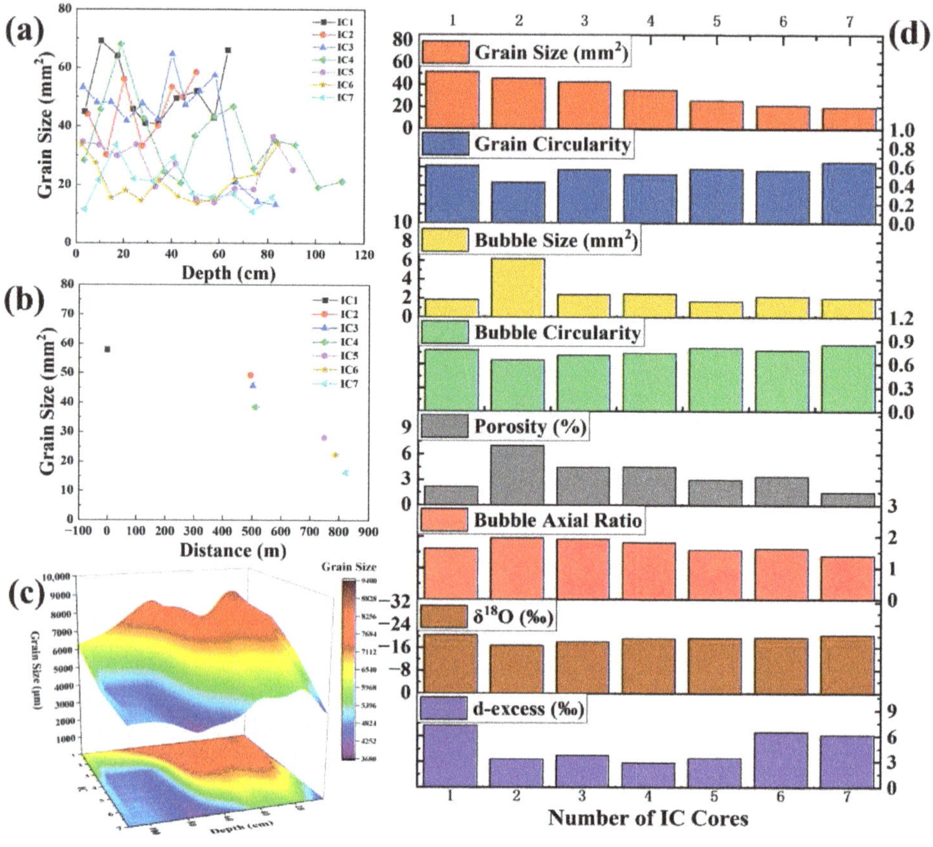

Figure 8. (**a**) Variations of IC grain sizes with depth; (**b**) variations of averaged grain sizes with the relative distances among different IC ice cores; (**c**) 3D smooth surface projection of grain size (μm), sample number and depth (cm). Grain size is the equivalent diameter corresponding to the grain average area. Smoothing method is adjacent-averaging, smoothing parameter is 0.05, total growth factor is 100; (**d**) averaged proxies of each IC core including grain size, grain circularity, bubble size, bubble circularity, porosity, bubble axial ratio, $\delta^{18}O$, and d-excess.

4.3. Depth and Spatial Variations of Bubble Size of IC Ice Cores

During the process of densification, snow and firn transform into ice and trap the air into bubbles. Blue ice areas at the margin of the ice sheet are usually in ablation zones, and the surface layers of the glacier are denuded in the summer season [61]. Variations of bubble size, circularity, axial ratio, and porosity with depth are shown in Figure 9. With the increase in depth, the bubble size decreases, and the bubble circularity increases.

According to the bubble size data of IC ice cores shown in Figure 8, the bubble sizes of IC1, IC3, and IC7 fluctuate around 2 mm^2, while the bubble size of the IC2 ice core is much larger than all the other six ice cores, up to 6.177 mm^2. It is shown in Figure 9 that at the shear margin of the Dalk Glacier, the bubbles become smaller and rounder, with increasing depth from the surface to 1 m-depth ice layer.

It is worth noting that this study is based on 2D thin sections for bubble analysis, which simplifies the 3D ice morphology. As a result, the original bubble characteristics in the 3D world are simplified into 2D profiles.

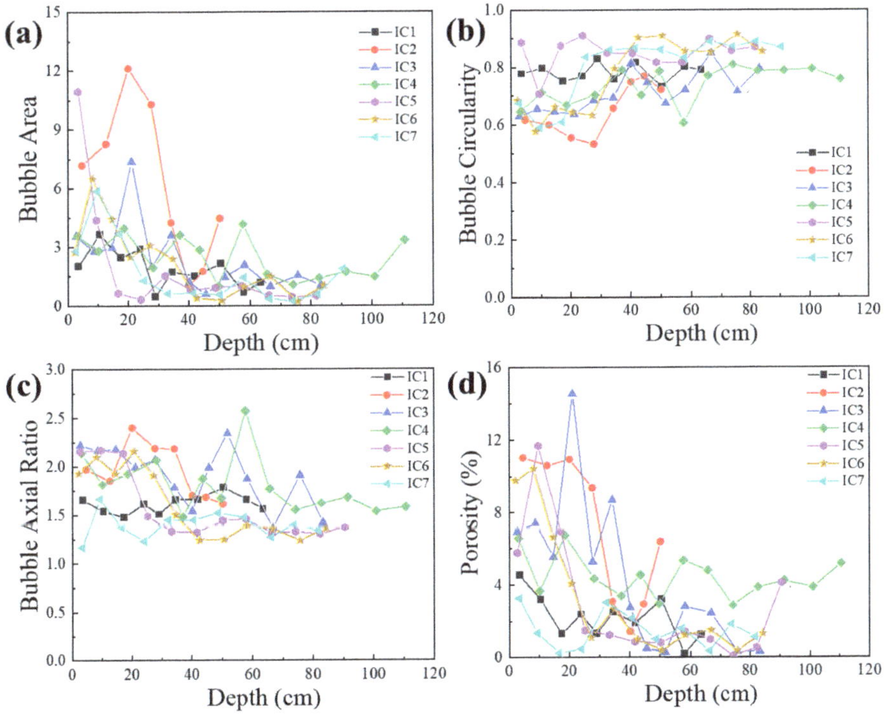

Figure 9. (**a**) Variations of bubble size with depth; (**b**) variations of bubble circularity with depth; (**c**) variations of bubble axial ratio with depth; (**d**) variations of porosity with depth.

4.4. The Mechanism of the Formation of Shallow Bubble Tunnels

The bubble morphology of the IC ice cores shows that there are strip-shaped bubble channels in the shallow ice. Moreover, compared with the IC1 sample, IC2–IC7 contains more bubbles with larger bubble areas and longer bubble long-axis lengths. From the microscopic images and bubble area statistics, it is clear that the bubble channels are mostly distributed at the location of ice crystal boundaries. There are two possible reasons for the formation of shallow bubble channels. Firstly, the temperature of the surface layer of blue ice increases due to solar radiation in summer. Melting and denudation of the topmost ice layers can occur, and the increase in surface temperature leads to the increase in ice grain size, movement, and connection of bubbles. Even refreezing of the surface ice occurs, leading to the formation of larger bubble channels. Secondly, as the glacier flows, the deeper blue ice is lifted to the surface by the subglacial topography, the bubble size increases to release pressure, and bubble channels form.

By comparing the studies on ice samples with melting and refreezing interference [36], it shows that the grain sizes are larger if there is melting and refreezing. As shown in Figure 8a, the grain sizes of the shallowest ice layers are not obviously larger than the deeper layers. As a result, ablation but no refreezing happens to surface ice layers in this region. In addition, microstructure and fabric studies of shear margin ice of the Taylor Glacier [35] and the subsequent dating and studies on blue ice [62–65] show that blue ice from deep layers of ice sheet flows towards the exposed bedrock and forms the Taylor Glacier shear margin. It can be inferred that in this study, the blue ice layer flows from the original accumulation area and is uplifted by the bedrock to the surface. In conclusion, under the influence of ablation and bubble pressure release, the ice surface is denudated and bubble channels form in shallow ice layers.

4.5. Characteristics and Formation Mechanisms of Basal Ice Cores

As shown in Figure 10, the grain size of OIW ice cores is smaller than that of IW ice cores, especially for OIW1–OIW6. This results from the different properties between the ice layers since IW and OIW ice cores were drilled at two different sites according to Figure 1. Combined with the fact that the grain circularity of OIW is also smaller than that of IW ice cores, it can be concluded that OIW ice cores experience much more basal sliding and the grains were sheared into small and elongated properties. Meanwhile, this finding also coincides with the SPO properties. As shown in Figure 6, the SPOs of bubbles are in the same direction with the SPOs of grains in OIW ice cores. It means that bubbles and grains of OIW ice cores are stretched in the same direction by the shear stress conducted by the bedrock during glacier movement. There is intense friction and shear stress between the OIW basal ice and the bedrock. Since the ice cores are drilled horizontally, the grains and bubbles are stretched in the direction of the ice flow.

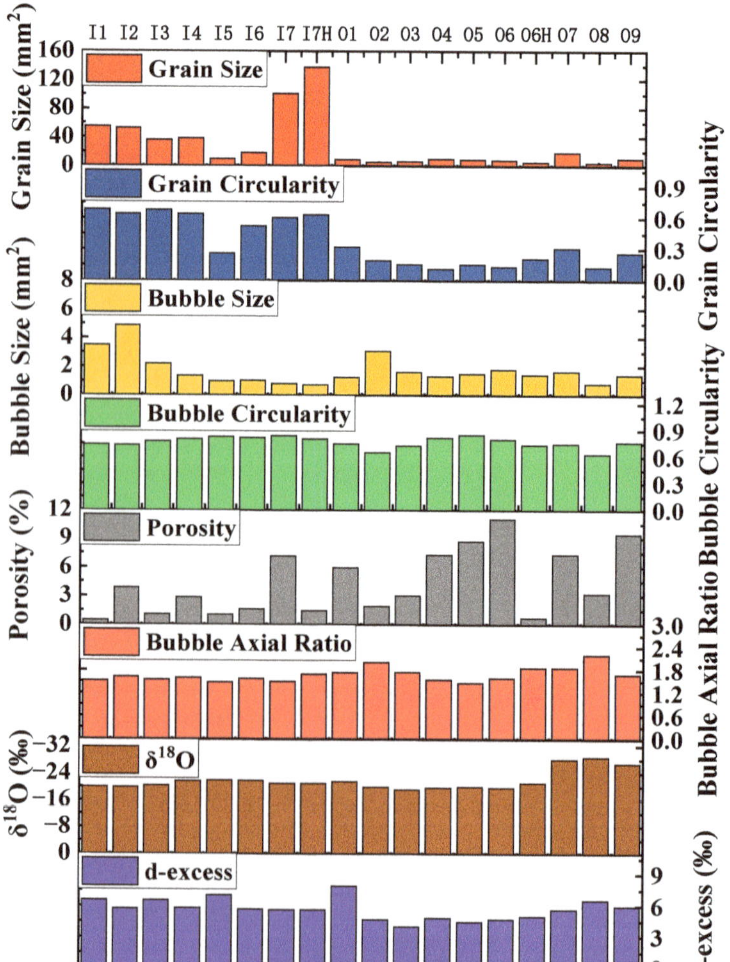

Figure 10. Grain and bubble characteristics of IW (I1-I7H) and OIW (O1–O9) ice cores. 'H' means horizontal thin sections, which cut perpendicularly to the drilling direction.

4.6. Correlations between Stable Water Isotopes and Grain Size

In studies of deep ice cores, grain size correlates with stable water isotopes in some cases to reveal layer information, such as folded ice layers [57]. For shallow ice cores at shear margins, conditions are more complicated due to ice flow and the influence of bedrock. The correlation between the IC1–IC7 ice core stable water isotope $\delta^{18}O$ and grain size with depth is shown in Figure 11. It can be seen from the figure that, except for IC7, the stable water isotope $\delta^{18}O$ has a weak negative correlation trend with the grain size of the other six ice cores. In this study, the grain size and stable water isotope of the ice cores fluctuate with depth. The seasonal/interannual variations of the original accumulation region are still preserved after the ice migrated to the surface shear margin. As shown in Figure 8d, the stable isotope $\delta^{18}O$ decreases from IC2 to IC7, which is in positive correlation with the decrease in grain size, contrary to the findings in depth variations. Combined with the information we have found in the previous sections, it can be concluded that bedrock topography and shear stress have a greater influence on grain size than stable isotope $\delta^{18}O$ among different ice cores over long distances. More studies are needed to further investigate the correlation between ice fabric and chemical proxies.

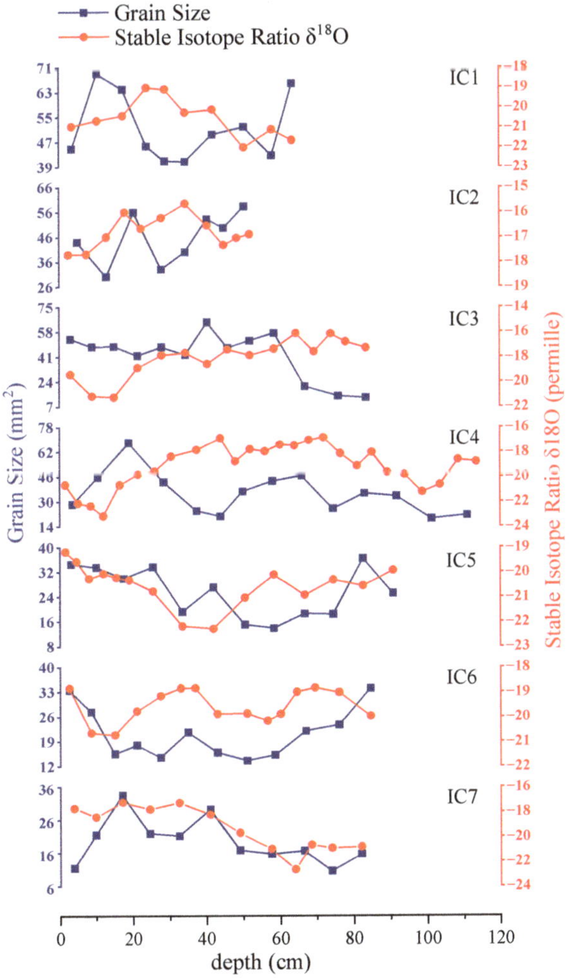

Figure 11. Correlations between ice grain sizes and stable water isotope $\delta^{18}O$.

5. Conclusions

In this study, we investigated the fabric, microstructure, and stable water isotope of ice cores from the shear margin of the Dalk Glacier, Antarctica. Based on the analysis and comparison of the depth and spatial variations of these proxies, we evaluated the properties and formation mechanisms of the ice layers in this region. The conclusions are as follows:

1. This study shows that the ice cores at the shear margin of the Dalk Glacier mostly have a cluster fabric. There exist rotations of the c-axis maximum and fabric evolution in the shallow ice layer at a depth of 1 m;
2. The formation mechanism of blue ice in this study region is that the ice from the original accumulation area flows to these sites, then it is lifted to the surface by the subglacial bedrock and shaped by the influence of summer ablation and denudation;
3. Spatial variations of ice grain size show that the grain size becomes smaller as the ice cores get closer to the exposed bedrock. This means that the blocking and squeezing of the bedrock topography have a significant influence on the ice microstructure;
4. The basal ice at the glacier shear margin was strongly affected by the intense friction of the bedrock during the glacial movement. The basal ice mostly shows a cluster fabric, with one or two c-axis maximums. Ice grains and bubbles have a very strong shear-stretching characteristic in one of the basal sites;
5. With the increase in depth, the grain size of shallow ice cores has a weak negative correlation with the stable water isotope. Bedrock topography and shear stress have a greater influence on grain microstructure among different ice cores over long distances at shear margins.

The conclusions of this paper can help understand the properties of shallow and basal ice at the shear margins of the East Antarctic ice sheet and provide a reference for improving the Antarctic ice flow model. Admittedly, there are limitations to this paper. The relative directions between each ice sample and the ice flow were not well recorded when drilling in the field. Studies on more proxies of deeper ice cores are needed in the future to better understand the properties of blue ice at the shear margin of the Antarctic ice sheet.

Author Contributions: Conceptualization, S.L., D.W. and G.S.; investigation, S.L., D.W., T.M. and C.A.; methodology, S.L.; data curation, S.L.; writing—original draft, S.L.; writing—review and editing, N.Z., G.S., T.M., H.M. and Y.L.; funding acquisition, N.Z., G.S., T.M. and H.M. All authors have read and agreed to the published version of the manuscript.

Funding: This research was funded by the National Natural Science Foundation of China (No. 42176232, 41876225, 42276243, 41922046) and the Fundamental Research Funds for the Central Universities (WK2080000161).

Data Availability Statement: The data presented in this study are available upon request from the first author.

Acknowledgments: The authors are grateful to the members of the 36th CHINARE members for technical support and assistance. The authors thank the editor and anonymous reviewers for their valuable comments and suggestions to this paper.

Conflicts of Interest: The authors declare no conflict of interest.

References

1. Willis, J.K.; Church, J.A. Regional Sea-Level Projection. *Science* **2012**, *336*, 550–551. [CrossRef] [PubMed]
2. Gregory, J.M.; White, N.J.; Church, J.A.; Bierkens, M.F.P.; Box, J.E.; den Broeke, M.R.; Cogley, J.G.; Fettweis, X.; Hanna, E.; Huybrechts, P.; et al. Twentieth-Century Global-Mean Sea Level Rise: Is the Whole Greater than the Sum of the Parts? *J. Clim.* **2013**, *26*, 4476–4499. [CrossRef]
3. Shepherd, A.; Fricker, H.A.; Farrell, S.L. Trends and connections across the Antarctic cryosphere. *Nature* **2018**, *558*, 223–232. [CrossRef] [PubMed]
4. Goodman, D.J.; Frost, H.J.; Ashby, M.F. The plasticity of polycrystalline ice. *Philos. Mag. A* **1981**, *43*, 665–695. [CrossRef]
5. Alley, R.B. Flow-law Hypotheses for Ice-Sheet Modeling. *J. Glaciol.* **1992**, *38*, 245–256. [CrossRef]
6. Faria, S.H.; Weikusat, I.; Azuma, N. The microstructure of polar ice. Part II: State of the art. *J. Struct. Geol.* **2014**, *61*, 21–49. [CrossRef]

7. Hunter, N.J.R.; Wilson, C.J.L.; Luzin, V. Crystallographic preferred orientation (CPO) patterns in uniaxially compressed deuterated ice: Quantitative analysis of historical data. *J. Glaciol.* **2022**, 1–12. [CrossRef]
8. Marshall, S.J. Recent advances in understanding ice sheet dynamics. *Earth Planet. Sci. Lett.* **2005**, *240*, 191–204. [CrossRef]
9. Aster, R.C.; Winberry, J.P. Glacial seismology. *Rep. Prog. Phys.* **2017**, *80*, 126801. [CrossRef]
10. Feldmann, J.; Reese, R.; Winkelmann, R.; Levermann, A. Shear-margin melting causes stronger transient ice discharge than ice-stream melting in idealized simulations. *Cryosphere* **2022**, *16*, 1927–1940. [CrossRef]
11. Hudleston, P.J. Progressive Deformation and Development of Fabric Across Zones of Shear in Glacial Ice. In *Energetics of Geological Processes: Hans Ramberg on His 60th Birthday*; Saxena, S.K., Bhattacharji, S., Annersten, H., Stephansson, O., Eds.; Springer: Berlin/Heidelberg, Germany, 1977; pp. 121–150.
12. Budd, W.F.; Jacka, T.H. A review of ice rheology for ice sheet modelling. *Cold Reg. Sci. Technol.* **1989**, *16*, 107–144. [CrossRef]
13. Duval, P.; Montagnat, M.; Grennerat, F.; Weiss, J.; Meyssonnier, J.; Philip, A. Creep and plasticity of glacier ice: A material science perspective. *J. Glaciol.* **2010**, *56*, 1059–1068. [CrossRef]
14. Montagnat, M.; Buiron, D.; Arnaud, L.; Broquet, A.; Schlitz, P.; Jacob, R.; Kipfstuhl, S. Measurements and numerical simulation of fabric evolution along the Talos Dome ice core, Antarctica. *Earth Planet. Sci. Lett.* **2012**, *357–358*, 168–178. [CrossRef]
15. Lipenkov, V.Y.; Barkov, N.I.; Duval, P.; Pimienta, P. Crystalline Texture of the 2083 m Ice Core at Vostok Station, Antarctica. *J. Glaciol.* **1989**, *35*, 392–398. [CrossRef]
16. Azuma, N.; Wang, Y.; Yoshida, Y.; Narita, H.; Hondoh, T.; Shoji, H.; Watanabe, O. Crystallographic analysis of the Dome Fuji ice core. In *Physics of Ice Core Records*; Hokkaido University Press: Sapporo, Japan, 2000; pp. 45–61.
17. Durand, G.; Svensson, A.; Persson, A.; Gagliardini, O.; Gillet-Chaulet, F.; Sjolte, J.; Montagnat, M.; Dahl-Jensen, D. Evolution of the texture along the EPICA Dome C ice core. *Low Temp. Sci.* **2009**, *68*, 91–105.
18. Weikusat, I.; Kipfstuhl, S.; Azuma, N.; Faria, S.H.; Miyamoto, A. Deformation Microstructures in an Antarctic Ice Core (EDML) and in Experimentally Deformed Artificial Ice. *Low Temp. Sci.* **2009**, *68*, 115–123.
19. Faria, S.H.; Weikusat, I.; Azuma, N. The microstructure of polar ice. Part I: Highlights from ice core research. *J. Struct. Geol.* **2014**, *61*, 2–20. [CrossRef]
20. Bamber, J.L.; Vaughan, D.G.; Joughin, I. Widespread Complex Flow in the Interior of the Antarctic Ice Sheet. *Science* **2000**, *287*, 1248–1250. [CrossRef] [PubMed]
21. Bennett, M.R. Ice streams as the arteries of an ice sheet: Their mechanics, stability and significance. *Earth-Sci. Rev.* **2003**, *61*, 309–339. [CrossRef]
22. Rignot, E.; Mouginot, J.; Scheuchl, B. Ice Flow of the Antarctic Ice Sheet. *Science* **2011**, *333*, 1427–1430. [CrossRef]
23. Echelmeyer, K.A.; Harrison, W.D.; Larsen, C.; Mitchell, J.E. The role of the margins in the dynamics of an active ice stream. *J. Glaciol.* **1994**, *40*, 527–538. [CrossRef]
24. Hruby, K.; Gerbi, C.; Koons, P.; Campbell, S.; Martín, C.; Hawley, R. The impact of temperature and crystal orientation fabric on the dynamics of mountain glaciers and ice streams. *J. Glaciol.* **2020**, *66*, 755–765. [CrossRef]
25. Kamb, W.B. Experimental Recrystallization of Ice Under Stress. In *Flow and Fracture of Rocks*; Geophysical Monograph Series; California Institute of Technology: Pasadena, CA, USA, 1972; pp. 211–241.
26. Bouchez, J.-L.; Duval, P. The Fabric of Polycrystalline Ice Deformed in Simple Shear: Experiments in Torsion, Natural Deformation and Geometrical Interpretation. *Textures Microstruct.* **1982**, *5*, 171–190. [CrossRef]
27. Jun, L.; Jacka, T.H.; Budd, W.F. Strong single-maximum crystal fabrics developed in ice undergoing shear with unconstrained normal deformation. *Ann. Glaciol.* **2000**, *30*, 88–92. [CrossRef]
28. Journaux, B.; Chauve, T.; Montagnat, M.; Tommasi, A.; Barou, F.; Mainprice, D.; Gest, L. Recrystallization processes, microstructure and crystallographic preferred orientation evolution in polycrystalline ice during high-temperature simple shear. *Cryosphere* **2019**, *13*, 1495–1511. [CrossRef]
29. Qi, C.; Prior, D.J.; Craw, L.; Fan, S.; Llorens, M.G.; Griera, A.; Negrini, M.; Bons, P.D.; Goldsby, D.L. Crystallographic preferred orientations of ice deformed in direct-shear experiments at low temperatures. *Cryosphere* **2019**, *13*, 351–371. [CrossRef]
30. Lawson, W.J.; Sharp, M.J.; Hambrey, M.J. The structural geology of a surge-type glacier. *J. Struct. Geol.* **1994**, *16*, 1447–1462. [CrossRef]
31. Harrison, W.D.; Echelmeyer, K.A.; Larsen, C.F. Measurement of temperature in a margin of Ice Stream B, Antarctica: Implications for margin migration and lateral drag. *J. Glaciol.* **1998**, *44*, 615–624. [CrossRef]
32. Barnes, P.R.F.; Wolff, E.W. Distribution of soluble impurities in cold glacial ice. *J. Glaciol.* **2004**, *50*, 311–324. [CrossRef]
33. Pettit, E.C.; Whorton, E.N.; Waddington, E.D.; Sletten, R.S. Influence of debris-rich basal ice on flow of a polar glacier. *J. Glaciol.* **2014**, *60*, 989–1006. [CrossRef]
34. Jackson, M.; Kamb, B. The marginal shear stress of Ice Stream B, West Antarctica. *J. Glaciol.* **1997**, *43*, 415–426. [CrossRef]
35. Samyn, D.; Svensson, A.; Fitzsimons, S. Dynamic implications of discontinuous recrystallization in cold basal ice: Taylor Glacier, Antarctica. *J. Geophys. Res.* **2008**, *113*, F03S90. [CrossRef]
36. Gerbi, C.; Mills, S.; Clavette, R.; Campbell, S.; Bernsen, S.; Clemens-Sewall, D.; Lee, I.; Hawley, R.; Kreutz, K.; Hruby, K. Microstructures in a shear margin: Jarvis Glacier, Alaska. *J. Glaciol.* **2021**, *67*, 1163–1176. [CrossRef]
37. Monz, M.E.; Hudleston, P.J.; Prior, D.J.; Michels, Z.; Fan, S.; Negrini, M.; Langhorne, P.J.; Qi, C. Full crystallographic orientation (c and a axes) of warm, coarse-grained ice in a shear-dominated setting: A case study, Storglaciären, Sweden. *Cryosphere* **2021**, *15*, 303–324. [CrossRef]

38. Hellmann, S.; Kerch, J.; Weikusat, I.; Bauder, A.; Grab, M.; Jouvet, G.; Schwikowski, M.; Maurer, H. Crystallographic analysis of temperate ice on Rhonegletscher, Swiss Alps. *Cryosphere* **2021**, *15*, 677–694. [CrossRef]
39. Thomas, R.E.; Negrini, M.; Prior, D.J.; Mulvaney, R.; Still, H.; Bowman, M.H.; Craw, L.; Fan, S.; Hubbard, B.; Hulbe, C.; et al. Microstructure and Crystallographic Preferred Orientations of an Azimuthally Oriented Ice Core from a Lateral Shear Margin: Priestley Glacier, Antarctica. *Front. Earth Sci.* **2021**, *9*. [CrossRef]
40. Bentley, C.R. Seismic-wave velocities in anisotropic ice: A comparison of measured and calculated values in and around the deep drill hole at Byrd Station, Antarctica. *J. Geophys. Res.* **1972**, *77*, 4406–4420. [CrossRef]
41. Kohnen, H.; Gow, A.J. Ultrasonic Velocity Investigations of Crystal Anisotropy in Deep Ice Cores from Antarctica. *J. Geophys. Res. -Ocean. Atmos.* **1979**, *84*, 4865–4874. [CrossRef]
42. Harland, S.R.; Kendall, J.M.; Stuart, G.W.; Lloyd, G.E.; Baird, A.F.; Smith, A.M.; Pritchard, H.D.; Brisbourne, A.M. Deformation in Rutford Ice Stream, West Antarctica: Measuring shear-wave anisotropy from icequakes. *Ann. Glaciol.* **2013**, *54*, 105–114. [CrossRef]
43. Smith, E.C.; Baird, A.F.; Kendall, J.M.; Martín, C.; White, R.S.; Brisbourne, A.M.; Smith, A.M. Ice fabric in an Antarctic ice stream interpreted from seismic anisotropy. *Geophys. Res. Lett.* **2017**, *44*, 3710–3718. [CrossRef]
44. Jordan, T.M.; Schroeder, D.M.; Elsworth, C.W.; Siegfried, M.R. Estimation of ice fabric within Whillans Ice Stream using polarimetric phase-sensitive radar sounding. *Ann. Glaciol.* **2020**, *61*, 74–83. [CrossRef]
45. Lutz, F.; Eccles, J.; Prior, D.J.; Craw, L.; Fan, S.; Hulbe, C.; Forbes, M.; Still, H.; Pyne, A.; Mandeno, D. Constraining Ice Shelf Anisotropy Using Shear Wave Splitting Measurements from Active-Source Borehole Seismics. *J. Geophys. Res. Earth Surf.* **2020**, *125*, e2020JF005707. [CrossRef]
46. Hellmann, S.; Grab, M.; Kerch, J.; Löwe, H.; Bauder, A.; Weikusat, I.; Maurer, H. Acoustic velocity measurements for detecting the crystal orientation fabrics of a temperate ice core. *Cryosphere* **2021**, *15*, 3507–3521. [CrossRef]
47. Rathmann, N.M.; Lilien, D.A.; Grinsted, A.; Gerber, T.A.; Young, T.J.; Dahl-Jensen, D. On the Limitations of Using Polarimetric Radar Sounding to Infer the Crystal Orientation Fabric of Ice Masses. *Geophys. Res. Lett.* **2022**, *49*, e2021GL096244. [CrossRef]
48. Shugar, D.H.; Jacquemart, M.; Shean, D.; Bhushan, S.; Upadhyay, K.; Sattar, A.; Schwanghart, W.; McBride, S.; de Vries, M.V.W.; Mergili, M.; et al. A massive rock and ice avalanche caused the 2021 disaster at Chamoli, Indian Himalaya. *Science* **2021**, *373*, 300–306. [CrossRef] [PubMed]
49. Smedley, A.R.D.; Evatt, G.W.; Mallinson, A.; Harvey, E. Solar radiative transfer in Antarctic blue ice: Spectral considerations, subsurface enhancement, inclusions, and meteorites. *Cryosphere* **2020**, *14*, 789–809. [CrossRef]
50. Muhuri, A.; Manickam, S.; Bhattacharya, A.; Snehmani. Snow Cover Mapping Using Polarization Fraction Variation With Temporal RADARSAT-2 C-Band Full-Polarimetric SAR Data Over the Indian Himalayas. *IEEE J. Sel. Top. Appl. Earth Obs. Remote Sens.* **2018**, *11*, 2192–2209. [CrossRef]
51. Tsai, Y.-L.S.; Dietz, A.J.; Oppelt, N.; Kuenzer, C. Remote Sensing of Snow Cover Using Spaceborne SAR: A Review. *Remote. Sens.* **2019**, *11*, 1456. [CrossRef]
52. Qiao, H.; Zhang, P.; Li, Z.; Liu, C. A New Geostationary Satellite-Based Snow Cover Recognition Method for FY-4A AGRI. *IEEE J. Sel. Top. Appl. Earth Obs. Remote Sens.* **2021**, *14*, 11372–11385. [CrossRef]
53. Wilson, C.J.L.; Russell-Head, D.S.; Kunze, K.; Viola, G. The analysis of quartz c-axis fabrics using a modified optical microscope. *J. Microsc.* **2007**, *227*, 30–41. [CrossRef]
54. Legland, D.; Arganda-Carreras, I.; Andrey, P. MorphoLibJ: Integrated library and plugins for mathematical morphology with ImageJ. *Bioinformatics* **2016**, *32*, 3532–3534. [CrossRef]
55. Schindelin, J.E.; Arganda-Carreras, I.; Frise, E.; Kaynig, V.; Longair, M.; Pietzsch, T.; Preibisch, S.; Rueden, C.T.; Saalfeld, S.; Schmid, B.; et al. Fiji: An open-source platform for biological-image analysis. *Nat. Methods* **2012**, *9*, 676–682. [CrossRef]
56. Arganda-Carreras, I.; Kaynig, V.; Rueden, C.; Eliceiri, K.W.; Schindelin, J.; Cardona, A.; Sebastian Seung, H. Trainable Weka Segmentation: A machine learning tool for microscopy pixel classification. *Bioinformatics* **2017**, *33*, 2424–2426. [CrossRef]
57. Montagnat, M.; Azuma, N.; Dahl-Jensen, D.; Eichler, J.; Fujita, S.; Gillet-Chaulet, F.; Kipfstuhl, S.; Samyn, D.; Svensson, A.; Weikusat, I. Fabric along the NEEM ice core, Greenland, and its comparison with GRIP and NGRIP ice cores. *Cryosphere* **2014**, *8*, 1129–1138. [CrossRef]
58. Dansgaard, W. Stable isotopes in precipitation. *Tellus* **1964**, *16*, 436–468. [CrossRef]
59. Wenk, H.R.; Christie, J.M. Comments on the interpretation of deformation textures in rocks. *J. Struct. Geol.* **1991**, *13*, 1091–1110. [CrossRef]
60. Young, T.J.; Schroeder, D.M.; Jordan, T.M.; Christoffersen, P.; Tulaczyk, S.M.; Culberg, R.; Bienert, N.L. Inferring Ice Fabric From Birefringence Loss in Airborne Radargrams: Application to the Eastern Shear Margin of Thwaites Glacier, West Antarctica. *J. Geophys. Res. Earth Surf.* **2021**, *126*, e2020JF006023. [CrossRef]
61. Cuffey, K.; Paterson, W. *The Physics of Glaciers*, 4th ed.; Butterworth-Heinemann: Oxford, UK, 2010; pp. 11–28.
62. Buizert, C.; Baggenstos, D.; Jiang, W.; Purtschert, R.; Petrenko, V.V.; Lu, Z.T.; Muller, P.; Kuhl, T.; Lee, J.; Severinghaus, J.P.; et al. Radiometric Kr 81 dating identifies 120,000-year-old ice at Taylor Glacier, Antarctica. *Proc. Natl. Acad. Sci. USA* **2014**, *111*, 6876–6881. [CrossRef]
63. Bauska, T.K.; Baggenstos, D.; Brook, E.J.; Mix, A.C.; Marcott, S.A.; Petrenko, V.V.; Schaefer, H.; Severinghaus, J.P.; Lee, J.E. Carbon isotopes characterize rapid changes in atmospheric carbon dioxide during the last deglaciation. *Proc. Natl. Acad. Sci. USA* **2016**, *113*, 3465–3470. [CrossRef]

64. Baggenstos, D.; Bauska, T.K.; Severinghaus, J.P.; Lee, J.E.; Schaefer, H.; Buizert, C.; Brook, E.J.; Shackleton, S.; Petrenko, V.V. Atmospheric gas records from Taylor Glacier, Antarctica, reveal ancient ice with ages spanning the entire last glacial cycle. *Clim. Past* **2017**, *13*, 943–958. [CrossRef]
65. Menking, J.A.; Brook, E.J.; Shackleton, S.A.; Severinghaus, J.P.; Dyonisius, M.N.; Petrenko, V.; McConnell, J.R.; Rhodes, R.H.; Bauska, T.K.; Baggenstos, D.; et al. Spatial pattern of accumulation at Taylor Dome during Marine Isotope Stage 4: Stratigraphic constraints from Taylor Glacier. *Clim. Past* **2019**, *15*, 1537–1556. [CrossRef]

water

MDPI

Article

Laboratory Studies on the Parametrization Scheme of the Melting Rate of Ice–Air and Ice–Water Interfaces

Zhijun Li [1], Qingkai Wang [1], Ge Li [2,*], Peng Lu [1], Zhiqun Wang [1] and Fei Xie [1]

1 State Key Laboratory of Coastal and Offshore Engineering, Dalian University of Technology, Dalian 116024, China; lizhijun@dlut.edu.cn (Z.L.); wangqingkai@dlut.edu.cn (Q.W.); lupeng@dlut.edu.cn (P.L.); 15522701687@163.com (Z.W.); xf_ice@mail.dlut.edu.cn (F.X.)

2 North China Sea Marine Forecasting Center of State Oceanic Administration, Qingdao 266000, China

* Correspondence: lige@ncs.mnr.gov.cn

Abstract: During the melt season, surface melting, bottom melting, and lateral melting co-occur in natural ice floes. The bottom melting rate is larger than the lateral melting rate, followed by the surface melting rate, and the smaller the size of an ice floe, the higher the lateral melting rate. To add the scale index of small-scale ice to the melting parametrization scheme, experiments on the melting process of sea ice and artificial fresh-water ice samples in the shape of a disc were carried out in a low-temperature laboratory, under conditions of no radiation, current, or wind, with controlled air and water temperatures. The variations of diameter, thickness, and mass of the ice discs were measured through the experiments. According to the experimental data, a new indicator was created using the ratio of the diameter to the thickness of an ice sample. Based on physical and statistical analyses, the relationships between the surface/bottom melting rates and temperature gradient were formulated. Additionally, the relationships among the lateral melting rate, temperature difference, and the ratio of the diameter to the thickness were also quantified. The equations can be applied to the melting parametrization scheme of ice for a range of diameters up to 100 m, which covers simulations of the energy and mass balance values of the Arctic sea ice and coastal freshwater ice during the summer melt season.

Keywords: sea ice; fresh-water ice; parametrization scheme; melting rate; laboratory study

Citation: Li, Z.; Wang, Q.; Li, G.; Lu, P.; Wang, Z.; Xie, F. Laboratory Studies on the Parametrization Scheme of the Melting Rate of Ice–Air and Ice–Water Interfaces. *Water* **2022**, *14*, 1775. https://doi.org/10.3390/w14111775

Academic Editor: Karl-Erich Lindenschmidt

Received: 5 May 2022
Accepted: 29 May 2022
Published: 1 June 2022

1. Introduction

With global warming, the annual warming trend of the Arctic temperature is 2–3 times that of the global temperature, which is also known as the Arctic amplification effect [1–3]. Sea ice changes in the Arctic are not only related to local climate, but also affect climate changes in the mid-latitudes of the Northern Hemisphere, such as by increasing extreme weather events in Eurasia and North America [4,5]. Sea ice models are effective tools to describe the evolution of sea ice which mainly include thermodynamic and dynamic processes [6,7].

In general, ice thickness distribution (ITD) and sea ice concentration are used to describe the state of sea ice in a given area in the sea ice models [8]. The World Meteorological Organization uses floe size (form of ice) in addition to these two terms to describe ice, and uses an egg code as the standard way to describe ice [9]. The floe size distribution (FSD) is also needed in the developments of sea ice models, especially in the simulation of a marginal ice zone (MIZ) [10]. In the empirical formula according to the observations, FSD generally could be represented by a power law [11]. To accurately describe FSD, lateral melting plays an important role in the contribution to the conservation formula of FSD. For example, in the Arctic summer, a rapid melting rate occurs not only in the bottom and surface, but also in the lateral interface between sea ice and ocean, especially in MIZ. The contribution of lateral melting to ice ablation and perimeter increases has a more important role than before, because of floe breaks [12]. The lateral melting rate and perimeter of floes often determine the magnitude of a net heat flux exchanged between the ocean and the lateral edge of sea ice [13]. The melting rate is sensitive to floe size due to the increase in the perimeter of the lateral ice edge when floe size is smaller than 30 m [14]. Due to the decrease in ice extent and thickness, the evolution of sea ice shows new characteristics different from previous cases. For larger floe sizes, the lateral melting of sea ice may also have an important contribution since the ice meltwater fronts accelerate the lateral mixing [15]. An active exploration of the lateral melting mechanism is necessary to understand Arctic sea ice melting during the summer.

Field investigations are important means of understanding the lateral melting physical processes of sea ice. In the programs of Arctic Ice Dynamics Joint Experiment and Marginal Ice Zone Experiment, researchers have investigated the relationship between floe size shape and lead heat flux [14,16]. Hwang et al. (2017) and Perovich and Jones (2014) have observed the evolution of floe size and found that lateral melting plays an important role in the variation of FSD [17,18]. Stern et al. (2018) are continuously conducting observational studies in which the seasonal cycle of the FSD is analyzed and lateral melting contributes quantitatively to the seasonal cycle [19].

At present, only a few field observations of the lateral melting of the Arctic sea ice have been conducted due to the limitations imposed by the polar field conditions and testing techniques. As a result, different techniques and methods were tested on stable lake ice in China and the acquired data were applied to the parametrization research of lateral melting [20,21]. These in situ experiments on lake ice have comprehensively investigated the multifactors of air temperature, solar radiation, and water temperature contributing to the lateral melting, and the limit in understanding the contributions of one particular factor to lateral melting. However, the effect of floe size on the lateral melting is still difficult to observe in in situ experiments due to the lack of effective techniques. In this paper, the laboratory experiments were conducted for the melting on the basal and the lateral interfaces of ice, improving the understanding of the lateral melting process, and then giving a parametrization scheme of the lateral melting rate varied with floe size.

2. Theoretical Foundation

The lateral melting of sea ice depends on the net heat flux from the ocean and atmosphere to lateral ice [22]. The net heat flux generally includes shortwave and longwave radiation, sensible heat flux, conductive heat flux, and latent heat flux [23]. The melting rate contributes to latent heat flux. In the literature (Bateson et al., 2020; Roach et al., 2018) [10,24], the lateral melting rate is expressed in a power law, which is:

$$W_{\text{lat}} = a\Delta T^{b} \tag{1}$$

where W_{lat} is the lateral melting rate of ice, ΔT is the difference of water temperature above the freezing point, and a and b are the statistical parameters based on observations.

Smith et al. (2022) used an equation for the ice surface, bottom, and lateral melting based on the sea ice concentration [25]:

$$\frac{d}{dt}(AH) = A\left(W_{\text{sur}} + W_{\text{bot}} + \frac{\pi H}{\mu L_D} W_{\text{lat}}\right) \tag{2}$$

where A is the sea ice concentration per unit area and ranges from 0–1, W_{sur} is the surface melting rate of ice, W_{bot} is the bottom melting rate of ice, H is the ice thickness, L_D is the floe diameter (300 m in the default set up of sea ice models), and μ is the geometric parameter representing the deviation of floe with a circular shape (μ = 0.66). W_{sur} and W_{bot} contribute to the variation of ice thickness H; and when H keeps constant, W_{sur} and W_{bot} can be neglected, and the equation can be written as:

$$H\frac{dA}{dt} = A\frac{\pi H}{\mu L_{\text{D}}} W_{\text{lat}} \tag{3}$$

In particular, ice with an equivalent diameter of less than 30 m is thought to be greatly affected by lateral melting [14]. In this paper, ice samples were used in the laboratory to model surface, bottom, and lateral melting of ice. The observed changes in the diameter were the result of the lateral melting on the surface of the samples, whereas the thickness variation data indicated the melting of the surface and the bottom, and the mass variation was an integral demonstration of the three melting processes. The Figure 1 schematic illustrates the heat flux and the thermal balance of the surface, bottom, and lateral interfaces during ice floe melting.

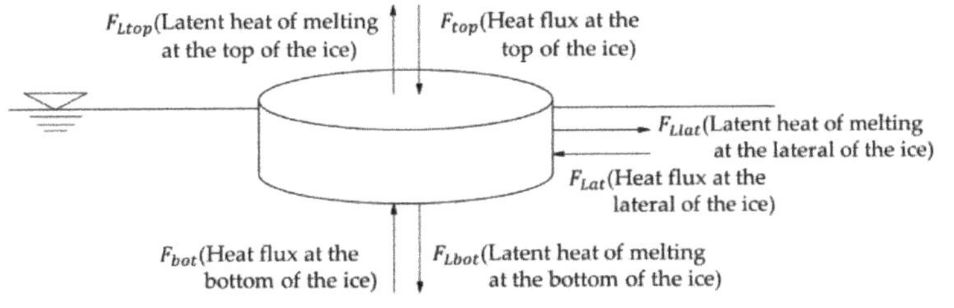

Figure 1. Thermal balance between the ice–air interface at the surface, between ice and water on the bottom, and the lateral surface of a floe.

Laboratory experiments are conducted in an ideal condition. As shown in Figure 1, the net heat flux mainly includes two parts. One is the latent heat flux resulting from melting, the other is heat flux from the environment (air or water) to ice. The latent heat flux is proportional to the melting rate, and the other heat flux is proportional to the gradient of temperature. The equation for the surface melting rate is:

$$W_{\text{sur}} = \left(\frac{\lambda_a}{\rho_i L}\right)\frac{\Delta T_a}{\Delta h_a} \tag{4}$$

The bottom melting rate is given by [26]:

$$W_{\text{bot}} = \left(\frac{\lambda_w}{\rho_i L}\right)\frac{\Delta T_w}{\Delta h_w} \tag{5}$$

where λ_a and λ_w are air and water thermal conductivity, respectively; ΔT_a and ΔT_w are the difference of temperatures between air and ice surface, and difference of temperatures between water and ice bottom, respectively; Δh_a and Δh_w are references to height and depth according to measurement positions; and L represents the latent heat.

3. Experimental Method

A water tank 2 m long, 1 m wide, and 0.8 m high was set up in the low-temperature laboratory. The tank was designed specially with a double-layer bottom, and a pump was installed at the lower layer to drive the water before tests and was stopped during test. The experimental water tank can hold water to a depth of 0.5 m. A schematic of the laboratory equipment system is shown in Figure 2. The melting process of the ice samples was affected jointly by the temperature of the air above the water tank and the water inside. So, the PT100 temperature probes were distributed along the direction of depth, from the air above the tank to the water in the tank. With a diameter of 5 mm, each temperature probe was precision-corrected in the factory to satisfy a temperature measurement accuracy of 0.1 °C. The air temperature probes were installed above the water surface at a height of 50 cm and 3 cm, respectively. For the water temperature measurements, probes were set up at depths of 0, 2, 4, 6, 8, 10, 20, 30, 40, and 50 cm. To reduce the interference caused by the small probe spacing, the adjacent two temperature probes were inserted vertically into the wood.

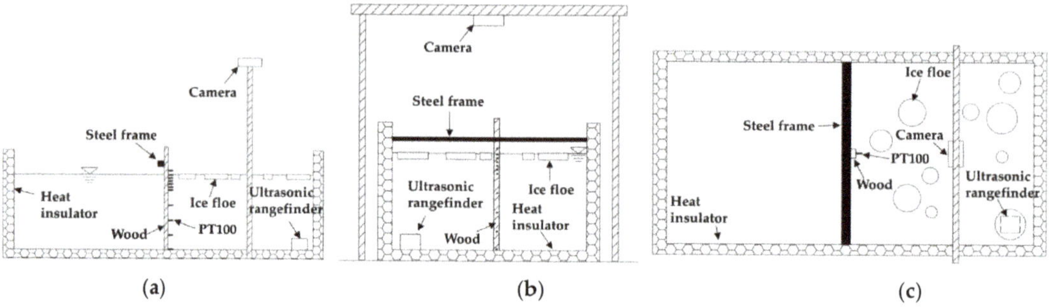

Figure 2. The laboratory equipment setup. (**a**) Front schematic; (**b**) lateral schematic; and (**c**) top schematic.

A COOLPIX P6000 camera, with 13.5 million pixels, was installed 0.7 m above the water surface at the tank center. Photos taken by the camera had the quality of each pixel corresponding to 0.5 mm × 0.5 mm. An under-ice ultrasonic range finder with high precision (±0.1 mm) was set on the bottom of the water tank. Both the camera and the ultrasonic range finder were used as aid technology in this experiment; that is, the measured data using the camera and ultrasonic range finder were not included into the analysis below but were used to validate the manual measurements during tests. The camera traced the diameter variation on the surface of the ice sample, and the ultrasonic range finder traced the level variation on the bottom of a specific ice sample.

In this experiment, two types of ice samples were used: natural sea ice from the Yellow Sea and artificial fresh-water ice. The sea ice was level ice with a thickness of about 30 mm that had been collected from the Xiajiahezi seashore, Dalian, during 4–6 January 2021, transported to the laboratory directly without large temperature variations and then stored at a low temperature. The sampling site was not far away from the laboratory so that large brine loss would not occur. The sea ice was then cut using a chain saw and lathe to diameters of 65, 90, 120, 150, and 190 mm. For the artificial fresh-water ice, constant temperature technology was used to maintain the lateral water temperature at the freezing point so that the ice would freeze unidirectionally. The unidirectionally frozen ice was processed using a lathe to diameters of 65, 90, 120, 150, and 190 mm, at a thickness of about 30 mm. The sea ice was melted for measuring salinity using a salinometer, and the mean salinity was 6.04 ppt, which is close to the typical salinity of sea ice in the Yellow Sea. The densities of sea ice and fresh-water ice were measured using the mass/volume method, and the mean densities were 800 kg/m^3 for sea ice and 900 kg/m^3 for fresh-water ice. The ice crystal structure was observed by first preparing 1 cm thick vertical sections using a band saw. These thick sections were attached to glass plates and cut further to an approximately 0.5 mm thickness. These thin sections were placed on a universal stage to observe the crystal structure under crossed polarized light, recorded by photography. Figure 3 shows the two crystal types. The sea ice was granular ice, and the fresh-water ice was columnar ice.

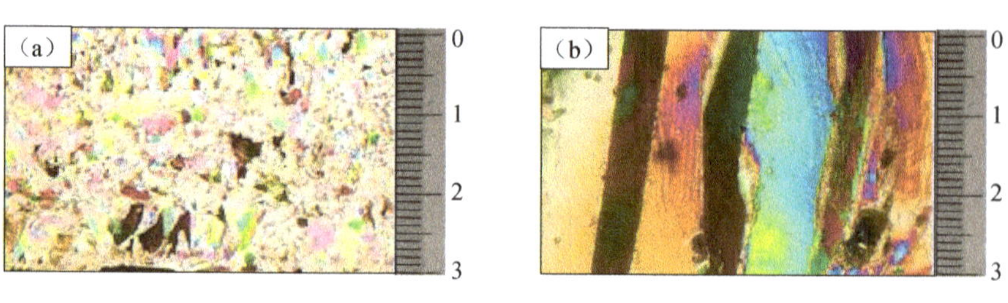

Figure 3. The crystals of the ice samples used in the experiments. (**a**) Yellow Sea ice (**b**) fresh-water ice.

By adjusting the air temperature in the low-temperature laboratory, the water temperature in the water tank was kept stable. In this experiment, the sea water was salty water prepared using sea salt, with a salinity (tested as 27.7 ppt) in accordance with the median value of the ocean salinity profile for the Arctic summer sea ice area, as reported in the literature [27]. The fresh water was tap water. Before the experiment, different preliminary water temperature ranges were set for the sea ice and the fresh-water ice experiments. Based on the temperature-controlling abilities of the laboratory, the water temperatures were set at −0.5, 0, 0.5, 1.0, and 1.5 °C in the sea ice tests and 0.5, 1.5, 2.0, and 3.0 °C in the fresh-water tests. Since the fresh water is simpler in phase composition than sea ice, the temperature groups were less in fresh-water tests and the test temperatures were set higher in fresh-water ice tests than in sea ice tests.

Before each experiment, ice samples were stored in the thermotank at the temperature of 1 °C lower than the freezing point for at least 24 h. When water temperature was measured close to the designed temperature, samples with five different diameters were placed in the water tank at the same time. Samples were labeled with different color stickers or numbers for identification. The variations in the diameter, maximum thickness, and mass of each sample were traced in the test. Both manual measurements and electro-test aided technology were carried out simultaneously during the melting process. At the beginning of each experiment, there was less melting of ice, so the lateral melting, thickness melting, and mass melting were measured directly after the samples were taken out of the water. In general, the measurement interval was set to about 1 h to minimize the influence of the manual testing on the experimental results. The maximum thickness and the surface diameter of the ice samples were measured by an electronic digital caliper (±0.01 mm), and the mass was measured using an electronic scale. The whole process was rapid and the samples were taken with caution so that no wave was produced and the temperatures of air, ice, and water would not be affected. A foam disc with a diameter of 10 cm was placed in the ice sample as a reference for photographing to calibrate the equivalent diameter on the surface of the ice samples at different moments determined using a sample area. Both the camera and ultrasonic sensor were set to record data at an interval of 10 min.

4. Results

Before the experiment, an electronic digital caliper was used to measure the diameter and thickness of each sample three times at different positions, respectively, based on which the average initial values of these two parameters were calculated. The initial size information of the two types of samples is listed in Table 1.

Throughout the experiment, no heat exchange was caused by solar radiation or wind convection in the low-temperature laboratory and only the indoor temperature was used as the driving factor to control the water–ice heat exchange. The designed preliminary water temperature was higher than the experimental water temperature to ensure that lateral and bottom melting occurred after the ice absorbed heat during the whole experiment. The temperature in the low-temperature laboratory was characterized by (1) temperature control fluctuations resulting from the refrigeration system, (2) a temperature increase "pulse" in the daily defrosting of the refrigeration system, and (3) a slight long-term temperature increase. Figure 4a presents the temperature changes at a position of 3 cm above the water surface during the experiment, where a sea ice sample with a designed preliminary water temperature of 1.0 °C is taken as an example.

Table 1. Preliminary parameters of the ice samples.

Specimen No.	Sea Ice					Fresh-Water Ice				
	1	2	3	4	5	1	2	3	4	5
Designed preliminary water temperature/°C	−0.5	−0.5	−0.5	−0.5	−0.5	0.5	0.5	0.5	0.5	0.5
Initial average diameter/mm	68.9	93.0	124.5	154.5	185.9	68.2	90.9	120.0	154.8	191.9
Initial thickness/mm	32.6	34.5	36.0	35.5	34.5	30.9	30.7	30.7	30.2	30.9
Designed preliminary water temperature/°C	0	0	0	0	0	1.5	1.5	1.5	1.5	1.5
Initial average diameter/mm	69.4	88.6	122.8	154.2	191.6	67.9	89.4	119.1	153.2	192.5
Initial thickness/mm	36.6	36.7	35.7	31.7	36.8	31.00	31.2	30.3	29.6	31.2
Designed preliminary water temperature/°C	0.5	0.5	0.5	0.5	0.5	2.0	2.0	2.0	2.0	2.0
Initial average diameter/mm	68.8	97.4	123.0	154.5	194.5	67.5	90.4	118.9	153.0	191.0
Initial thickness/mm	31.9	41.9	36.0	36.1	39.5	30.9	30.4	30.6	30.6	31.5
Designed preliminary water temperature/°C	1.0	1.0	1.0	1.0	1.0	3.0	3.0	3.0	3.0	3.0
Initial average diameter/mm	69.6	88.1	124.4	152.8	193.8	67.6	89.5	120.0	153.2	191.0
Initial thickness/mm	38.8	37.3	35.7	40.5	38.1	30.9	30.2	30.5	30.9	30.6
Designed preliminary water temperature/°C	1.5	1.5	1.5	1.5	1.5					
Initial average diameter/mm	67.0	94.9	126.9	155.9	193.3					
Initial thickness/mm	28.8	32.9	35.3	32.5	34.1					

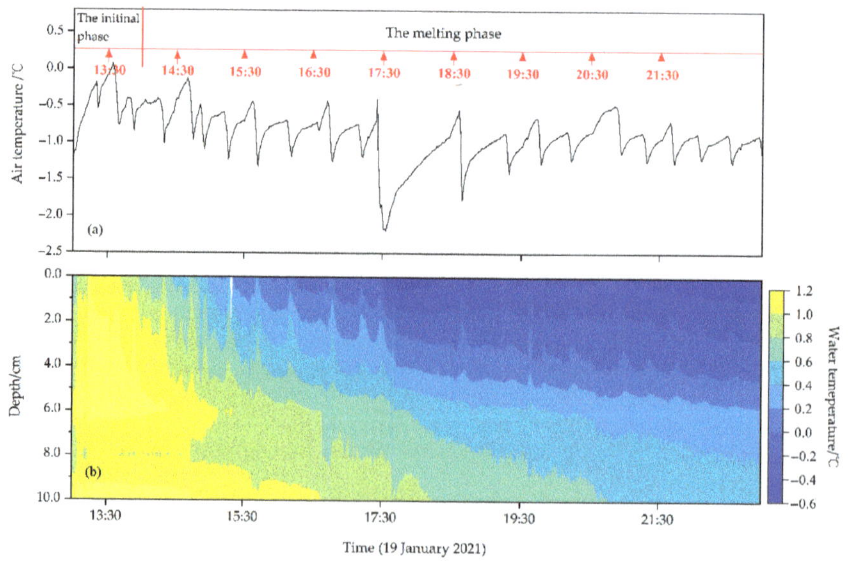

Figure 4. The (**a**) air and (**b**) water temperature variation processes of the tested samples of sea ice at a designed preliminary water temperature of 1.0 °C.

The experiment was conducted from 13:30 to 21:30 on 19 January. Before the experiment, the water in the tank was stirred constantly to uniformly reduce the water temperature to 1.0 °C. The air temperature at a distance of 3 cm from the water surface was kept below 1.0 °C for a period of time to stabilize the water temperature. Next, the sample was placed into the tank. Thus, the temperature in the laboratory was reduced due to the heat absorption of the ice sample. However, the operation of the refrigerator guaranteed a regular fluctuation of 0.4 °C. Finally, a temperature "pulse" caused by the short-term defrosting was observed at 17:30.

The sample melted after absorbing heat from the water, leading to a temperature reduction from the water surface to the depth of ice immersion. The temperature of the water at the same depth also experienced a decrease with the adjustment of air temperature. However, the vertical temperature gradient of the water under the ice gradually increased with depth, which could be attributed to the heat absorption in the ice-melting process. Taking the sea ice sample with the designed preliminary water temperature of 1.0 °C as an example, Figure 4b records the changes in water temperature, including changes in the section of the example experiment, where the water temperature decreased with the melting of the sample. The water and air temperatures first remained stable and then increased after the ice sample had completely melted.

As seen in Figure 4, large changes occurred in the frequency and amplitude of the air temperature, whereas the water temperature remained stable during the experiment. According to the color gradation diagram of the water temperature in Figure 4b, the water was uniformly stirred before 14:00 and its temperature reached the designed preliminary water temperature of about 1.0 °C, indicated by the yellow color. After all the ice samples were placed in the tank, the surface water temperature reduced quickly, while the temperature of the lower water reduced slowly. Abrupt changes in water temperature were observed before 18:00, and the whole experiment displayed a clear vertical water temperature gradient. Moreover, the temperature of the water from a depth of 2 to 4 cm underwent a higher rate of reduction owing to the ice thickness of 3 cm.

A typical picture of a melting ice sample is shown in Figure 5. The diameter, maximum thickness, and mass changes of the sea ice samples with different diameters with a designed preliminary water temperature of 1.0 °C are shown in Figure 6. The changes in the surface diameter during the ice-melting process are displayed in Figure 6a. The diameter reduced gradually in the initial stage due to the hysteresis of heat absorption. The reduction rate then increased with time. Figure 6b depicts the thickness changes in the samples. As the experiment continued, the sample first melted rapidly and then slowly, eventually reaching a stable state. The melting rates of the samples with different diameters were basically the same. According to Figure 6c, the overall mass of the sample decreased in the experiment. At the beginning, due to the large pores of the unclosed bubbles, salty water infiltrated the sample when it was first placed into the water, resulting in an increased sample mass.

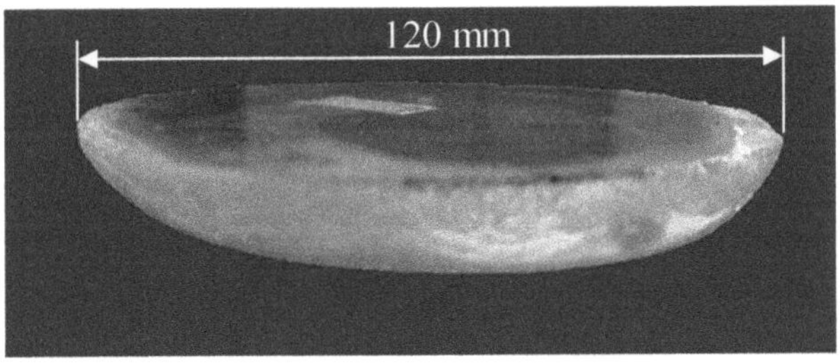

Figure 5. The lateral shape of a melted ice sample.

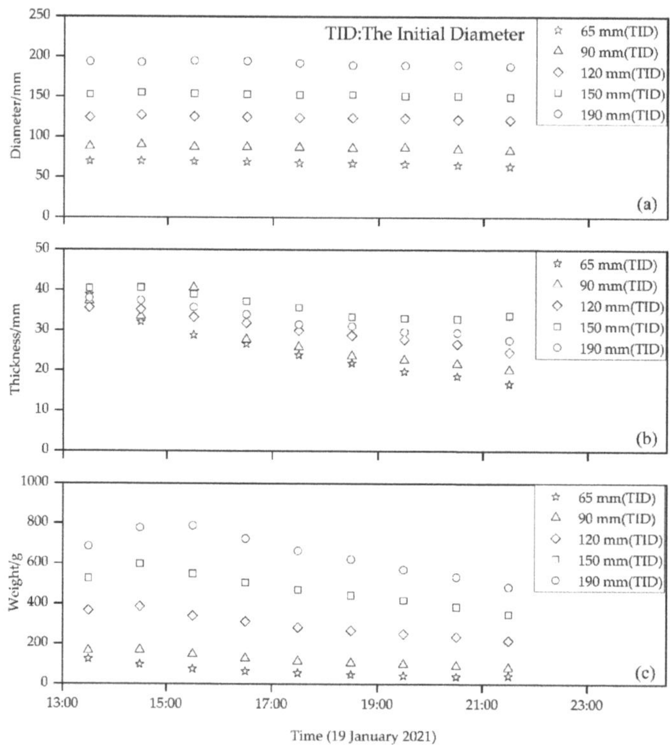

Figure 6. (**a**) Diameter, (**b**) thickness, and (**c**) mass variability of the tested sea ice samples at the designed preliminary water temperature of 1.0 °C.

Similar results were obtained from experiments conducted on the other four groups of sea ice and three groups of fresh-water ice samples in water with different initial preliminary water temperatures. Therefore, they are not presented in detail here but shown in an online dataset [28].

Furthermore, the lateral melting of the sample was affected by the temperature difference in the water at different depths. The amount of melting of the sample varied with water depth. In general, the surface and bottom of the ice melted the most in nature. However, solar radiation and wind were not present in the low-temperature laboratory, where water temperature adjustment was completely based on air temperature, which led to a consistent and large vertical temperature gradient in the water, as shown in Figure 4b. For this reason, along the depth below water surface, the water–ice temperature difference and the amount of melting rose, showing a decreasing trend (Figure 6), which agrees with reports from the literature [21,29]. Other findings included a high lateral melting rate of the sample underwater, a small lateral melting rate of the ice surface, and a protruding edge of the ice upper surface. The ice specimen became thinner in the later stage and lost its round shape. This phenomenon was fully consistent with the observation that the melt rate of sea ice is higher than that of fresh-water ice.

5. Parametrization Schemes for the Melting Rate at the Ice–Water Interface

According to the literature [30], for Arctic sea ice, the bottom melting rate is higher than the surface melting rate, followed by the lateral melting rate. In the low-temperature laboratory without solar radiation or wind, the surface of the specimen melted more slowly than the bottom surface. The melting rates were higher than those of Arctic sea ice in the actual situation since the laboratory had a larger vertical temperature gradient than the Arctic, but its rate was still lower than that of the bottom melting for the sample in our experiments. In the melting experiment, the ice floes became thinner during the melting process. In the data analysis, the thickness changes of the disc sample and the measured average densities of the sea ice and fresh-water ice samples (800 kg/m^3 and 900 kg/m^3, respectively) were used to estimate distances from the sample surface to the temperature probe in air and from the sample bottom to the temperature probe in water. At the temperature where the disc sample achieved thermal equilibrium, sea ice was a mixture containing liquid brine, in which pure ice crystals accounted for the majority. From a physicochemical perspective, the freezing and melting points of these crystals are different [31], and the melting temperature is a range value. It was assumed that the minimum temperature was slightly lower than the freezing point of sea water, namely, −1.5 °C, and the maximum temperature was slightly lower than the melting point of pure ice crystals, namely, −0.2 °C. Fresh-water ice was considered pure-water ice with the same freezing and melting points of 0 °C. In this way, the air temperature gradient on the sample surface and the vertical water temperature gradient on the sample bottom were calculated. The melting point of fresh-water ice is higher than sea ice. Therefore, the temperature differences between air and sea ice samples were higher than those between air and fresh-water ice samples, leading to higher surface melting rates of sea ice samples.

Given the high frequency of fluctuations in surface air temperature, the surface melting accounted for a small proportion of thickness melting. Thus, based on Equations (4) and (5), the thickness melting rate was decomposed into the surface and bottom melting rates. Due to the short time period, the temperature fluctuations during the defrosting period and the moment when the refrigeration stopped failed to cause rapid melting or freezing of the sample surface. However, in individual cases, the calculated temperature gradient values were negative (i.e., exothermic and ice freeze) and hence eliminated. Statistical analysis was conducted on the surface and the bottom melting rates, respectively, and the corresponding results are depicted in Figures 7 and 8.

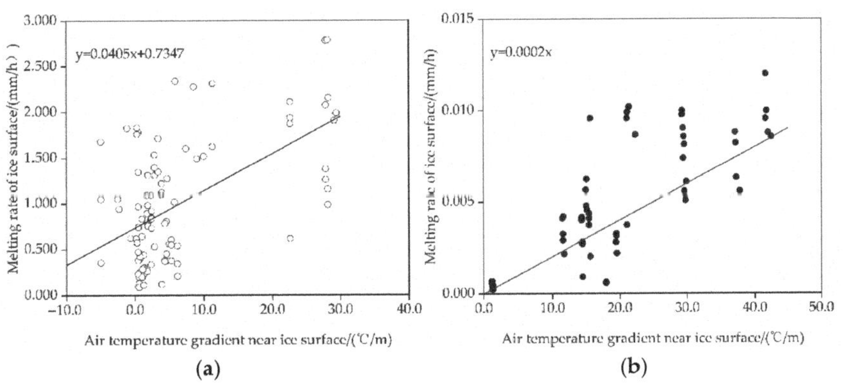

Figure 7. The statistical relationship between the surface melting rate and air temperature gradient. (**a**) Sea ice (**b**) fresh-water ice.

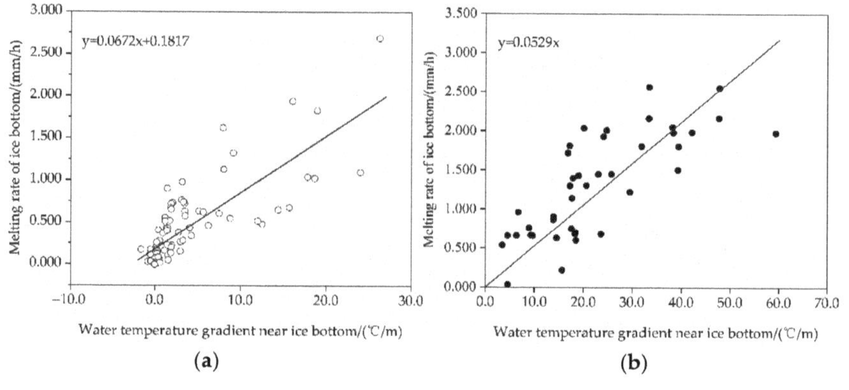

Figure 8. The statistical relationship between the bottom melting rate and vertical water temperature gradient. (**a**) Sea ice (**b**) fresh-water ice.

According to Figure 7, the empirical parametrization scheme of the surface melting rate of sea ice and fresh-water ice can be expressed by Equations (6) and (7), respectively, and those of the bottom melting rate are expressed by Equations (8) and (9), respectively. These can be applied to the melting process of coastal fresh water or low-salinity ice in most of the Arctic Ocean that is covered by summer sea ice and several estuaries of the Arctic Ocean, such as the Mackenzie River flowing into the Chukchi Sea, the Lena River flowing into the East Siberian Sea, and the Ob River and Yenisei River flowing into the Kara Sea [32]. Notably, the laboratory was subject to high frequency and large amplitude temperature fluctuations, resulting in large data dispersion. For this reason, the parametrization confidence of the surface melting rate might be reduced. Thus, it was suggested that Equations (6) and (7) should be modified based on experimental and field results obtained in the future in a controlled and stable environment. Long-term observational data from the Arctic can be directly used to verify Equations (8) and (9).

$$W_{sur,sea} = 0.0405\frac{\Delta T_a}{\Delta h_a} + 0.7347 \quad r = 0.5542 \tag{6}$$

$$W_{sur,fresh} = 0.0002\frac{\Delta T_a}{\Delta h_a} \quad r = 0.7654 \tag{7}$$

$$W_{bot,sea} = 0.0672\frac{\Delta T_w}{\Delta h_w} + 0.1817 \quad r = 0.7989 \tag{8}$$

$$W_{bot,fresh} = 0.0529\frac{\Delta T_w}{\Delta h_w} \quad r = 0.8185 \tag{9}$$

where the subscripts sea and fresh denote sea ice and fresh-water ice, respectively; $W_{sur,sea}$, $W_{sur,fresh}$, $W_{bot,sea}$, and $W_{bot,fresh}$ are in mm/h, respectively; and r is the correlation coefficient. The equations were empirically established using regression analysis of the test data, and thus the dimensions were not considered.

Equations (6)–(9) are specific manifestations of Equations (4) and (5). Therefore, under natural conditions, as for the interface heat flux obtained based on meteorological and marine hydro-meteorological environmental elements, Equations (6)–(9) can be used to estimate the bottom and surface melting rates of the coastal fresh-water ice in the sea ice area and estuaries of the Arctic Ocean. Moreover, according to the thermodynamic theory, the heat flux on the ice surface and bottom is determined by the temperature gradient, and the above parametrized expressions are not limited by the floe size.

The target of manual measurements in the experiment was to observe the diameter changes in the sample surface, since it was difficult to measure the dimension of the sample in other depths below surface (as shown in Figure 5). Thus, in the statistical analysis of the experimental results focused on lateral melting, the difference between the water temperature at the water surface and the thermal equilibrium melting point of the ice sample were taken into consideration. To demonstrate the effect of the floe size on the lateral melting rate, the ratio of the diameter (L_D) to the thickness (H) of the ice sample was introduced into the parametric scheme as a new index, thereby achieving consistency between test results obtained in the laboratory and those measured on site. Previous investigation showed that the lateral melting is important for floes with a diameter of less than 30 m [14], and thus, the laboratory parametrization scheme for ice within a 10 m diameter can be first expanded to that within 30 m through the geometric similarity criterion. With warming in the Arctic, the lateral melting might be enhanced. Moreover, the proportions of surface melting, bottom melting, and lateral melting for onsite ice floes, as reported in the literature [30], can be extended to applications where the diameter of the ice floe is within 100 m using the geometric similarity criterion. When the ice diameter is greater than 100 m, lateral melting occurs as well, but in small quantities. Hence, the lateral melting parametric scheme including the floe size effect is more applicable to the marginal area of the ice floes in the Arctic Ocean.

In the manual measurements of the sea ice and fresh-water ice samples, data corresponding to the negative temperature gradient and instantaneous temperature gradient changes were eliminated first. Afterwards, a total of 86 and 58 measured data were obtained for sea ice and fresh-water ice samples, respectively. Finally, the expression of the predictive empirical parametrization scheme was fitted with the results shown in Figure 9.

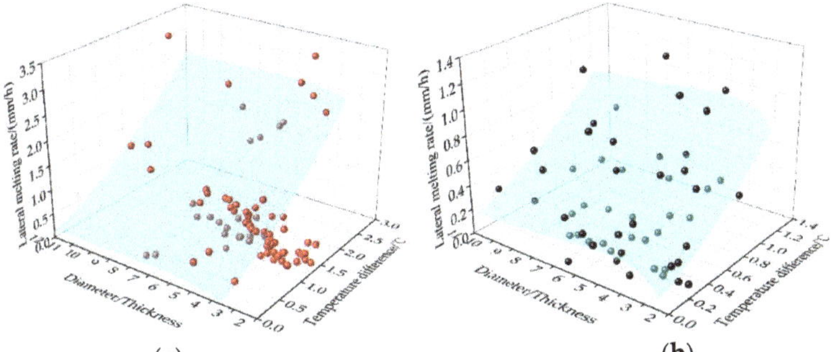

Figure 9. The statistical relationship between the lateral melting rate and the temperature difference, and the ratio of the diameter and the sample thickness. (**a**) Sea ice (**b**) fresh-water ice.

According to the fitted results in Figure 9a, the parametrization scheme for the lateral melting of sea ice can be expressed as:

$$W_{lat,sea} = 0.534\Delta T^{1.337} + \frac{2.637}{L_D/H} - \frac{9.145}{(L_D/H)^2} \quad (200 \geq L_D/H \geq 6) \quad r = 0.8273 \tag{10}$$

According to the result in Figure 9b, the parametrization scheme for the lateral melting of fresh-water ice is:

$$W_{lat,fresh} = 0.573\Delta T^{1.326} + \frac{2.361}{L_D/H} - \frac{5.246}{(L_D/H)^2} \quad (200 \geq L_D/H \geq 6) \quad r = 0.6323 \tag{11}$$

When only the scale parameters of ice are considered, the larger the L_D, the greater the L_D/H, and the lower the melting rate of the ice floes. This is consistent with the results reported in previous studies [13,14,25]. In nature, ice diameters range widely from kilometers to meters, while their thicknesses are only at the meter level. However, the proportion of melt in ice thickness is much larger than in lateral surface [30]. If diameter was selected as the scale parameter of ice and introduced into the parametric scheme, it failed to reflect the overall loss of ice; however, L_D/H is an ideal parameter. The first item in Equations (10) and (11) has also been used by international scholars [14,16,33]. The latter two demonstrate that the lateral melting rate changes with L_D/H. The contributions of L_D/H can be reflected by the latter two items only when L_D/H is lower than 200. The case for the ratio <6 can only be achieved in a laboratory without dynamic conditions, but it is rare in nature. Because surface, bottom, and lateral melting occurred simultaneously in the ice, the L_D/H effect could not be expressed by the monotonic function. In this study, the function $y = a/x + b/x^2$ was adopted, where the case of $x < 6$ was excluded. This mathematical expression conveys two meanings. First, the lateral melting rate can be ignored when the ice floe is small. Second, under the same water temperature difference, the larger the ice, the smaller the lateral melting rate. When the floe size is large, in spite of the lateral melt that occurs, the sea ice melts completely in thickness because the size is much smaller in thickness than in diameter.

Based on the experimental results using sea ice, a parametrization scheme for the lateral melting of Arctic sea ice was obtained through Equation (10), where the coefficients of the first item are 0.534 and 1.337, respectively. Coefficient 1.337 is close to 1.36, as recommended in the literature [18]. Coefficient 0.534, which is subject to environmental conditions and the selected units, should be corrected based on future observation data from the polar regions. Coefficients 0.573 and 1.326 in Equation (11) are on the same level but slightly less than those for sea ice, indicating that the fresh-water ice had no brine channels, but a higher density and lower melting rate. A similar phenomenon has been reported in the literature [29]. Likewise, this equation is mainly applicable to the coastal ice in the estuaries of the Arctic Ocean.

6. Conclusions and Discussion

Surface, bottom and lateral melting occurs simultaneously in ice. Regardless of the heat source responsible for melting ice in nature, the water temperature at the ice–water interface is the ultimate driver. Therefore, the parametrization for ice melting can be realized by the construction of relationships between the melting rate at the ice–water interface and the water temperature or the temperature difference. Water temperature changes caused by other natural factors can be considered as peripheral products of such parametrization. The parametrization scheme for ice floe melting formulated based on laboratory experiments is characterized as follows:

(1) The surface melting rate of ice is determined by the gradient between the near-surface air temperature and temperature of the ice surface, and the bottom melting rate depends on the gradient between the near-bottom water temperature and the temperature of the ice bottom. Notably, the surface melting rate is lower than the bottom one under the same temperature gradient. This seems to hold even when the air temperature gradient during the ice melting season is much higher than the vertical water temperature gradient. Despite the scattered data in the parametrization scheme and the large deviation of sea ice from fresh-water ice in this paper, the results still confirm the feasibility of the parametrization scheme for surface melting as continuously verified by laboratory research with high-precision temperature control and field data. Furthermore, the theoretical analysis indicates that parametrization schemes for surface and bottom melting rates are not affected by the diameter of ice.

(2) The lateral melting rate is determined by the difference between the temperature at the ice–water interface and the water temperature at different depths, as well as the ratio of the diameter (L_D) to the thickness (H) of ice. This ratio can reflect the negative correlation between the ice floe diameter and the lateral melting rate. When only the scale parameters of ice are considered, the larger the L_D, the greater the L_D/H, and the lower the melting rate of the ice floes. Because of the same reason, the large near-surface diameter and the small bottom diameter during the ice melting season can be explained.

(3) The lateral melting rates of fresh-water ice and sea ice are close to those in the parametric schemes reported internationally. The two parametrization schemes formulated in this study can be used for the numerical simulation of the lateral melting of sea ice with a diameter of <100 m in the marginal seas of the Arctic Ocean with low water salinity (e.g., Chukchi Sea and East Siberian Sea) and the melting of coastal fresh-water ice with a diameter of <100 m in the Arctic Ocean (e.g., the estuaries of Lena River and Mackenzie River).

(4) Given the complexity of sea ice melting during the Arctic summer, more experiments are still essential in the future to respond to different situations in the Arctic, even though the lateral sea ice melting mechanism developed in the present study is effective. Furthermore, for laboratory experiments, the technology of simultaneously observing the surface and bottom melting of ice needs to be explored.

Admittedly, there are limitations in this paper, especially because only the effects of air and water temperatures on ice melting were investigated, but the other environmental drivers including wind, wave, and solar radiation were not included under the restriction of the experimental apparatus. Therefore, more laboratory experiments are still required in the future with more specially designed devices that can resemble solar radiation, generate waves with small heights, and measure the ice lateral surface automatically. More temperature probes are also needed to give a complete air–ice–water temperature profile. The control of air temperature needs to be more precise with less fluctuation to reduce the influence on ice melting.

Author Contributions: Conceptualization, Z.L. and G.L.; investigation, Z.W., F.X. and Q.W.; writing—original draft preparation, Z.L. and P.L.; writing—review and editing, Q.W. All authors have read and agreed to the published version of the manuscript.

Funding: This research was supported by the National Natural Science Foundation of China (41922045, 41906198, 41876213, 41807501), the National Key Research and Development Program of China (2018YFA0605901), the Liaoning Revitalization Talents Program (XLYC2007033), and the Fundamental Research Funds for the Central Universities (DUT21RC3086, DUT20GJ206).

Data Availability Statement: The experiment data are available at the website (https://doi.org/10.5281/zenodo.6517801) (accessed on 4 May 2022).

Acknowledgments: The authors thank the editor and anonymous reviewers for their valuable comments and suggestions to this paper.

Conflicts of Interest: The authors declare no conflict of interest.

References

1. Comiso, J.C. Large decadal decline of the Arctic multiyear ice cover. *J. Clim.* **2012**, *25*, 1176–1193. [CrossRef]
2. Screen, J.A.; Deser, C.; Smith, D.M.; Zhang, X.D.; Blackport, R.; Kushner, P.J.; Oudar, T.; Mccusker, K.E.; Sun, L.T. Consistency and discrepancy in the atmospheric response to Arctic sea-ice loss across climate models. *Nat. Geosci.* **2018**, *11*, 155–163. [CrossRef]
3. Dekker, E.; Bintanja, R.; Severijns, C. Nudging the Arctic ocean to quantify sea ice feedbacks. *J. Clim.* **2019**, *32*, 2381–2395. [CrossRef]
4. Vihma, T. Effects of arctic sea ice decline on weather and climate: A review. *Surv. Geophys.* **2014**, *35*, 1175–1214. [CrossRef]
5. Onarheim, I.H.; Eldevik, T.; Smedsrud, L.H.; Stroeve, J.C. Seasonal and regional manifestation of Arctic sea ice loss. *J. Clim.* **2018**, *31*, 4917–4932. [CrossRef]
6. Stroeve, J.; Notz, D. Changing state of arctic sea ice across all seasons. *Environ. Res. Lett.* **2018**, *13*, 103001. [CrossRef]
7. Notz, D.; SIMIP Community. Arctic sea ice in CMIP6. *Geophys. Res. Lett.* **2020**, *47*, e2019GL086749. [CrossRef]
8. Thorndike, A.S.; Rothrock, D.A.; Maykut, G.A.; Colony, R. The thickness distribution of sea ice. *J. Geophys. Res.* **1975**, *80*, 4501–4513. [CrossRef]
9. Milaković, A.-S.; Schütz, P.; Piehl, H.; Ehlers, S. A method for estimation of equivalent-volume ice thickness based on WMO egg code in absence of ridging parameters. *Cold Reg. Sci. Technol.* **2018**, *155*, 381–395. [CrossRef]
10. Bateson, A.W.; Feltham, D.L.; Schröder, D.; Hosekova, L.; Ridley, J.K.; Aksenov, Y. Impact of sea ice floe size distribution on seasonal fragmentation and melt of Arctic sea ice. *Cryosphere* **2020**, *14*, 403–428. [CrossRef]
11. Toyota, T.; Takatsuji, S.; Nakayama, M. Characteristics of sea ice floe size distribution in the seasonal ice zone. *Geophys. Res. Lett.* **2006**, *33*, L02616. [CrossRef]

12. Morison, J.H.; Mcphee, M.G.; Maykut, G.A. Boundary layer, upper ocean, and ice observations in the Greenland Sea Marginal Ice Zone. *J. Geophys. Res.* **1987**, *92*, 6987–7011. [CrossRef]
13. Horvat, C.; Tziperman, E.; Campin, J.M. Interaction of sea ice floe size, ocean eddies, and sea ice melting. *Geophys. Res. Lett.* **2016**, *43*, 8083–8090. [CrossRef]
14. Steele, M. Sea ice melting and floe geometry in a simple ice-ocean model. *J. Geophys. Res.* **1992**, *97*, 17729–17738. [CrossRef]
15. Lu, K.; Weingartner, T.; Danielson, S.; Winsor, P.; Dobbins, E.; Martini, K.; Statscewich, H. Lateral mixing across ice meltwater fronts of the Chukchi Sea shelf. *Geophys. Res. Lett.* **2015**, *42*, 6754–6761. [CrossRef]
16. Maykut, G.A.; Perovich, D.K. The role of shortwave radiation in the summer decay of a sea ice cover. *J. Geophys. Res.* **1987**, *92*, 7032–7044. [CrossRef]
17. Hwang, B.; Wilkinson, J.; Maksym, T.; Graber, H.C.; Schweiger, A.; Horvat, C.; Perovich, D.K.; Arntsen, A.E.; Stanton, T.P.; Ren, J.C.; et al. Winter-to-summer transition of Arctic sea ice breakup and floe size distribution in the Beaufort Sea. *Elem. Sci. Anthr.* **2017**, *5*, 40. [CrossRef]
18. Perovich, D.K.; Jones, K.F. The seasonal evolution of sea ice floe size distribution. *J. Geophys. Res. Ocean.* **2014**, *119*, 8767–8777. [CrossRef]
19. Stern, H.L.; Schweiger, A.J.; Stark, M.; Zhang, J.L.; Steele, M.; Hwang, B. Seasonal evolution of the sea-ice floe size distribution in the Beaufort and Chukchi seas. *Elem. Sci. Anthr.* **2018**, *6*, 48. [CrossRef]
20. Wang, Q.K.; Li, Z.J.; Cao, X.W.; Yan, L.H. Analysis of measured thermodynamic melting rate of lateral interface between ice and water. *South-to-North Water Transf. Water Sci. Technol.* **2016**, *14*, 81–86. (In Chinese) [CrossRef]
21. Wang, Q.K.; Fang, H.; Li, Z.J.; Zu, Y.H.; Li, G.Y. Field investigations on lateral and bottom melting of lake ice and thermodynamic analysis. *J. Hydraul. Eng.* **2018**, *49*, 1207–1215. (In Chinese) [CrossRef]
22. Zhao, J.; Gao, G.; Jiao, Y. Warming in Arctic intermediate and deep waters around Chukchi Plateau and its adjacent regions in 1999. *Sci. China Ser. D Earth Sci.* **2005**, *48*, 1312–1320. [CrossRef]
23. Parkinson, C.L.; Washington, W.M. A large-scale numerical model of sea ice. *J. Geophys. Res.* **1979**, *84*, 311–337. [CrossRef]
24. Roach, L.A.; Horvat, C.; Dean, S.M.; Bitz, C.M. An Emergent Sea Ice Floe Size Distribution in a Global Coupled Ocean-Sea Ice Model. *J. Geophys. Res. Ocean.* **2018**, *123*, 4322–4337. [CrossRef]
25. Smith, M.M.; Holland, M.; Light, B. Arctic sea ice sensitivity to lateral melting representation in a coupled climate model. *Cryosphere* **2022**, *16*, 419–434. [CrossRef]
26. Chen, X.D.; Wang, A.L.; Høyland, K.; Ji, S.Y. Study of the influence of sea ice salinity on the thermodynamics. *Mar. Sci. Bull.* **2019**, *38*, 38–46. (In Chinese) [CrossRef]
27. Zhao, J.P.; Shi, J.X.; Jin, M.M.; Li, C.L.; Jiao, Y.T.; Lu, Y. Water mass structure of the Chukchi Sea during ice melting period in the Summer of 1999. *Adv. Earth Sci.* **2010**, *25*, 154–162. (In Chinese) [CrossRef]
28. Li, Z.J.; Wang, Q.K.; Lu, P.; Wang, Z.Q.; Xie, F. Laboratory Studies on the Melting Rate of Ice-Air and Ice-Water Interfaces. Zenodo [Data Set]. Available online: https://doi.org/10.5281/zenodo.6517801 (accessed on 4 May 2022). [CrossRef]
29. Ai, R.B.; Xie, T.; Liu, B.X.; Zhao, L.; Fang, H. An experimental study on parametric scheme of lateral melting rate of ice layer based on temperature. *Haiyang Xuebao* **2020**, *42*, 150–158. (In Chinese) [CrossRef]
30. Tsamados, M.; Feltham, D.; Petty, A.; Schroeder, D.; Flocco, D. Processes controlling surface, bottom and lateral melt of Arctic sea ice in a state of the art sea ice model. *Philos. Trans. R. Soc. A. Math. Phys. Eng. Sci.* **2015**, *373*, 20140167. [CrossRef]
31. Overduin, P.P.; Kane, D.L.; van Loon, W.K.P. Measuring thermal conductivity in freezing and thawing soil using the soil temperature response to heating. *Cold Reg. Sci. Technol.* **2006**, *45*, 8–22. [CrossRef]
32. Ahmed, R.; Prowse, T.; Dibike, Y.; Bonsal, B.; O'Neil, H. Recent trends in freshwater influx to the Arctic Ocean from four major Arctic-Draining rivers. *Water* **2020**, *12*, 1189. [CrossRef]
33. Perovich, D.K. On the Summer Decay of a Sea Ice Cover. Ph.D. Thesis, University of Washington, Seattle, WA, USA, 1983.

water

MDPI

Article

Experimental Investigation of Uniaxial Compressive Strength of Distilled Water Ice at Different Growth Temperatures

Yujia Zhang , Zuoqin Qian, Song Lv *, Weilong Huang, Jie Ren , Ziwei Fang and Xiaodong Chen

School of Naval Architecture, Ocean and Energy Power Engineering, Wuhan University of Technology, Wuhan 430063, China

* Correspondence: lvsong@whut.edu.com; Tel.: +86-027-8658-2035

Abstract: The existence of ice in nature will threaten the safety of navigation and water operations in cold regions. In order to improve the knowledge system of ice strength, the uniaxial compressive strength of distilled water ice grown at different temperatures is studied in this paper. Distilled water ice samples grown at −5 °C, −10 °C, −15 °C, −20 °C, −25 °C, −30 °C and −35 °C are prepared in the cryogenic laboratory. The density and grain size are measured. The uniaxial compressive strength tests are carried out at −10 °C. The stress-strain curves and the mechanical properties and failure modes of ice are obtained by loading along the vertical direction in the strain rate range of 10^{-6} s^{-1} to 10^{-2} s^{-1}. It is found that the uniaxial compressive strength of ice is a power function of strain rate and a linear relationship with the −1/2 power of grain size. Combined with the relationship between strength and grain size and the relationship between grain size and temperature, it is deduced that the peak compressive strength has a logarithmic relationship with the growth temperature. In addition, it shows that the growth temperature affects the strength of ice by controlling the grain size.

Keywords: distilled water ice; uniaxial compressive strength; growth temperature of ice; strain rate; grain size

Citation: Zhang, Y.; Qian, Z.; Lv, S.; Huang, W.; Ren, J.; Fang, Z.; Chen, X. Experimental Investigation of Uniaxial Compressive Strength of Distilled Water Ice at Different Growth Temperatures. *Water* **2022**, *14*, 4079. https://doi.org/10.3390/w14244079

Academic Editors: Zhijun Li, Tomasz Kolerski, Xiaohong Shi, Zhengyong Zhang, Li Zhou and Frédéric Frappart

Received: 12 November 2022
Accepted: 11 December 2022
Published: 14 December 2022

1. Introduction

Ice can be regarded as a composite material with a complex internal structure [1]. Its mechanical properties have been widely studied [2,3]. The driving force of the research lies in the development of the oil and natural gas industry, the construction and operation of hydropower equipment and the design of ice breakers, offshore platforms and other structures. The purpose is to predict the maximum ice load acting on these structures and ensure the safety of maritime operations in cold regions. In icy waters, different failure modes such as compression, buckling and radial cracking will occur in the ice interaction with ships, oil and gas platforms and coastal buildings [4,5]. These failure modes mainly involve the mechanical properties of ice, such as compressive strength, bending strength and shear strength [6–8]. Compression is one of the main failure modes when ice interacts with vertical structures, and the ice compressive strength is a key parameter in the calculation of ice load [9–11]. Moreover, the ice compressive strength is also an important factor affecting the dynamic process and deformation characteristics of ice at the geophysical scale [12].

In the compressive strength test, the loading rate that determines the strain rate is an important external factor affecting the strength of ice. Generally, the mechanical properties and failure modes of ice under different strain rates are also different [13]. Jones [14] carried out compression tests on single crystal ice in the temperature range as low as −90 °C and discussed the stress dependence of strain rate. Wu [15] studied the dynamic response of lake ice and distilled water ice under dynamic uniaxial compression. The uniaxial compressive strength shows positive strain rate sensitivity in the selected high strain rate range 60–800 s^{-1}. Kim indicated that the uniaxial compressive strength of ice is almost constant when the strain rate is high enough [16]. Timco and Frederking [17] studied the mechanical properties of freshwater ice and summarized the quantitative expressions of compressive strength and strain rate of ice in a ductile regime. Bonath, Sinha and Chen also discussed the relationship between strain rate and stress for different types of ice [18–20]. A series of studies have shown that the maximum compressive strength of ice appears in a special strain rate range, under which ice performs a transition state from ductility to brittleness, which is called the ductile-to-brittle transition regime. Qi's results show that the ductile-to-brittle transition regime appears in the strain rate range of 10^{-4} s^{-1} to 10^{-3} s^{-1} [21]. However, some studies found that the compressive strength reaches a maximum at a larger strain rate. Deng [22] found that the ductile-to-brittle transition regime of ice at the test temperature of -18 °C was in the range of 10^{-4} s^{-1} to 10^{-2} s^{-1} through a uniaxial compression test. Schulson [23] obtained the peak compressive strength at the strain rate of 10^{-2} s^{-1}.

The crystallization process of water will be different under different external conditions such as pressure and temperature, resulting in different structures of ice crystals. Crystals with the same structure are affected by the environment, which leads to diverse grain sizes [24,25]. Schulson [26] measured the uniaxial compressive strength of granular ice with different grain sizes and obtained that the brittle strength is related to the power exponent of grain diameter. Iliescu [27] investigated the relationship between the grain structure and the mechanical behavior of seasonal lake ice and river ice. The grain structure and the resulting mechanical properties change with the thickness of the ice. Cole's research results are different from others, which show that the compressive strength of ice increases with the increase in particle size at a low strain rate [28]. Moreover, during the formation of ice, air may not be discharged in time, so it is limited to forming bubbles inside. Due to the salt content of seawater, there may also be brine cells in sea ice, forming ice with different porosity, density and salinity. Many studies have shown that natural defects such as bubbles have a significant impact on the uniaxial compressive strength of the ice, and the higher the air content, the lower the compressive strength of ice [29–31]. Han [32] found that the uniaxial compressive strength of Arctic summer sea ice decreases linearly with the increase in porosity. Timco and Frederking [33] indicated that the uniaxial compressive strength in the vertical direction will be lost when the porosity of sea ice reaches 200‰ and the uniaxial compressive strength in the horizontal direction will be lost when the porosity reaches 270‰.

Ice temperature is another important factor affecting the uniaxial compressive strength of ice [34,35]. Arakawa [36] conducted uniaxial compressive strength tests on polycrystalline ice and discussed the relationship between strength and temperature and strain rate. With the decrease in temperature and the increase in strain rate, the structural strength of ice increases systematically. Wu [37] found that the compressive strength of ice increases with the decrease in temperature, but when the temperature decreases to a certain extent, the strength almost remains unchanged. However, the research on the effect of temperature on compressive strength mainly focuses on the test temperature at present. In the process of ice growth, the heat exchange with the atmosphere and water will also affect the strength of ice. The strength in the early stage of ice formation in cold waters greatly affects the safety of ship navigation and structures. In this study, the freezing process of distilled water ice and the preparation of test samples were carried out in the cryogenic laboratory. The density, crystal structure and grain size of ice condensed at different growth temperatures were measured. The relationship between the compressive strength at the initial stage of freezing and the growth temperature was systematically discussed under laboratory conditions. At present, there are few studies on the effect of growth temperature on the compressive strength of ice. The purpose of this study is to summarize the relationship between the compressive strength of ice and growth temperature and conduct further supplementary research on the mechanical properties of ice.

2. Methods

2.1. Preparation of Distilled Water Ice and Test Ice Samples

Distilled water ice was prepared in a cryogenic laboratory. The lowest ambient temperature of the cryogenic laboratory can be reduced to −40 °C. The temperature control accuracy is 0.1 °C. In order to study the effect of growth temperature on ice structural strength, the temperature of the cryogenic laboratory during ice making was set to −5 °C, −10 °C, −15 °C, −20 °C, −25 °C, −30 °C and −35 °C, respectively. The foam boxes with a length of 600 mm, a width of 450 mm and a depth of 300 mm were selected as the condensation tanks. The reason is that when water condensed in the box, the heat exchange intensity of the part close to the box was large, resulting in the growth of ice from the edge of the box to the middle. The smaller thermal conductivity of plastic foam leads to a smaller growth trend from the edge, which ensures that ice grows from top to bottom as in the natural environment. When sampling, an electric chain saw was used for cutting and taking out the distilled water ice in the middle of the box. Ice preparation and cryogenic laboratory temperature control panel are shown in Figure 1.

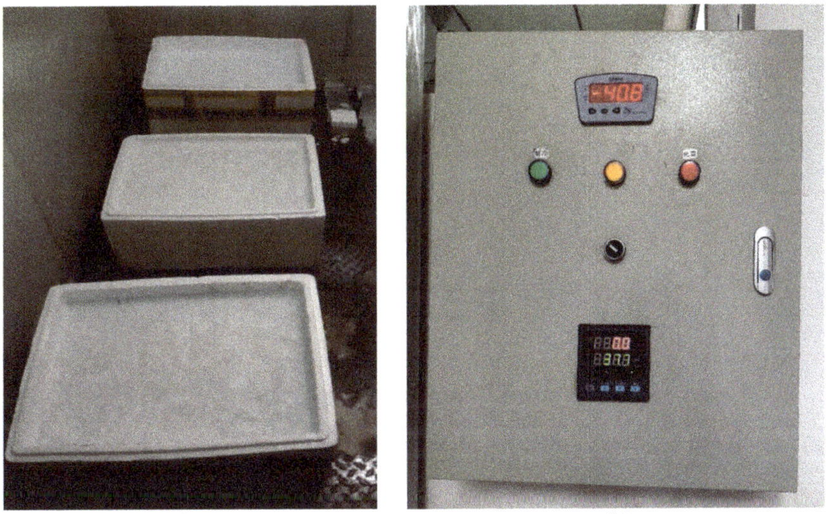

(a) (b)

Figure 1. Ice preparation and cryogenic laboratory temperature control panel. (**a**) Plastic foam condensation tanks filled with distilled water in a cryogenic laboratory. (**b**)Temperature control panel.

Preparation of the samples was also carried out in the cryogenic laboratory. According to the recommendations of the IAHR (International Association for Hydro-Environment Engineering and Research) [38], the upper and lower section dimensions of the ice uniaxial compressive strength test specimen are square with a side length of 7–10 cm (or circle with a diameter of 7–10 cm), and the length of the specimen is 2.5 times of the side length. The test samples were uniformly processed into 7 cm × 7 cm × 17.5 cm standard cuboids by a cutting machine. In order to prevent the samples from sticking together and weathering due to the continuous blowing of the laboratory, they were packed in plastic bags. The prepared samples needed to be kept at a constant temperature in the incubator for more than 24 h before the test, so that the samples can fully reach thermal equilibrium. In order to avoid the influence of ambient temperature on compressive strength and better study the relationship between growth temperature and compressive strength, the storage temperature and test temperature were both set to −10 °C. Test sample preparation is shown in Figure 2.

Figure 2. Test sample preparation.

2.2. Physical Properties Measurement

2.2.1. Ice Density Measurement and Porosity Calculation

In this study, the ice density was measured by the mass volume method. Standard cuboid samples with a size of 10 cm × 10 cm × 5 cm were processed by a cutting machine, of which 5 cm is along the direction of ice thickness. The exact length, width and thickness of the samples were measured by the vernier caliper many times and the volumes were calculated. The mass of the samples was weighed with an electronic scale, and the ratio of mass to volume is the ice density. Ice porosity is the sum of brine volume fraction ν_b and air volume fraction ν_a. Cox has derived empirical formulas for the volume fraction of air and brine from the temperature, salinity and density of sea ice as follows [39]:

$$\nu_a = 1 - \frac{\rho}{\rho_i} + \rho S \frac{F_2(T)}{F_1(T)} \tag{1}$$

$$\nu_b = \frac{\rho S}{F_1(T)} \tag{2}$$

$$\rho_i = 0.917 - 1.403 \times 10^{-4} T \tag{3}$$

where ρ is ice density (g/cm^3); S is salinity (‰); T is ice temperature (°C); ρ_i is the density of pure ice (g/cm^3); $F_1(T)$ and $F_2(T)$ are cubic polynomials about temperature. The research object of this study is distilled water ice with a salinity of 0, thus the porosity is the air volume fraction. Equation (1) can be simplified as:

$$\nu_T = \nu_a = 1 - \frac{\rho}{\rho_i} \tag{4}$$

2.2.2. Ice Crystal Structure Measurement

There are two types of ice crystal structures: granular and columnar ice. The ice crystal structure was observed under orthogonally polarized light through the thin ice slice pasted on the glass. The observation device is shown in Figure 3.

Figure 3. Ice crystal structure observation device.

The steps of preparing thin ice slices in the cryogenic laboratory are as follows: (1) After taking out the ice sample, slice the ice surface vertically and horizontally from different positions. The ice sections are 5–10 cm in length with a thickness of about 1–2 cm. Flatten one side of the slice gradually with a planer so that it can fit seamlessly with the clean glass. (2) Heat the prepared glass slice in a water bath and stick the flat surface of the ice section onto the glass slice slightly higher than 0 °C. Leave it for a few minutes to let the ice section freeze firmly with the glass. (3) Thin the slice to a thickness of less than 1 mm, which is convenient for distinguishing the crystal boundary. (4) Mark the slices, put them separately into clean plastic bags and keep them closed to prevent sublimation. The procedures of ice crystal structure observation under crossed polarized light are as follows: (1) Place the instrument in the cryogenic laboratory to reach room temperature. (2) Turn on the power, observe from the upper polarizer and rotate both upper and lower polarizer at the same time until the polarizers reach an angle of 90 ° to make it completely opaque. (3) Place and fix the cut ice crystal observation slice on the observation platform, and place the ruler on the ice slice. (4) Adjust the rotating disc until the ice crystal shows the clearest state for observation and take photos. The total area of the ice section can be measured by the ruler, and the total number of grains can also be obtained. The grain size is expressed by the equivalent circle diameter of the average crystal area on the section. The average grain diameter is calculated as follows:

$$D_g = 2\sqrt{S/n\pi} \quad (5)$$

where D_g is the average grain diameter (mm); S is the area of the ice section (mm^2); n is the number of grains in the section.

2.3. Uniaxial Compression Test

2.3.1. Test Devices

The main loading device required for the test is the WDW-100E electronic universal testing machine produced by Jinan Chuance Testing Equipment Co., Ltd. The testing machine is driven by a servo motor and the maximum test force is 100 kN. Previous studies have shown that the smaller the stiffness of the testing machine, the smaller the strain rate obtained, which affects the test results [40,41]. The parameters of the testing machine are as follows: displacement measurement accuracy is ±0.5%, test force measurement accuracy is ±0.5%, displacement resolution is 0.03 μm, the deformation resolution is 0.001 mm and the indication error of displacement and deformation is ±0.5%. In order to create a low-temperature test environment, DWC-40 low-temperature test chamber is installed between the beam and the bottom plate of the testing machine, and the temperature control accuracy can reach 0.1 °C to ensure strict control of the test temperature. In order to ensure that the upper indenter fits well with the surface of the test sample, springs are installed at the upper indenter. The test machine and testing zone and low-temperature test chamber are shown in Figures 4 and 5.

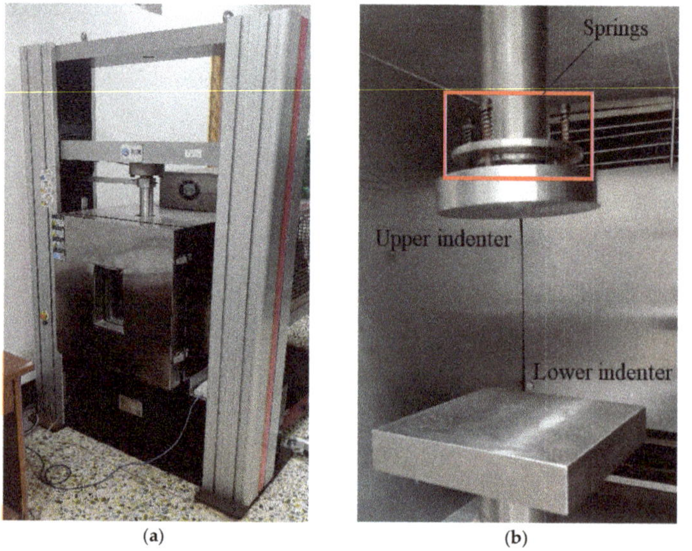

(a) (b)

Figure 4. Test devices. (**a**) WDW-100E electronic universal testing machine. (**b**) Testing zone.

Figure 5. Low-temperature test chamber.

2.3.2. Test Principle and Procedure

In the uniaxial compression test, the load is recorded by the pressure sensor, and the displacement of the beam and indenter is recorded by the laser displacement sensor. The load-time curve and displacement-time curve can be obtained in real time during each test. Combined with the cross-sectional area and initial length of the sample, the real-time stress and strain can be calculated, and then the stress-strain curve can be drawn. Generally speaking, the strength of materials is defined as the ability of materials to resist failure under the action of external forces, that is, the stress corresponding to the failure of materials. However, due to the particularity of ice materials, cracks occur and continue to develop during the external force loading process until it is completely destroyed. Therefore, it is difficult to judge the exact starting point of failure and the stress condition of this point cannot be monitored in the experiment. From the perspective of engineering, the uniaxial compressive strength of ice is represented by the maximum compressive stress measured in the test regardless of the real-time development of damage inside the ice during loading. The calculation method of uniaxial compressive strength is given as follows:

$$\sigma_c = \frac{F_{max}}{A} \tag{6}$$

where F_{max} is the maximum load on the load time curve; A is the original cross-sectional area of the sample. The strain rate $\dot{\varepsilon}$ (s^{-1}) is controlled by adjusting the displacement rate $\dot{X}$ (mm/s) of the indenter. It can be calculated by combining the original length L_0 (mm) of the sample as follows:

$$\dot{\varepsilon} = \frac{\dot{X}}{L_0} \tag{7}$$

In this study, the loading direction of the test is parallel to the core axis. The test procedure is as follows: (1) Set the low-temperature test chamber to the temperature required for the test and wait for it to complete the internal cooling. (2) After measuring the length, width and height of the ice sample, seal the sample with a plastic bag, and then put it into the low-temperature test chamber for constant temperature. The constant temperature time should be more than 24 h to ensure that the interior of the sample also reaches the temperature required for the test. (3) Place the sample at the center of the lower indenter to ensure that the geometric axis of the specimen is in the same line with the centerline of the testing machine. (4) Turn on the control and data acquisition program of the testing machine. After placing the sample, adjust the position of the upper indenter so that it just contacts the upper surface of the sample, then keep the temperature constant for a period of time to reduce the temperature disturbance during the manual placing of the sample. (5) Set the test parameters including sample sizes, loading speed, data acquisition frequency and turn on the testing machine for loading. (6) Stop loading immediately after the sample is destroyed and take photos. (7) Take out the damaged sample, clean up the residual fragments and prepare for the next test.

3. Results

Distilled water ice was prepared by controlling different growth temperatures in the cryogenic laboratory. The ice crystal structure was observed. The ice density and grain size were measured. Uniaxial compressive strength tests of distilled water ice grown at several temperatures were carried out at −10 °C. The loading direction was the same vertical direction as the natural growth direction of ice crystals. The loading speed was selected in the range of 0.01 mm/min to 105 mm/min and the strain rate range was 10^{-6} s^{-1} to 10^{-2} s^{-1} calculated by Equation (7).

3.1. Results of Uniaxial Compressive Strength Test

3.1.1. Deformation Types

Figure 6 shows the stress-strain curves of the uniaxial compressive strength test and the deformation types of distilled water ice.

It can be seen that although the strength of ice samples at different growth temperatures is different, the mechanical behavior and failure trend are almost the same. At a low strain rate, the uniaxial compressive strength increases linearly with the increase in strain. When the strength reaches the maximum value, the stress decreases with the increase in strain, and finally tends to a constant value, and the curve tends to be horizontal. The mechanical behavior of ice is ductile. Its failure mode is that with the progress of test loading, micro-cracks appear in the ice body of the sample and gradually expand with the progress of the test. When the cracks develop to a certain extent, the ability of the ice body to bear external loads decreases, but it will not completely lose its compressive capacity. At a high strain rate, the uniaxial compressive strength increases linearly with the increase in strain and then decreases rapidly after reaching the peak value. The mechanical behavior of ice is brittle. Its failure mode is that large cracks penetrating the upper and lower surfaces of the sample are directly generated in the ice during the loading process, and the sample is directly split into several pieces completely losing its compressive capacity. From the test results, the test duration in the high strain rate range is short, and the longitudinal displacement and deformation of the specimen are relatively small.

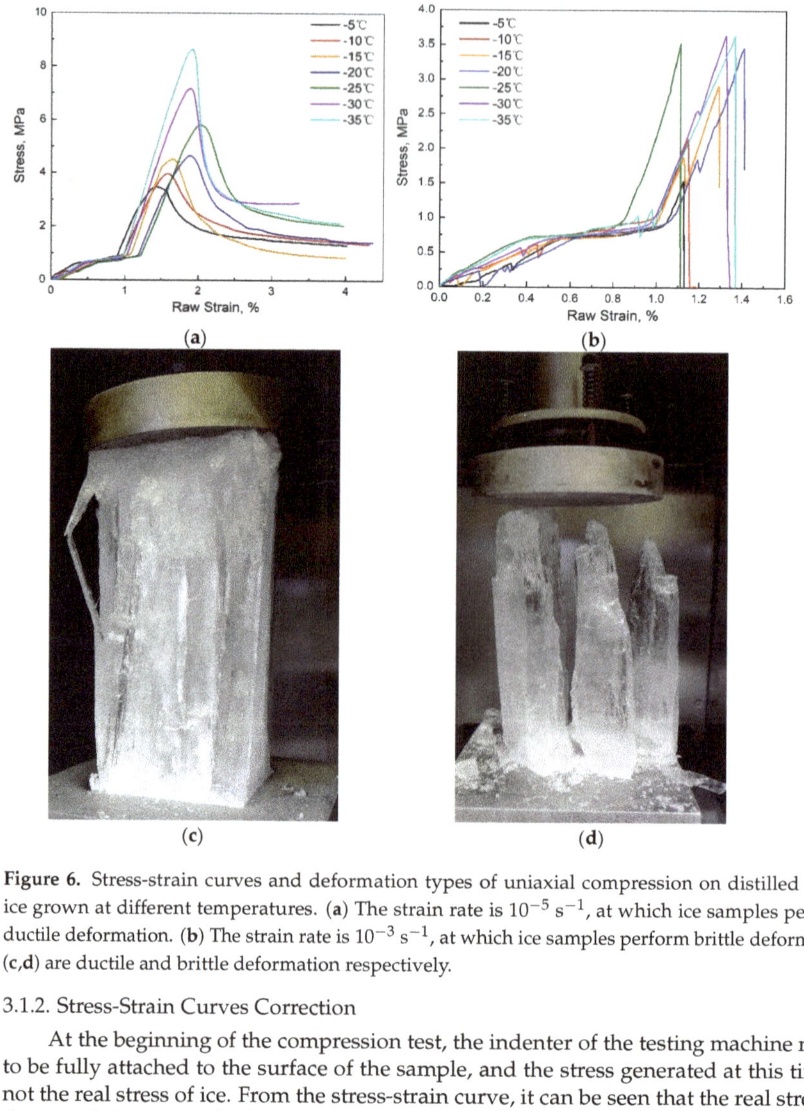

Figure 6. Stress-strain curves and deformation types of uniaxial compression on distilled water ice grown at different temperatures. (**a**) The strain rate is 10^{-5} s^{-1}, at which ice samples perform ductile deformation. (**b**) The strain rate is 10^{-3} s^{-1}, at which ice samples perform brittle deformation. (**c**,**d**) are ductile and brittle deformation respectively.

3.1.2. Stress-Strain Curves Correction

At the beginning of the compression test, the indenter of the testing machine needs to be fully attached to the surface of the sample, and the stress generated at this time is not the real stress of ice. From the stress-strain curve, it can be seen that the real stress of the sample is obtained when the raw strain reaches about 1%, so the stress-strain curve needs to be modified. The correction method refers to reference [36], which suggests using tangent modulus to calculate the true zero point of the stress-strain curve. The tangent modulus of a point on the stress-strain curve is the ratio of the stress change to the strain change near the point. Taking the test results of growth temperature −10 °C and strain rate

10^{-5} s^{-1} as an example, Figure 7 shows the correction method of the stress-strain curve when ice shows ductile behavior at a low strain rate.

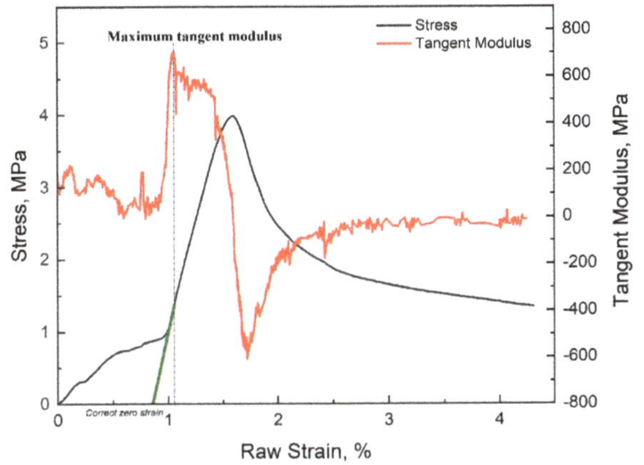

Figure 7. Correction of stress-strain curve in ductile regime.

When the raw strain reaches 0.93%, the upper surface of the sample is fully attached to the upper indenter, and the tangent modulus rises rapidly from 0, reaching the maximum value of 740 MPa at 1.04% raw strain. The point corresponding to the maximum tangent modulus is given on the curve. At this point, take the maximum tangent modulus as the slope as the auxiliary line, and the intersection with the x-axis is the correct zero strain, that is, 0.85% raw strain. When the tangent modulus curve reaches the maximum value, it decreases rapidly with the increase in strain. When the tangent modulus decreases to 0, the stress reaches the maximum value of 3.98 MPa, which is defined as the maximum compressive strength under this condition, and the corresponding corrected strain is 0.72%. After reaching the peak, the stress and tangent modulus decrease rapidly with the increase in strain. The tangent modulus reaches the minimum value at 0.85% of the corrected strain and then gradually rises, and the corresponding stress reduction rate also gradually decreases. Finally, the stress tends to be constant, and the tangent modulus tends to be 0. When ice shows brittle behavior at a high strain rate, the correction method of the stress-strain curve is the same as that of the ductile regime, as shown in Figure 8. See Table 1 for the maximum compressive strength σ_{max} and corresponding corrected strain ε_{max} in some typical tests.

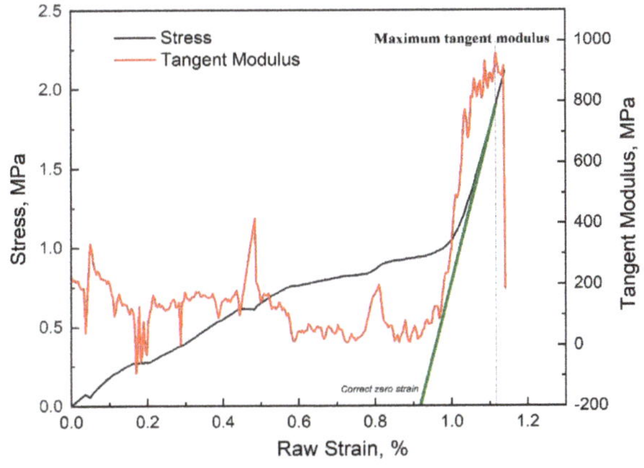

Figure 8. Correction of stress-strain curve in brittle regime.

Table 1. Typical Experimental conditions and corrected results.

T (°C)	$\dot{\varepsilon}$ (s^{-1})	σ_{max} (MPa)	ε_{max} (%)
−5	1×10^{-6}	1.95	0.51
	1×10^{-5}	3.52	0.64
	1×10^{-4}	5.88	0.71
	1×10^{-3}	1.54	0.18
	1×10^{-2}	1.30	0.26
−10	1×10^{-6}	2.08	0.49
	1×10^{-5}	3.98	0.72
	1×10^{-4}	6.39	0.71
	1×10^{-3}	2.11	0.22
	1×10^{-2}	1.75	0.27
−15	1×10^{-6}	1.82	0.40
	1×10^{-5}	4.51	0.67
	1×10^{-4}	6.08	0.54
	1×10^{-3}	2.91	0.29
	1×10^{-2}	2.29	0.27
−20	1×10^{-6}	4.61	0.63
	1×10^{-5}	4.67	0.79
	1×10^{-4}	5.49	0.77
	1×10^{-3}	3.46	0.36
	1×10^{-2}	1.75	0.48
−25	1×10^{-6}	4.34	0.73
	1×10^{-5}	5.81	0.87
	1×10^{-4}	5.13	0.52
	1×10^{-3}	3.52	0.27
	1×10^{-2}	1.22	0.34
−30	1×10^{-6}	3.07	0.83
	1×10^{-5}	7.17	0.94
	1×10^{-4}	7.45	0.87
	1×10^{-3}	3.64	0.38
	1×10^{-2}	1.23	0.31
−35	1×10^{-6}	4.70	1.09
	1×10^{-5}	8.65	0.99
	1×10^{-4}	7.02	0.74
	1×10^{-3}	3.63	0.48
	1×10^{-2}	2.18	0.22

3.2. Physical Properties and Crystal Structure of Distill Water Ice

The density of distilled water ice grown at different growth temperatures was measured in the low-temperature laboratory. The density range is 900–920 kg/m^3 increasing with the decrease in temperature. The ice density measured in this study is relatively large because ice is frozen from distilled water without salt and other impurities. Because there is almost no disturbance of air and water flow, which makes the ice crystal growth more uniform and stable, the icing process is static. In this way, the bubble content is low. The average porosity of distilled water ice grown at different temperatures is calculated by Equations (3) and (4) combined with the average density of ice samples. The results are shown in Table 2.

Figure 9 shows the horizontal and vertical crystal structure of the ice sample. From the figure, it can be seen that the distilled water ice frozen in the laboratory is a typical columnar structure. The reason is that the temperature in the laboratory is almost constant during icing, and the growth rate of ice crystals is slow, which gives ice crystals enough time to develop. Each ice crystal can only grow vertically downward due to the restriction of surrounding ice crystals during growth, forming columnar ice crystals.

Table 2. Measured average density and porosity of ice samples at different growth temperatures.

Growth Temperature (°C)	Average Density (kg/m^3)	Average Porosity (‰)
−5	906	12.75
−10	907	12.42
−15	915	4.47
−20	916	4.14
−25	918	2.72
−30	919	2.40
−35	919	3.16

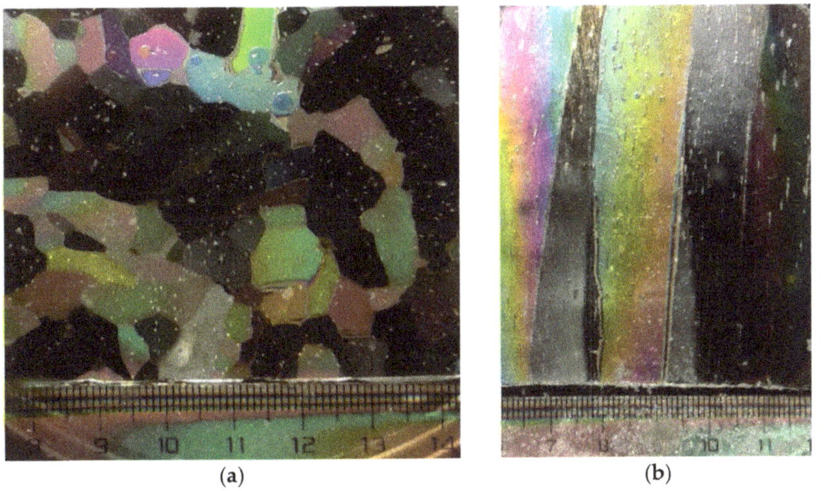

Figure 9. Thin sections of distilled water ice grown at −15 °C photographed in polarized light. (**a**) is the horizontal section at the depth of 0 cm (top) and (**b**) is the vertical section of 8–16 cm.

Figure 10 shows the variation curves of average grain size with depth. The average grain size of distilled water ice frozen under laboratory conditions ranges from 2 to 7 mm. The greater the depth, the slower the growth rate, and the larger the average grain size. The average grain size decreases roughly with the decrease in growth temperature.

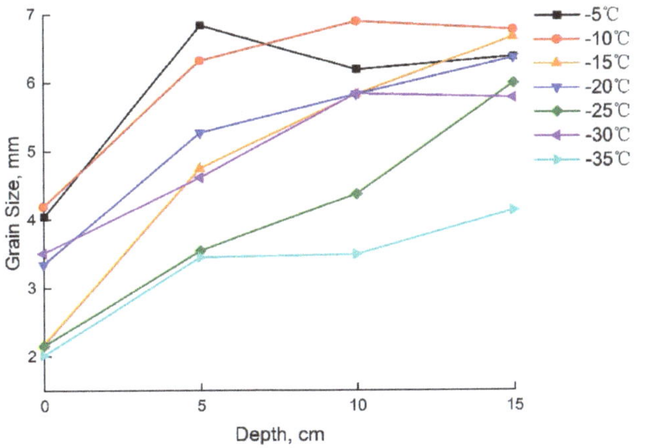

Figure 10. Grain size vs. depth curves of distilled water ice grown at different temperatures.

4. Discussion

4.1. Stress vs. Strain Rate

In Section 3, ice shows different mechanical behaviors in different strain rate ranges, so the uniaxial compressive strength of ice can correspond to the strain rate. Figure 11 shows the curves of the uniaxial compressive strength of distilled water ice with strain rate in the double logarithmic coordinate system.

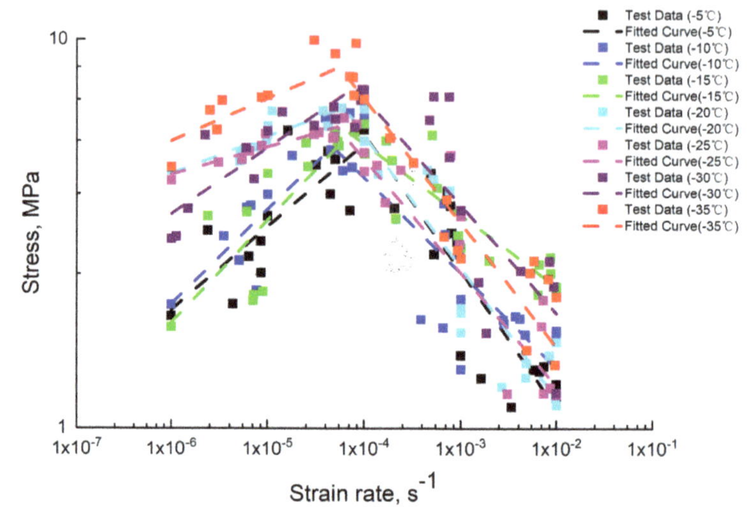

Figure 11. Test data of uniaxial compressive strength versus strain rate and fitted curves for distilled water ice samples grown at different temperatures.

In the range of strain rate from 10^{-6} s^{-1} to 10^{-2} s^{-1}, the uniaxial compressive strength first increases and then decreases. The strain rate range can be divided into ductile and brittle regimes according to the mechanical behavior of ice. The uniaxial compressive strength of ice reaches a maximum at the ductile-to-brittle transition point. The uniaxial compressive strength of ice at a strain rate higher than 10^{-2} s^{-1} was not obtained due to the limitation of test conditions. Some researchers have studied the uniaxial compressive strength of ice under a high strain rate. Test results from Jones [42] show that the uniaxial compressive strength of freshwater ice and low salinity sea ice does not decrease in the strain rate range of 10^{-1} s^{-1} to 101 s^{-1} but continues to increase. Wu et al. [17] conducted uniaxial compression tests on distilled water ice and lake ice in the range of strain rate 80 s^{-1} to 600 s^{-1} and the uniaxial compressive strength changed little.

The relationship between uniaxial compressive strength and strain rate of ice in the ductile regime should conform to Glen's law [43] as shown in Equation (8).

$$\dot{\varepsilon} = B\sigma_{c,v}^{n} \tag{8}$$

where B and n are empirical coefficients. In order to keep the equation consistent in dimension, Equation (8) is transformed into Equation (9) as follows:

$$\frac{\sigma_{c,v}}{\sigma_1} = B\left(\frac{\dot{\varepsilon}}{\dot{\varepsilon_1}}\right)^n \tag{9}$$

where σ_1 = 1 MPa, $\dot{\varepsilon_1}$ = 1 s^{-1}. The fitting curves of the relationship between uniaxial compressive strength and strain rate in the ductile regime are shown in the dashed lines in Figure 11 and the empirical coefficient and judgment coefficient are given in Table 3.

Table 3. The fitting coefficients and coefficients of determination R^2 of uniaxial compressive strength tests on distilled water ice.

T_g (°C)	Ductile Regime			Brittle Regime			σ_{cp} (MPa)
	B	n	R^2	B	n	R^2	
−5	38.798	0.214	0.58	0.225	−0.35	0.78	5.09
−10	60.236	0.243	0.70	0.460	−0.246	0.76	5.34
−15	71.611	0.264	0.77	0.991	−0.184	0.77	5.75
−20	16.837	0.095	0.73	0.248	−0.34	0.88	6.70
−25	12.001	0.071	0.75	0.332	−0.293	0.85	5.96
−30	38.262	0.172	0.71	0.553	−0.277	0.63	7.55
−35	24.741	0.109	0.60	0.370	−0.32	0.89	8.50

There is no unified conclusion about the relationship between uniaxial compressive strength and strain rate in a brittle regime. It can be seen from Figure 11 that within the strain rate range of this study, the relationship between uniaxial compressive strength and strain rate is still approximate to the power function. Therefore, Equation (9) is still used for fitting, and the fitting curves are shown in Figure 11. The empirical coefficient and judgment coefficient are also given. The fitting curve is in good agreement with the test data. The peak compressive strength of distilled water ice σ_{cp} at different growth temperatures can be calculated through the fitting function, which is also listed in Table 3.

4.2. *Stress vs. Grain Size*

From Section 3, it can be seen that the grain size tends to decrease with the decrease in growth temperature. Grain size is another key factor affecting ice strength. Cole concluded from the tests that the stress of ice decreases with the increase in grain size [28]. This conclusion can be explained by dislocation theory, that is, for the same material, the smaller the grain diameter, the more the grain boundary; the greater the obstacle to dislocation movement, the greater the resistance to material deformation; and the higher the macro strength. Schulson [26] summarized the functional relationship between compressive strength and grain size after studying the brittle compression failure of ice, as shown in Equation (10):

$$\sigma_c = \sigma_0 + k_c d^{-p} \tag{10}$$

where σ_c is compressive strength, d is grain size, σ_0 and k_c are material constants and $p = 1/2$. Other researchers [26,28,44] mainly focused on the brittle regime to study the relationship between stress and grain size. In this paper, the relationship between the peak compressive strength of ice and grain size is analyzed in a larger range of strain rates (including ductile and brittle regime). Table 4 shows the average grain size of distilled water ice grown at different growth temperatures.

Table 4. Result of grain size measurement.

T_g (°C)	Average Grain Size (mm)	σ_{cp} (MPa)
−5	5.86	5.09
−10	6.05	5.34
−15	4.85	5.75
−20	5.20	6.70
−25	4.02	5.96
−30	4.93	7.55
−35	2.37	8.50

Combining the peak compressive strength given in Section 4.1, the relationship between the two is fitted with Equation (10). The test results and fitting curves are shown in Figure 12, and the empirical coefficients and judgment coefficients obtained by fitting this study with other studies are given in Table 5.

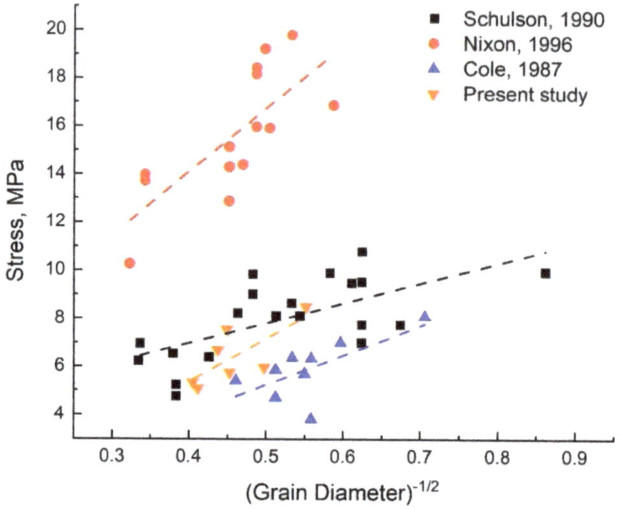

Figure 12. Stress versus grain size (data from Schulson (1990) [26], Nixon (1996) [44], Cole (1987) [28] and present study) and fitted curves.

Table 5. Regression and correlation coefficients for stress vs. grain size fitting function.

Items	Schulson, 1990	Nixon, 1996	Cole, 1987	Present Study
σ_0	3.67	3.52	−0.89	−1.81
k_c	8.27	26.4	12.25	17.91
R^2	0.44	0.56	0.45	0.54

There is a good correlation between stress and the $-1/2$ power of grain size, and the trend of the results obtained in the present study is roughly the same as that of other studies. From the numerical point of view, the maximum stress in this study is roughly the same as the test results of Schulson. The stress obtained by Nixon is relatively large, while the result obtained by Cole is smaller. The reason is that the sample storage and test of this study and Schulson are carried out at −10 °C. The test temperatures of Nixon and Cole are −50 °C and −5 °C, respectively. The lower the test temperature, the greater the compressive strength of the ice. Many previous studies have also confirmed this conclusion [34–37].

4.3. Stress vs. Growth Temperature

The uniaxial compressive strength of ice is closely related to temperature. Many previous studies on temperature and strength mainly discussed the influence of test temperature on strength. There is little research on the relationship between growth temperature and compressive strength at present. The test temperature in this study was kept constant at −10 °C in order to avoid the influence. The ice samples used in this study are distilled water ice, with no influence of salt and other solid impurities. The porosity of distilled water ice sample is small as shown in Section 3.2, so the influence of bubbles on strength can also be ignored, which can more directly study the influence of growth temperature on the peak compressive strength. It can be seen from the test results that the peak value of uniaxial compressive strength appears at −35 °C, which is 8.50 MPa calculated by using the fitting function.

Three functions are selected for data fitting referring to the relationship between test temperature and strength in previous studies. Yu [45] found that there is a linear relationship between the compressive strength of river ice and ice temperature in the range of −5 to −30 °C as Equation (11):

$$\sigma = A|\theta| + B \quad (11)$$

Li [46] proposed the logarithmic relationship between the peak uniaxial compressive strength of Bohai Sea Ice and ice temperature as Equation (12):

$$\sigma_{c,max} = A \ln|\theta| + B \quad (12)$$

Zhang [8] measured the shear and tensile strength of freshwater ice and sea ice at different temperatures and obtained the highest third-order polynomial relationship between shear/tensile strength and ice temperature as Equation (13):

$$\sigma = A\theta^3 + B\theta^2 + C\theta + D \quad (13)$$

where σ is ice strength, θ is ice temperature, A, B, C, D are fitting coefficients. The peak uniaxial compressive strength σ_{cp} and ice growth temperature T_g (T_g replaces θ in the original equation) in this study are fitted by the three functions above, as shown in Figure 13. The empirical coefficient and judgment coefficient are listed in Table 6.

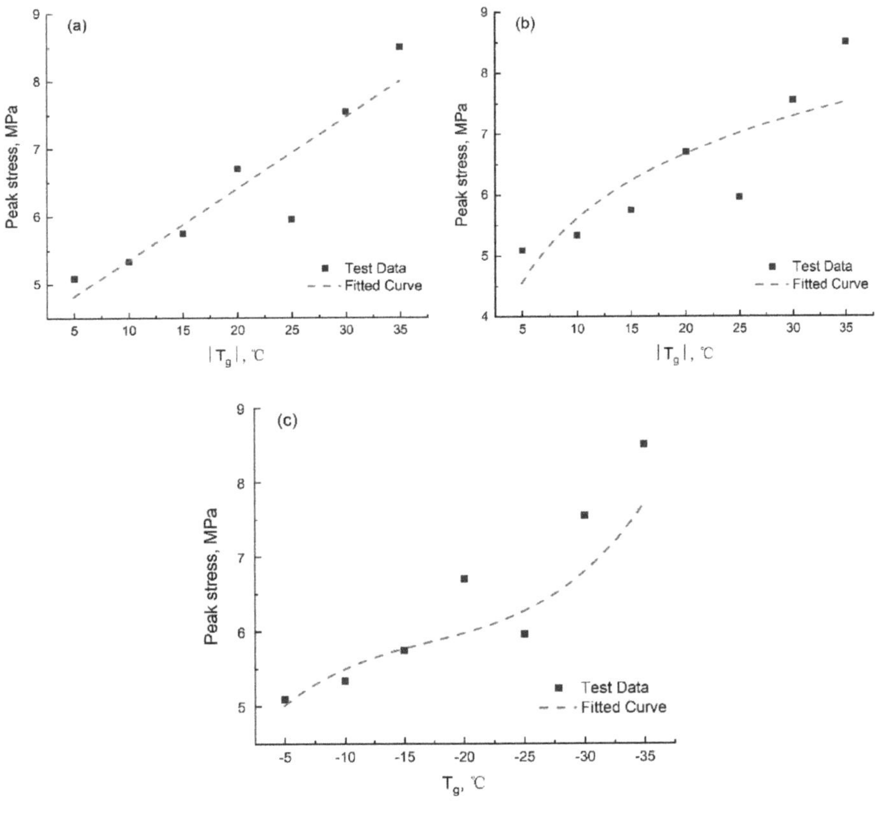

Figure 13. Peak stress versus growth temperature and curves fitted with different function types. Solid squares are test data. Dashed lines are fitted curves. (**a**–**c**) are linear, logarithmic and polynomial function respectively.

Table 6. Regression and correlation coefficients for peak stress vs. growth temperature fitting functions.

Function Type	Fit Coefficients				R^2
	A	B	C	D	
Linear	0.106	4.290	/	/	0.85
Logarithmic	1.527	2.096	/	/	0.71
Polynomial	−0.0002	−0.0103	−0.217	4.16	0.91

The correlation of the three functions is good according to the fitting results, but the most appropriate function should be selected from the perspective of physical facts. From the above, it can be seen that the growth temperature affects the grain size. It is found that the growth temperature has a logarithmic or linear relationship with the $-1/2$ power of the grain size, as shown in Figure 14.

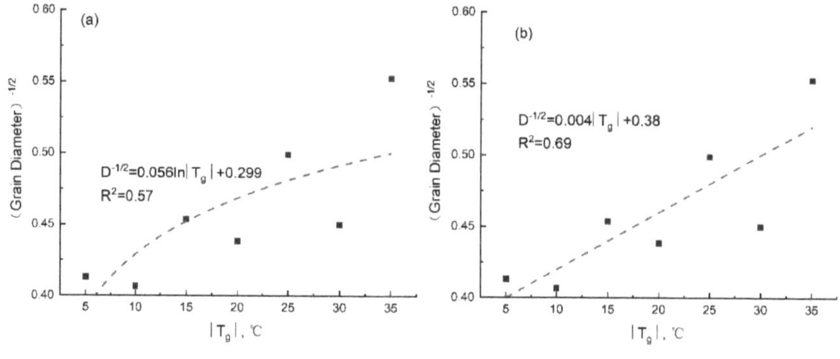

Figure 14. Grain size versus growth temperature and curves fitted with different function types. Solid squares are test data. Dashed lines are fitted curves. (**a**) is a logarithmic function and (**b**) is a linear function. Fitting equations and correlation coefficients are given.

The relationship between peak uniaxial compressive stress and growth temperature can be obtained by substituting the relationship between growth temperature and grain size into Equation (10) as follows:

$$\sigma_{cp} = A_1|T_g| + B_1 \tag{14}$$

$$\sigma_{cp} = A_2 \ln|T_g| + B_2 \tag{15}$$

Finally, the form of Equation (15) is selected as the relationship between the peak uniaxial compressive strength and the growth temperature. The reason is that the freezing process occurs below 0 °C. When the growth temperature is equal to 0 °C, there is no ice formation, that is, there will be no strength, which is inconsistent with the physical fact that the strength is close to zero when the temperature is near the freezing point. The logarithmic function conforms to the requirements of both physics and mathematics [46]. By deducing and summarizing the relationship between the peak uniaxial compressive strength and the growth temperature, it is found that the growth temperature affects the strength of ice by controlling the grain size of ice crystal when other factors are ignored. Finally, the relationship between peak uniaxial compressive stress and growth temperature is given as follows:

$$\sigma_{cp} = 1.527 \ln|T_g| + 2.096 \tag{16}$$

This conclusion improves the research system of ice strength. It is also conducive to guiding the estimation and judgment of ice strength in engineering applications in cold regions. Finally, it provides a theoretical basis and technical support for ships to choose appropriate routes and ice-breaking schemes, as well as the design of maritime structures and risk avoidance.

5. Conclusions

In this study, uniaxial compressive strength tests of distilled water ice at different growth temperatures were carried out. The conclusions based on experimental results and comparison with other research work are given as follows:

1. The growth temperatures were set at −5 °C, −10 °C, −15 °C, −20 °C, −25 °C, −30 °C and −35 °C. The ice crystal structure in this study is columnar. Ice density ranges from 900 kg/m^3 to 920 kg/m^3 and increases with decreasing temperature. Grain size increases with the increases in ice depth, and the average grain size range is 2–7 mm, which roughly decreases with the decrease in temperature.
2. The uniaxial compressive strength of ice at different strain rates is obtained, which first increases with the increase in strain rate, and then decreases with the increase in strain rate after reaching the peak value. The results show that ice is ductile at a low strain rate and brittle at a high strain rate. The relationship between uniaxial compressive strength and strain rate is a power function.
3. Compared with previous studies, it is found that the peak compressive strength of ice gradually increases with the decrease in grain size. The results show that the relationship between the peak compressive strength and the—1/2 power of grain size is a linear function.
4. Referring to the previous research, three functional forms of linear, logarithmic and polynomial have been proposed. The experimental results of this study are fitted in three functions to reveal the mathematical relationship. In addition, by summarizing the relationship between growth temperature and grain size, combined with mathematical requirements and physical facts, it is finally determined that the relationship between ice peak compressive strength and growth temperature conforms to the logarithmic function. This result explains that the growth temperature affects the compressive strength of ice by controlling the grain size.

The conclusions of this paper can help guide the safe navigation of ships and the design of offshore marine structures in cold regions.

Author Contributions: Conceptualization, Y.Z. and Z.Q.; methodology, Y.Z.; software, J.R.; validation, Y.Z., Z.Q. and S.L.; formal analysis, Y.Z.; investigation, Z.F.; resources, X.C.; data curation, Y.Z.; writing—original draft preparation, Y.Z.; writing—review and editing, S.L.; visualization, W.H.; supervision, Z.Q.; project administration, Z.Q. and Y.Z.; funding acquisition, Z.Q. All authors have read and agreed to the published version of the manuscript.

Funding: This research was funded by the National Natural Science Foundation of China, grant number 51676144.

Institutional Review Board Statement: Not applicable.

Informed Consent Statement: Not applicable.

Data Availability Statement: Not applicable.

Acknowledgments: The authors would like to thank the Wuhan University of Technology (WUT).

Conflicts of Interest: The authors declare no conflict of interest.

References

1. Overland, J.E.; McNutt, S.L.; Salo, S.; Groves, J.; Li, S. Arctic sea ice as a granular plastic. *J. Geophys. Res.-Oceans.* **1998**, *103*, 21845–21867. [CrossRef]
2. Schulson, E.M. Brittle failure of ice. *Eng. Fract. Mech.* **2001**, *68*, 1839–1887. [CrossRef]
3. Snyder, S.A.; Schulson, E.M.; Renshaw, C.E. Effects of prestrain on the ductile-to-brittle transition of ice. *Acta. Mater.* **2016**, *108*, 110–127. [CrossRef]

4. Kjerstad, Ø.K.; Metrikin, I.; Løset, S.; Skjetne, R. Experimental and phenomenological investigation of dynamic positioning in managed ice. *Cold Reg. Sci. Technol.* **2015**, *111*, 67–79. [CrossRef]
5. Montewka, J.; Goerlandt, F.; Kujala, P.; Lensu, M. Towards probabilistic models for the prediction of a ship performance in dynamic ice. *Cold Reg. Sci. Technol.* **2015**, *112*, 14–28. [CrossRef]
6. Sinha, N.K. Uniaxial compressive strength of first-year and multi-year sea ice. *Can. J. Civ. Eng.* **1984**, *11*, 82–91. [CrossRef]
7. Saeki, H.; Ozaki, A.; Kubo, Y. Experimental study on flexural strength and elastic modulus of sea ice. In Proceedings of the 6th international Conference on Port and Ocean Engineering under Arctic Conditions, Quebec City, QC, Canada, 27–31 July 1981.
8. Zhang, Y.; Zhang, Y.; Guo, R.; Cui, B. Method for Testing Shear and Tensile Strengths of Freshwater/Seawater Ice. *Water* **2022**, *14*, 1363. [CrossRef]
9. Lainey, L.; Tinawi, R. The Mechanical Properties of Sea Ice—A compilation of available data. *Can. J. Civ. Eng.* **1984**, *11*, 884–923. [CrossRef]
10. Richter, M.J.A. US Research in Ice Mechanics: 1987–1990. *Cold Reg. Sci. Technol.* **1992**, *20*, 231–246. [CrossRef]
11. Moslet, P.O. Field Testing of Uniaxial Compression strength of columnar sea ice. *Cold Reg. Sci. Technol.* **2007**, *48*, 1–14. [CrossRef]
12. Weertman, J. Creep deformation of ice. *J. Annu. Rev. Earth Planet. Sci.* **1983**, *11*, 215–240. [CrossRef]
13. Timco, G.W.; Weeks, W.F. A review of the engineering properties of sea ice. *Cold Reg. Sci. Technol.* **2010**, *60*, 107–129. [CrossRef]
14. Jones, S.J.; Glen, J.W. The mechanical properties of single crystals of pure ice. *J. Glaciol.* **1969**, *54*, 463–473. [CrossRef]
15. Wu, X.; Prakash, V. Dynamic strength of distill water and lake water ice at high strain rates. *Int. J. Impact. Eng.* **2015**, *76*, 155–165. [CrossRef]
16. Kim, H.; Keune, J.N. Compressive strength of ice at impact strain rates. *J. Mater. Sci.* **2007**, *42*, 2802–2806. [CrossRef]
17. Timco, G.W.; Frederking, R.M.W. Comparative strengths of fresh water ice. *Cold Reg. Sci. Technol.* **1982**, *6*, 21–27. [CrossRef]
18. Bonath, V.; Edeskär, T.; Lintzén, N.; Fransson, L.; Cwirzen, A. Properties of ice from first-year ridges in the Barents Sea and Fram Strait. *Cold Reg. Sci. Technol.* **2019**, *168*, 102890. [CrossRef]
19. Sinha, N.K. Constant strain- and stress-rate compressive strength of columnar-grained ice. *J. Mater. Sci.* **1982**, *17*, 785–802. [CrossRef]
20. Chen, A.C.T.; Lee, J. Large-Scale Ice Strength Tests at Slow Strain Rates. *J. Offshore Mech. Arct.* **1988**, *110*, 302–306. [CrossRef]
21. Qi, S.; Schulson, E.M. The effect of temperature on the ductile-to-brittle transition in columnar ice. In Proceedings of the 14th International Symposium on ice, Potsdam, NY, USA, 27–31 July 1998.
22. Deng, K.; Feng, X.; Tian, X.; Hu, Y. Experimental research on compressive mechanical properties of ice under low strain rates. *Mater. Today Commun.* **2020**, *24*, 101029. [CrossRef]
23. Schulson, E.M.; Buck, S.E. The ductile-to-brittle transition and ductile failure envelopes of orthotropic ice under biaxial compression. *Acta. Metall. Mater.* **1995**, *43*, 3661–3668. [CrossRef]
24. Baker, R.W. The influence of ice-crystal size on creep. *J. Glaciol.* **1978**, *21*, 485–500. [CrossRef]
25. Jones, S.J.; Chew, H.A.M. Effect of sample and grain size on the compressive strength of ice. *Ann. Glaciol.* **1983**, *4*, 129–132. [CrossRef]
26. Schulson, E.M. The brittle compressive fracture of ice. *Acta Metall. Mater.* **1990**, *38*, 1963–1976. [CrossRef]
27. Iliescu, D.; Baker, I. The structure and mechanical properties of river and lake ice. *Cold Reg. Sci. Technol.* **2007**, *48*, 202–217. [CrossRef]
28. Cole, D.M. Strain-rate and Grain-size Effects in Ice. *J. Glaciol.* **1987**, *33*, 274–280. [CrossRef]
29. Potter, R.S.; Cammack, J.M.; Braithwaite, C.H.; Church, P.D.; Walley, S.M. A study of the compressive mechanical properties of defect-free, porous and sintered water-ice at low and high strain rates. *Icarus* **2020**, *351*, 113940. [CrossRef]
30. Ouyang, Q.; Qi, C.; Lian, J.; Zhao, X. Dynamic Compressive Strength and Failure of Natural Lake Ice Under Moderate Strain Rates at Near Melting Point Temperature. *Lat. Am. J. Solids. Stru.* **2017**, *14*, 1669–1694. [CrossRef]
31. Kovacs, A. Estimating the full-scale flexural and compressive strength of first-year sea ice. *J. Geophys. Res-Oceans.* **1997**, *102*, 8681–8689. [CrossRef]
32. Han, H.; Li, Z.; Huang, W.; Lu, P.; Lei, R. The uniaxial compressive strength of the Arctic summer sea ice. *Acta. Oceanol. Sin.* **2015**, *34*, 129–136. [CrossRef]
33. Timco, G.W.; Frederking, R.M.W. Compressive strength of sea ice sheets. *Cold Reg. Sci. Technol.* **1990**, *17*, 227–240. [CrossRef]
34. Shazly, M.; Prakash, V.; Lerch, B.A. High strain-rate behavior of ice under uniaxial compression. *Int. J. Solids Struct.* **2009**, *46*, 1499–1515. [CrossRef]
35. Farid, H.; Saeidi, A.; Farzaneh, M. Prediction of failure in atmospheric ice under triaxial compressive stress. *Cold Reg. Sci. Technol.* **2017**, *138*, 46–56. [CrossRef]
36. Arakawa, M.; Maeno, N. Mechanical strength of polycrystalline ice under uniaxial compression. *Cold Reg. Sci. Technol.* **1997**, *26*, 215–229. [CrossRef]
37. Wu, X.; Prakash, V. Dynamic compressive behavior of ice at cryogenic temperatures. *Cold Reg. Sci. Technol.* **2015**, *118*, 1–13. [CrossRef]
38. Schwarz, J.; Frederking, R.; Gavrillo, V.; Petrov, I.G.; Hirayama, K.I.; Mellor, M.; Tryde, P.; Vaudrey, K.D. Standardized testing methods for measuring mechanical properties of ice. *Cold Reg. Sci. Technol.* **1981**, *4*, 245–253. [CrossRef]
39. Cox, G.F.N.; Weeks, W.F. Equations for Determining the Gas and Brine Volumes in Sea-Ice Samples. *J. Glaciol.* **1983**, *29*, 306–316. [CrossRef]

40. Sinha, N.K.; Frederking, R.M.W. Effect of test system stiffness on strength of ice. In Proceedings of the Fifth International Conference on Port and Ocean Engineering under Arctic Conditions, Trondheim, Norway, 13–18 August 1979.
41. Timco, G.W.; Frederking, R.M.W. A procedure to account for machine stiffness in uni-axial compressive compression tests. In Proceedings of the International Symposium on Ice (IAHR), Hamburg, Germany, 27–31 August 1984.
42. Jones, S.J. High strain-rate compression tests on ice. *J. Phys. Chem. B.* **1997**, *101*, 6099–6101. [CrossRef]
43. Glen, J.W. The creep of polycrystalline ice. *P. Roy. Soc. A-Math. Phy.* **1955**, *228*, 519–538. [CrossRef]
44. Nixon, W.A. Wing crack models of the brittle compressive failure of ice. *Cold Reg. Sci. Technol.* **1996**, *24*, 41–55. [CrossRef]
45. Yu, T.; Wang, J.; Du, F.; Zhang, L. Experimental research on ice disaster in Huma River. *J. Nat. Disaster.* **2007**, *16*, 43–48. (In Chinese) [CrossRef]
46. Li, Z.; Zhang, L.; Lu, P.; Leppäranta, M.; Li, G. Experimental study on the effect of porosity on the uniaxial compressive strength of sea ice in Bohai Sea. *Sci. China Technol. Sci.* **2011**, *54*, 2429–2436. [CrossRef]

water

Article

Numerical Simulation and Experimental Study on Single Point Blasting of Ice Jam of Heilongjiang River Based on ANSYS/LSDYNA

Dong Liu [1,2,3], Changlei Dai [1,2,3,*], Chenggang Yu [4] and Yaodong Ma [1,2,3]

1 Institute of Groundwater in Cold Region, Heilongjiang University, Harbin 150080, China
2 School of Hydraulic & Electric Power, Heilongjiang University, Harbin 150080, China
3 China and Russian Cold Region Hydrology and Water Conservancy Engineering Joint Laboratory, Heilongjiang University, Harbin 150080, China
4 Daxing'anling Hydrology Bureau, Daxing'anling 165000, China
* Correspondence: daichanglei@126.com; Tel.: +86-132-5866-2595

Abstract: Based on the thickness of the common ice layer in the Heilongjiang Basin, explosions featuring different methods of firing, different thicknesses of the common ice layer, and different distances between the explosive and the ice are hereby simulated using ANSYS/LSDYNA software. The most suitable blasting scheme after field test verification is obtained, and it is concluded through the field blasting test comparison and verification that the central ignition produces the best blasting effect and the largest blasting volume and radius when the explosion is fired under the ice. A larger ice thickness indicates a worse explosive blasting effect, and the number of explosive packs should be used reasonably to achieve the best blasting effect. An ice blast does not finish at once, but experiences energy transfer in the form of waves. The blasting effect is the best in the case of a distance between the explosive and the ice of about L = 150 cm. The shallower the depth of the river is, the more powerful the blast becomes and the more obvious the blast volume becomes. The ice layer is blasted when the maximum temperature on the day of blasting is stable and positive, which is when the texture of the ice sheet maintains its hardness and high efficiency.

Keywords: blasting under ice; dynamic simulation; blasting volume; parameter analysis

Citation: Liu, D.; Dai, C.; Yu, C.; Ma, Y. Numerical Simulation and Experimental Study on Single Point Blasting of Ice Jam of Heilongjiang River Based on ANSYS/LSDYNA. *Water* **2022**, *14*, 2808. https://doi.org/10.3390/w14182808

Academic Editors: Zhijun Li, Tomasz Kolerski, Xiaohong Shi, Zhengyong Zhang and Li Zhou

Received: 3 August 2022
Accepted: 5 September 2022
Published: 9 September 2022

1. Introduction

River ice is an integral component of the hydrological and hydraulic regimes of cold region rivers and affects an extensive portion of the global hydrologic system, particularly in the Northern Hemisphere [1]. It regulates numerous hydrologic, geomorphic, biological, and chemical process in rivers, and provides important socio-economic and ecological benefits [2]. However, river ice can also cause extensive damage to infrastructure, navigation, and hydroelectric production, and can result in detrimental effects to river ecology and water quality [3].

The formation of river ice is a rather complex physical process [4,5], which involves the interaction of hydrodynamic, mechanical and thermal forces, and is also affected by meteorological and hydrological conditions as well as river topography features [6]. There are four main types of river ice formation. In turbulent flow, water flows so fast that the flow can carry away the melting heat so quickly that ice can form at only the tiniest cooling temperature. Ice crystals (known as ice flowers) are generated as many tiny disks and continue to form ice [7]. In calm water, the ice that first appears also takes the shape of a small disk, and the disk shape will develop into a dendritic shape when a critical size is reached. This dendritic ice develops rapidly across the surface of the water until an entire lake or pond is covered with it, often overnight. In addition, the surface is covered with an extremely thin but uniform layer of smooth and unbroken ice. The newly fallen snow is light and loose, with a density ranging from 0.1 to 0.3 kg/m^{-3}. The tension of edges of the crystalline and snow water vapor at the edge is higher than that inside, and the molecules evaporate towards the edges and condense inward, so the crystallization tends to be round [8]. A solid fine substance, known as snow ice, is formed during the deformation, refreezing, recrystallization, and other processes. Recrystallization makes the space between particles more compact and tighter; the density is between 0.82~0.84 kg/m, and the air gap is squeezed out and becomes an impermeable object, i.e., the conversion of snow to ice. Affected by the gravity and centrifugal force, the surface water flows to the concave bank, and the bottom water flows to the convex bank at the curved reach of the river, forming a transverse circulation. The ice flowers floating on the water surface gather towards the concave bank under the action of transverse circulation. When the mechanical force caused by the flow turbulence is less than the bond force between the ice flowers, the ice flowers stick together and result in blockage, and such ice flowers fill the whole river, thereby giving rise to an iceberg [9,10].

River ice impacts local river hydraulics in many different ways [11]. Similarly, in some small river systems, freezing of riverbeds may block groundwater inflows, which may also generate flooding through the aufeis [12]. Ice jams can also change river channel properties and flow paths through ice scouring and erosion [13]. Therefore, it is of great significance to study the blasting of river ice in order to open the river.

Heilongjiang is located in the northernmost part of China, as shown in Figure 1, and its geographical location is 108°20′–141°20′ E and 42°00′–55°45′ N [14]. This section belongs to the Sino–Russian Boundary River, where there is an ice dam of a certain size every 3 years on average, and where there are also serious melting ice floods. However, Heilongjiang is a remote province subject to underdeveloped transportation and communication, as well as backward observation equipment and technology, and relatively little research has been conducted on its ice jam flood disaster prevention and reduction. As the boundary river between China and Russia, the occurrence of ice disasters cannot be controlled, prevented or mitigated by engineering measures (such as the construction of hydraulic structures). Instead, only effective non-engineering measures can be used to prevent the occurrence of ice dams or smoothly break the ice dam [15,16]. Given that the measured ice sheet thickness can reach up to 1.8 m before the upper reach of the Heilongjiang River is opened up, that the ice sheet is of high strength, and that the snow cover on the ice sheet is thick, the measured data show the difficulty of achieving the destruction and movement of large ice structures by non-engineering measures (such as ice surface treatment, mechanical destruction of the ice sheet, cutting of the ice sheet, icebreaking by hovercraft, and river trepanning). In this case, exploding the ice sheet or ice dam using explosives has become the most effective way to prevent or alleviate ice disasters in the Heilongjiang River.

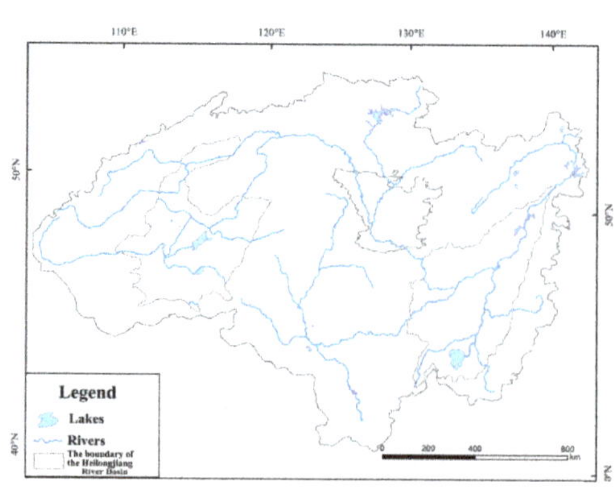

Figure 1. Location diagram of Heilongjiang Basin.

Although there was a record regarding icebreaking by explosives 100 years ago [17], the research is limited by the ice conditions of the river and the surrounding environment of the river, and the measurement is difficult and risky, heavily impeding the research progress [18,19]. Mellor from the United States Army Corps of Engineers obtained a series of explosion data for more than 10 years from the infraglacial blasting experiment and explored the relationship among blasting pit radius, ice sheet thickness and explosive dosage using the regression analysis [20,21]. In China, He Bingshun et al. [22] put forward the icebreaking principle of using the icebreaking and cannon firing technology by dropping a bomb with an aircraft. Based on the freezing layer with thick ice of the Yellow River and Inner Mongolia as the study object, Liang Xiangqian et al. [23,24] studied the vibration velocity of explosive blasting and the change in shock wave pressure in water under two icebreaking methods, i.e., ice contact energy-collected explosion and underwater explosion. Tong Zheng et al. [25] investigated the crack and pressure distribution of ice blocks after blasting by simulating the field blasting and numerical situation of the Yellow River ice run. In terms of the blasting technique, Duan Yuansheng et al. [26] summarized the parameter indexes (such as the arrangement form of explosive holes and the proper channel width of blasting the ice sheet for ice prevention) via engineering practice. Yan Huaitang et al. [27] deduced the relationship between the length of extended explosive charge and the amount of explosives used for icebreaking via blasting under the ice sheet by studying the properties of the explosives. It can be observed from the above research results of infraglacial blasting in the river that the average thickness of the ice sheet is mainly 0.5 m and below, and that the measured maximum ice thickness of the Heilongjiang River can reach as high as 1.8 m. To this end, the field test of infraglacial blasting in the Mohe section in the upper reach of the Heilongjiang River was hereby studied pursuant to the characteristics of the Heilongjiang River (thick ice sheet, solid ice layer, and sub-glacial water depth) to obtain the effective methods and techniques of ice prevention blasting and establish the theoretical relationship between blasting pit radius as well as explosive dosage, ice sheet thickness, and water depth under ice, thereby providing a scientific basis and theoretical support for realizing efficient icebreaking by explosives. Therefore, it is of great significance to develop a blasting method with a good icebreaking effect to prevent the ice run in the Heilongjiang River. Additionally, the Huma section in the Heilongjiang River Basin is one of the main reaches where ice damage occurs, and ice damage of different degrees happens at the moment of closure and ice breakup every year. Therefore, the river at the Huma section of the Heilongjiang River main stream during the freeze-up period is hereby taken as the study object, and numerical simulation of ice explosions is hereby carried out for the river during the freeze-up period under different ice thicknesses and explosive charges, so as to provide the theoretical basis for flood control and disaster reduction, and develop a series of ice explosion methods in the Heilongjiang River Basin [28].

2. Models and Methods

2.1. Selection of Calculation Software

In recent years, blasting engineering has seen much progress due to the use of numerical simulation technology and effective numerical models [29]. ANSYS/LSDYNA (The software version: ANSYS 2021 R1; Creator and the location: John Swanson, New York, NY, USA; Software Sourced: Livermore Software Technology Company, Canonburg, PA, USA.) is the main international explicit dynamic analysis program; it uses ANSYS to simulate the establishment of environmental characteristics and an ANSYS/LSDYNA built-in solver (LS-RUN) to solve K files. Data analysis can be performed using LS-PrePost (The software version: LS-PrePost 4.7; Creator and the location: John Swanson, New York, NY, USA; Software Sourced: Livermore Software Technology Company, Canonburg, PA, USA.), a post-processing software developed by LSDYNA. In the Explosion Analysis Module, ANSYS/LSDYNA not only provides the corresponding explosive material models and equations of state but also simulates the propagation of shock waves and the instantaneous state responses of structures [30]. Therefore, ANSYS/LSDYNA software was selected for simulation studies in this study.

2.2. Software Simulation Process

(1) Model establishment: This step mainly comprises model building, meshing, element-type selection, definition of material parameters, creation of parts, definition of (symmetry) constraints and (non-reflection) boundary conditions, energy settings, solution time control, and output K files.
(2) K-file check: In this step, a K file is opened and checked in LS-PrePost. K files contain material parameter verification, equation of state, Lagrange algorithm, initial conditions for detonation, ignition method, and coupling method.
(3) Solution process: In this step, the modified K file is imported into the LS-RUN solver to solve. Then, the resulting file is saved and used to record the blasting area, blast radius, and other results to draw a corresponding cloud map [31].
(4) Result post-processing viewing: In this step, the LS-PrePost post-processing software is used to open a D3plot in the result file for the viewing and analysis of the solution results.

A numerical model of ice blasting involves the blasting and underwater explosion process of an ice material model, reflecting the coupling between air, explosives, water, and ice, and the whole change process is dynamic, complex, and highly nonlinear. According to universities and research institutes such as the National University of Defense Technology and Dalian University of Technology, ANSYS/LSDYNA procedures are often used for explicit dynamic analysis [32].

2.3. Model Building

2.3.1. The Theory of the Model

In accordance with classification as per the coordinate system, two kinds of methods in the nonlinear continuum mechanics, i.e., the Lagrangian method (mainly for solid mechanics) and Euler's method (mainly for fluid mechanics), have been developed for numerical calculation [33]. For the Lagrangian method, the mesh is fixed to the object, and will then move or deform with the object, which is used for calculation to intuitively and naturally handle the free surface and boundary surface of the object and accurately describe the stress history of different materials. However, in the case of the occurrence of the large deformation, the mesh will suffer from serious distortion or aberration along with the object, which decreases the calculation accuracy or even causes calculation termination. For Euler's method, the mesh is fixed to the space, and its shape always remains unchanged, which is suitable for calculating the problem of large deformation. However, considering the difficulty of using Euler's method to describe the moving boundary or the boundary surface between different substances, the tracking and processing of the boundary surface using some special methods is necessarily important, which is one of the main directions of the current development of Euler's method.

Given that both the Lagrangian method and Euler's method are provided with their respective advantages and disadvantages, a batch of problems that cannot be solved merely by either of them can be solved by combining the two organically, fully absorbing their advantages and overcoming their disadvantages. The Arbitrary Lagrangian and Eulerian method (referred to as ALE method for short) is thus produced for this very purpose.

For the ALE method, the mesh can move arbitrarily in space, namely moving independently of the object coordinate system and the space coordinate system, when the moving boundary of the object can be accurately described, and the reasonable shape of the mesh can be maintained by specifying the appropriate motion form of the mesh. In this case, the ALE method has wide applicability in the case of simulating the impact, explosion, and other large deformation problems, and is of practical value to numerical simulation of the underground explosion dynamic process.

The nonlinear continuum mechanics method and the method of Gauss theorem can be used for deducing the governing equation under ALE description.

The following is the mass conservation equation.

$$\frac{\partial \rho}{\partial t} l_{\xi} + \frac{\partial \rho}{\partial x_i} + \rho \frac{\partial v_i}{\partial x_i} = 0 \tag{1}$$

In the formula, l_{ξ} is the distance from the reference point ξ to the origin, v_i is the material flow velocity of the particle X, ρ is the density of the material, x_i is the moving distance of the particle X, and ∂ is the time.

The formula below is the momentum equation.

$$\rho \frac{\partial v_i}{\partial t} l_{\xi} + \rho c_j \frac{\partial v_i}{\partial x_j} = \frac{\partial \sigma_{ji}}{\partial x_j} + \rho f_i \tag{2}$$

In the formula, c_j is the grid migration velocity, σ_{ji} is the Cauchy stress tensor, and f_i is the body force per unit mass.

The following formula is the equation of energy conservation.

$$\rho \frac{\partial e}{\partial t} l_{\xi} + \rho c_i \frac{\partial e}{\partial x_i} = \sigma_{ij} \frac{\partial v_i}{\partial x_j} - \frac{\partial Q_i}{\partial x_i} \tag{3}$$

In the formula, e is the mass specific internal energy and Q_i is the heat flux.

The model consisted of four parts: air, ice, explosives, and water. The air model size was 600 cm × 600 cm × 100 cm, the size of the ice model was 600 cm × 600 cm × X cm (X values are 100, 130, and 160 cm), the size of the water body model was 600 cm × 600 cm × 200 cm (in order to ensure greater realism, a non-reflective model was used to prevent the reflected waves at the bottom of the river from affecting the blasting results), and the dynamite model size was 10 cm × 10 cm × 10 cm, as shown in Figure 2.

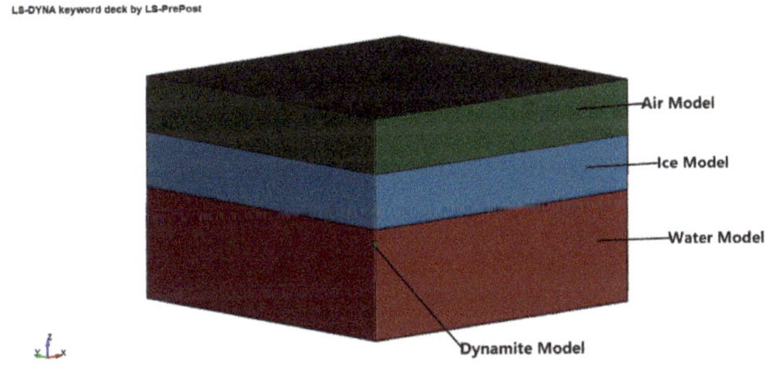

Figure 2. Numerical model of ice blast.

(1) Meshing was set to a 5 cm × 5 cm × 5 cm grid.
(2) All material units were set using SOLID 164.
(3) After the symmetry of the blasting model, calculation convenience, and calculation time reductions were set, a 1/4 three-dimensional model was established. The YOZ and XOZ planes were symmetrical planes that were set as symmetry boundaries, and symmetric normal constraints were applied.
(4) To simulate infinite air basin and the infinite water basin, the left, back, and bottom boundaries of the model were set to non-reflective boundary. The bottom boundary was set to the free surface.
(5) The ignition positions were determined as being located at the top, center, and bottom of the explosive center axis.
(6) Numerical simulation units were calculated using cm-g-μs.

(7) Considering the duration of the entire blasting process, the model blasting solution time in this paper was set to 1000 μs.

2.3.2. Model Parameters and Algorithms

The main model materials in this study were ice, explosives, water, and air. The ice body was considered with the Lagrange algorithm, and the air, explosives, and water were considered with the ALE algorithm, that is, the multi-substance Arbitrary Lagrange–Euler algorithm [34–36]. The model calculation used the multi-substance ALE fluid–structure interaction algorithm so that the mesh did not have serious deformations [37] that were too different from actual situations. The ice body was considered with an isotropic elastic fracture model (MAT_ISOTROPIC_ELASTIC_FAILURE), and the main model parameters are shown in Table 1.

Table 1. Mechanical property parameters of ice materials.

Density/ $(g{\cdot}cm^{-3})$	Shear Modulus/GPa	Yield Stress/MPa	Plastic Hardening Modulus/GPa	Volume Modulus/GPa	Plastic Failure Strain	Truncation Pressure/MPa
0.9	2.2	2.12	4.26	5.26	0.35	−4

Explosive parameter was set as a TNT explosive parameter, the model was set to adopt the (MAT_HIGH_EX-PLOSIVE_BURN) high-energy explosive model, and the explosion pressure equation of state was calculated using the JWL (Jones–Wilkins–Lee) equation of state [38], which is expressed as follows:

$$P = A(1 - \frac{\omega}{R_1 V})e^{-R_1 V} + B(1 - \frac{\omega}{R_2 V})e^{-R_2 V} + \frac{\omega E_0}{V} \quad (4)$$

In the above equation, E_0 is the initial internal energy, P is the pressure, V is the relative volume, and A, B, R_1, R_2, and ω are material constants.

TNT material parameters are shown in Table 2.

Table 2. Material parameters of TNT.

Density/ $(g{\cdot}cm^{-3})$	$D/(m{\cdot}s^{-1})$	P_{CJ}/GPa	A/GPa	B/GPa	R_1	R_2	ω	$E_0/(J/m^{-3})$	V_0
1.63	6930	21	373.8	3.747	4.15	0.90	0.35	7.0×10^9	1

The water material model was set to adopt the (MAT_NULL) material model [39], as shown in Table 3.

Table 3. Parameters of water materials.

Density/$(g{\cdot}cm^{-3})$	C	α	γ_0	S_1	S_2	S_3	$E_0/(J/m^{-3})$
1.02	0.165	0	0.35	1.92	−0.096	0	0

Note: C is the intercept of the shear-compression wave velocity curve, γ_0 is a constant, α is the volume correction amount, and S_1, S_2, and S_3 are constants. The air modeling was set to use a (MAT_NULL) material model, with the material parameters shown in Table 4.

Table 4. Air material parameters.

Density/$(g{\cdot}cm^{-3})$	C_0	C_1	C_2	C_3	C_4	C_5	C_6	$E_0/(J/m^{-3})$	V_0
0.00125	0	0	0	0	0.4	0.4	0	2.5×10^5	1

In the table, C_0, C_1, C_2, C_3, C_4, C_5, and C_6 are constants.

2.3.3. Introduction to Working Conditions

In this article, simulations of the working conditions were established with three variables: the distance of the explosives from the ice surface, the ignition location, and the thickness of the ice layer.

Working condition 1:

(1) The upper end of the explosive is ignited, the ice thickness is 100 cm, and the upper end of the explosive is 20 cm from the ice.
(2) Ignition occurs at the center of the explosive, the thickness of the ice layer is 100 cm, and the upper end of the explosive is 20 cm from the ice layer.
(3) Ignition occurs at the bottom of the explosive, the thickness of the ice layer is 100 cm, and the upper end of the explosive is 20 cm from the ice layer.

Working condition 2:

(1) Ignition occurs at the center of the explosive, the thickness of the ice layer is 100 cm, the upper end of the explosive is 20 cm from the ice layer, and the explosive mass is 3.2 kg.
(2) Ignition occurs at the center of the explosive, the thickness of the ice layer is 100 cm, the upper end of the explosive is 20 cm from the ice layer, and the explosive mass is 4.8 kg.
(3) Ignition occurs at the center of the explosive, the thickness of the ice layer is 100 cm, the upper end of the explosive is 20 cm from the ice layer, and the explosive mass is 6.4 kg.
(4) Ignition occurs at the center of the explosive, the thickness of the ice layer is 100 cm, the upper end of the explosive is 20 cm from the ice layer, and the explosive mass is 8.0 kg.

Working condition 3:

(1) Ignition occurs at the center of the explosive, the thickness of the ice layer is 100 cm, and the upper end of the explosive is 30 cm from the ice layer.
(2) Ignition occurs at the center of the explosive, the thickness of the ice layer is 100 cm, and the upper end of the explosive is 40 cm from the ice layer.
(3) Ignition occurs at the center of the explosive, the thickness of the ice layer is 100 cm, and the upper end of the explosive is 50 cm from the ice layer.
(4) Ignition occurs at the center of the explosive, the thickness of the ice layer is 100 cm, and the upper end of the explosive is 60 cm from the ice layer.
(5) Ignition occurs at the center of the explosive, the thickness of the ice layer is 100 cm, and the upper end of the explosive is 70 cm from the ice layer.
(6) Ignition occurs at the center of the explosive, the thickness of the ice layer is 100 cm, and the upper end of the explosive is 80 cm from the ice layer.
(7) Ignition occurs at the center of the explosive, the thickness of the ice layer is 100 cm, and the upper end of the explosive is 90 cm from the ice layer.
(8) Ignition occurs at the center of the explosive, the thickness of the ice layer is 100 cm, and the upper end of the explosive is 100 cm from the ice layer.
(9) Ignition occurs at the center of the explosive, the thickness of the ice layer is 100 cm, and the upper end of the explosive is 110 cm from the ice layer.
(10) Ignition occurs at the center of the explosive, the thickness of the ice layer is 100 cm, and the upper end of the explosive is 120 cm from the ice layer.
(11) Ignition occurs at the center of the explosive, the thickness of the ice layer is 100 cm, and the upper end of the explosive is 130 cm from the ice layer.
(12) Ignition occurs at the center of the explosive, the thickness of the ice layer is 100 cm, and the upper end of the explosive is 140 cm from the ice layer.
(13) Ignition occurs at the center of the explosive, the thickness of the ice layer is 100 cm, and the upper end of the explosive is 150 cm from the ice layer.
(14) Ignition occurs at the center of the explosive, the thickness of the ice layer is 100 cm, and the upper end of the explosive is 160 cm from the ice layer.

(15) Ignition occurs at the center of the explosive, the thickness of the ice layer is 100 cm, and the upper end of the explosive is 170 cm from the ice layer.

3. Results

3.1. Analysis of the Influence of Ignition Location on Ice Sheet

The established ice blasting model was used to analyze condition 1 regarding the impact of different ignition methods on the blast volume of ice bodies when the bottom was set as a non-reflective surface. The equivalent pressure distribution plot is shown in Figure 3.

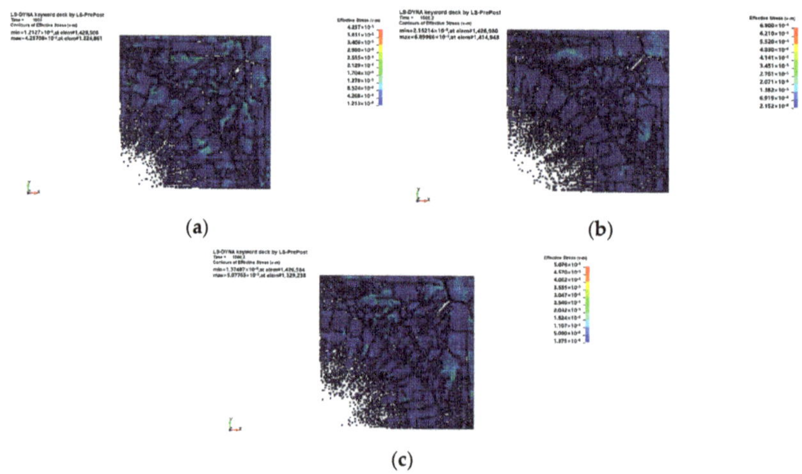

Figure 3. (**a**) Top ignition; (**b**) central ignition; (**c**) bottom ignition.

As can be seen from the figure, the maximum pressure of the equivalent pressure cloud at the center ignition was 6.89966×10^{-5} GPa, the maximum pressure of the bottom ignition was 5.07765×10^{-5} GPa, and the minimum pressure of the top ignition was 4.25708×10^{-5} GPa. In addition, analysis of its pressure distribution showed that the center ignition explosion was generated by the most concentrated pressure distribution, the top of the ignition explosion was generated by the most dispersed pressure distribution, and the bottom of the ignition explosion was generated by the pressure distribution between these. Furthermore, the analysis of pressure distribution showed that the center ignition pressure distribution was the most concentrated, the top ignition pressure distribution was the most dispersed, and the bottom ignition was between the two.

The time variation of ice blast volume for different ignition modes is shown in Figure 4.

An analysis of the blast volume in time at different locations of ignition showed that there was no effect of ignition at different locations at the beginning of blasting, there was a significant volume change from 49.2 μs on, and the remaining volume after blasting was 1.1×10^7 cm^3 at the center ignition, 1.13×10^7 cm^3 at the bottom ignition, and 1.22×10^7 cm^3 at the top ignition. The volume of blasting caused by the center ignition was the largest, followed by the bottom ignition, and the volume of blasting caused by the top ignition was the smallest.

The change in kinetic energy at the moment of detonation for different ignition methods is shown in Figure 5.

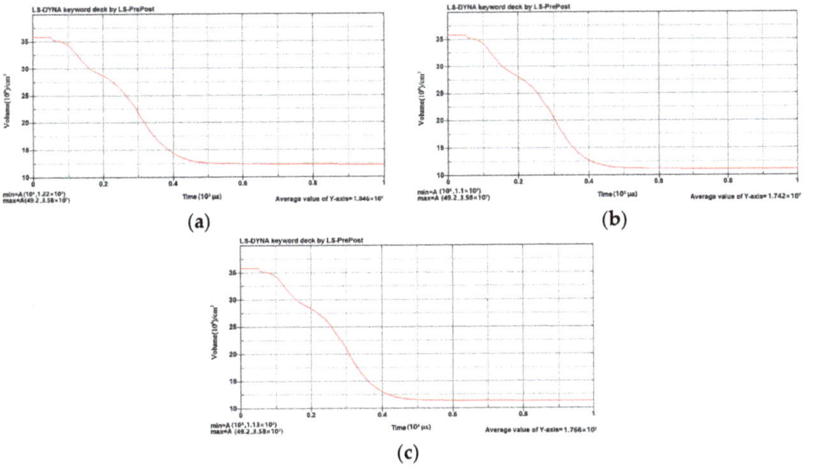

Figure 4. (**a**) Top ignition; (**b**) central ignition; (**c**) bottom ignition.

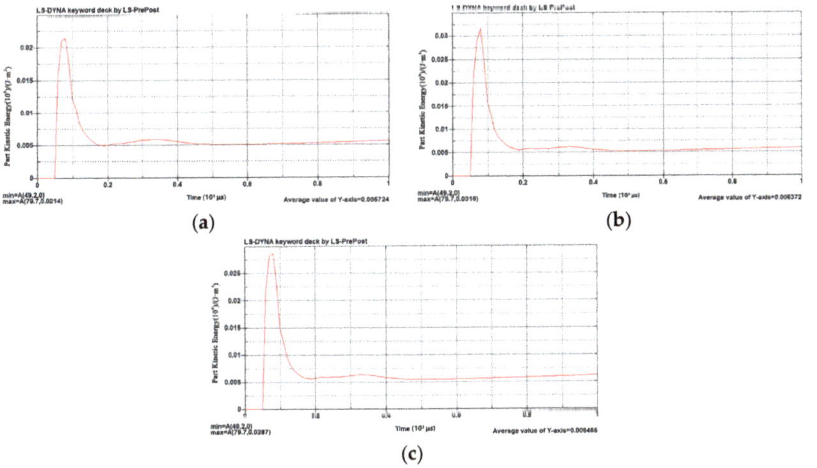

Figure 5. (**a**) Top ignition; (**b**) central ignition; (**c**) bottom ignition.

The figure shows that all three ignition methods resulted in a maximum kinetic energy at 79.7 μs following the maximum initial kinetic energy directly at 49.2 μs. The maximum initial kinetic energy was reached by wave transmission in a short time, and the maximum initial kinetic energy was 3.16×10^7 J/m^{-3} at the center ignition, 2.87×10^7 J/m^{-3} at the bottom ignition, and 2.14×107 J/m^{-3} at the top ignition.

The relationships between ignition position and ice blasting volume, final maximum pressure, and instantaneous kinetic energy are shown in Table 5.

Table 5. Relationship between ignition position and ice blasting volume, final maximum pressure, and instantaneous kinetic energy.

Ignition Mode	Final Volume V_{End} (cm^3)	Maximum Pressure $P_{explosion}$ (GPa)	Maximum Instantaneous Kinetic Energy (J/m^{-3})	Maximum Blast Radius (cm)
Top ignition	1.22×10^7	4.25708×10^{-5}	2.14×10^7	341.5
Center ignition	1.10×10^7	6.89966×10^{-5}	3.16×10^7	390.4
Bottom end ignition	1.13×10^7	5.07765×10^{-5}	2.87×10^7	363.8

3.2. Analysis of the Influence of Explosive Masses on Ice Sheet

The established ice blasting model was used to analyze simulated condition 2 regarding the effect of different explosive masses on the ice blasting volume at center ignition. Equivalent pressure distributions are shown in Figure 6.

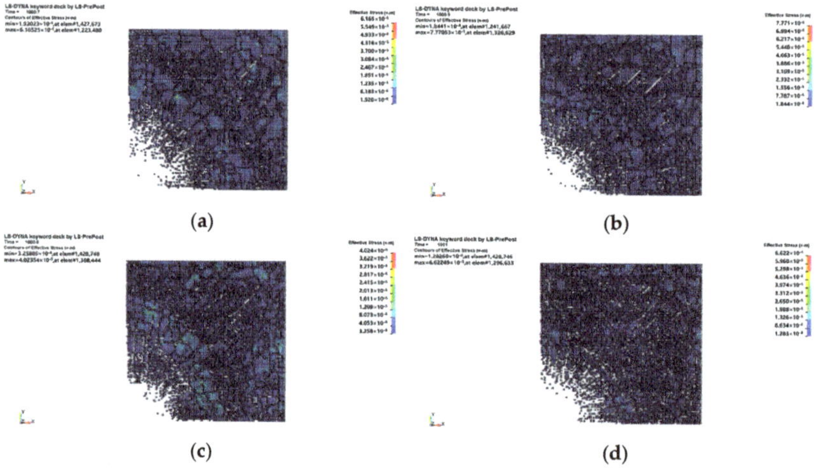

Figure 6. (**a**) Blasting with 3.2 kg TNT; (**b**) blasting with 4.8 kg TNT; (**c**) blasting with 6.4 kg TNT; (**d**) blasting with 8.0 kg TNT.

From the figure, it can be seen that the maximum pressure of the equivalent pressure cloud was 6.89966×10^{-5} GPa when the explosive mass was 1.6 kg, the maximum pressure of the equivalent pressure cloud was 6.16525×10^{-5} GPa when the explosive mass was 3.2 kg, the maximum pressure of the equivalent pressure cloud was 7.77053×10^{-5} GPa when the explosive mass was 4.8 kg, the maximum pressure of the equivalent pressure cloud was 4.02354×10^{-5} GPa when the explosive mass was 6.4 kg, and the maximum pressure of the equivalent pressure cloud was 6.62249×10^{-5} GPa when the explosive mass was 8.0 kg. The maximum pressure of the equivalent pressure changes in waves.

The time variations of the ice blast volume for different explosive masses are shown in Figure 7.

The instantaneous kinetic energy changes for different explosive masses are shown in Figure 8.

The relationships between explosive mass and ice blasting volume, maximum pressure at the end, instantaneous kinetic energy, and average kinetic energy are shown in Table 6.

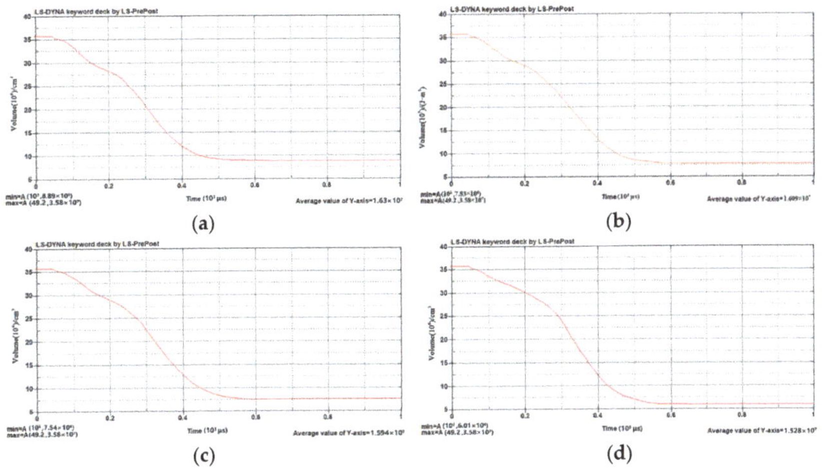

Figure 7. (**a**) Blasting with 3.2 kg TNT; (**b**) blasting with 4.8 kg TNT; (**c**) blasting with 6.4 kg TNT; (**d**) blasting with 8.0 kg TNT.

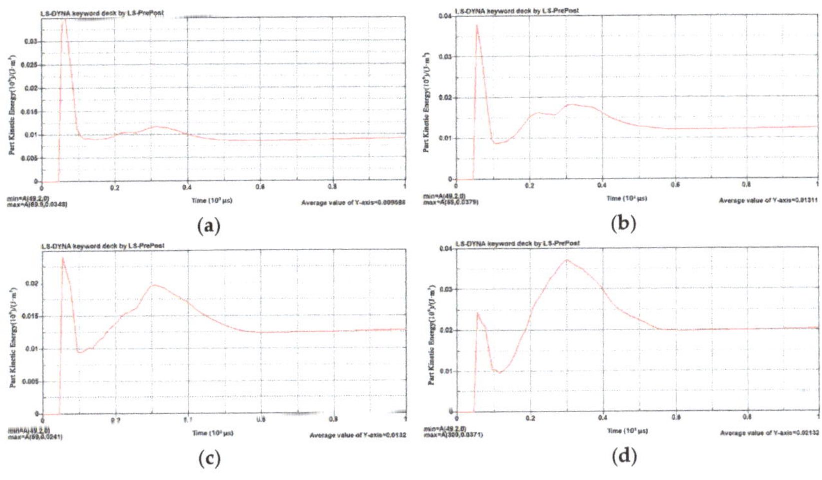

Figure 8. (**a**) Blasting with 3.2 kg TNT; (**b**) blasting with 4.8 kg TNT; (**c**) blasting with 6.4 kg TNT; (**d**) blasting with 8.0 kg TNT.

Table 6. Relationship between ice thickness and ice blasting volume, maximum pressure at the end, instantaneous kinetic energy, and average kinetic energy.

Explosive Mass (kg)	Volume Burst Rate (%)	Volume Burst Rate for Every 1.6 kg Increase (%)	Maximum Pressure P_{burst} (GPa)	Maximum Kinetic Energy (J/m^{-3})	Average Kinetic Energy (J/m^{-3})
1.6	69.44		6.89966×10^{-5}	3.16×10^{7}	6.372×10^{6}
3.2	75.31	5.87	6.16525×10^{-5}	3.48×10^{7}	9.586×10^{6}
4.8	78.25	2.94	7.77053×10^{-5}	3.79×10^{7}	1.311×10^{7}
6.4	79.06	0.81	4.02354×10^{-5}	2.41×10^{7}	1.320×10^{7}
9.0	83.31	4.25	6.62249×10^{-5}	3.71×10^{7}	2.132×10^{7}

From the above table, the volume of explosive blast is found to increase gradually as the mass of explosive increases. However, in the third column of the table, the blast volume is not found to increase linearly when the explosive mass increases from 1.6 kg to 6.4 kg. Here, the blast volume is found to increase gradually and more slowly, but when the explosive mass reaches 8 kg, a significant increase in blast volume can be observed. Therefore, an increase in blasting efficiency is assumed to be possible by increasing the number of explosive packages. However, as the number of individual explosive packages increases, the blasting effect will gradually approach its limit. This limit can only be exceeded by using a large quantity of explosives. When observing the maximum kinetic energy when blasting different masses of explosives, it was found that the maximum kinetic energy (3.16×10^7 J/m^{-3}) was generated when the mass of explosives was 4.8 kg. As the explosive mass increased, the average kinetic energy of the ice layer was gradually increasing. However, the average kinetic energy generated by the explosion of 6.4 kg of explosives did not exhibit a significant change compared to the average kinetic energy generated by the explosion of 4.8 kg of explosives.

3.3. Analysis of the Effect of Explosive Location on Ice Blasting Model

The established ice blasting model was used to analyze simulated working condition 3 regarding the effect of different positions of explosives on the blasting volume of the ice body at the center ignition and an ice thickness of 100 cm. The stress effects are shown in Figure 9a–o.

The simulation results regarding the effects of different distances between explosives and the ice model on the ice blasting under working condition 3 are shown in Table 7.

Table 7 shows that at different distances between the explosive and ice model, the blasting effects were also different. Figures 10–12 illustrate the relationships between different distances between the explosive and ice model and ice blasting volume, maximum pressure at the end, instantaneous kinetic energy, and average kinetic energy.

Table 7. Relationship between explosive ice distance and ice blasting volume, maximum pressure at the end, instantaneous kinetic energy, and average kinetic energy.

Distance from Ice (cm)	Volume at the End V_0 (cm^3)	Maximum Pressure P Explosion (GPa)	Maximum Kinetic Energy (J/m^{-3})	Average Kinetic Energy (J/m^{-3})
20	1.10×10^7	6.89966×10^{-5}	3.16×10^7	6.372×10^6
30	8.84×10^6	6.48349×10^{-5}	4.84×10^7	8.786×10^6
40	7.96×10^6	4.64357×10^{-5}	5.55×10^7	9.816×10^6
50	7.56×10^6	1.37037×10^{-4}	5.00×10^7	1.080×10^7
60	7.13×10^6	3.87005×10^{-5}	3.71×10^7	1.107×10^7
70	6.76×10^6	4.12181×10^{-5}	3.33×10^7	1.145×10^7
80	6.11×10^6	5.45746×10^{-5}	4.19×10^7	1.302×10^7
90	5.56×10^6	6.85551×10^{-5}	5.64×10^7	1.489×10^7
100	5.23×10^6	7.15330×10^{-5}	6.77×10^7	1.661×10^7
110	5.00×10^6	6.24793×10^{-5}	7.63×10^7	1.787×10^7
120	4.91×10^6	5.82885×10^{-5}	8.05×10^7	1.959×10^7
130	4.62×10^6	5.95254×10^{-5}	8.19×10^7	2.084×10^7
140	4.67×10^6	6.85479×10^{-5}	8.34×10^7	2.139×10^7
150	4.63×10^6	5.87875×10^{-5}	8.52×10^7	2.229×10^7
160	4.58×10^6	6.88285×10^{-5}	8.53×10^7	2.207×10^7
170	4.87×10^6	7.06496×10^{-5}	8.00×10^7	2.126×10^7

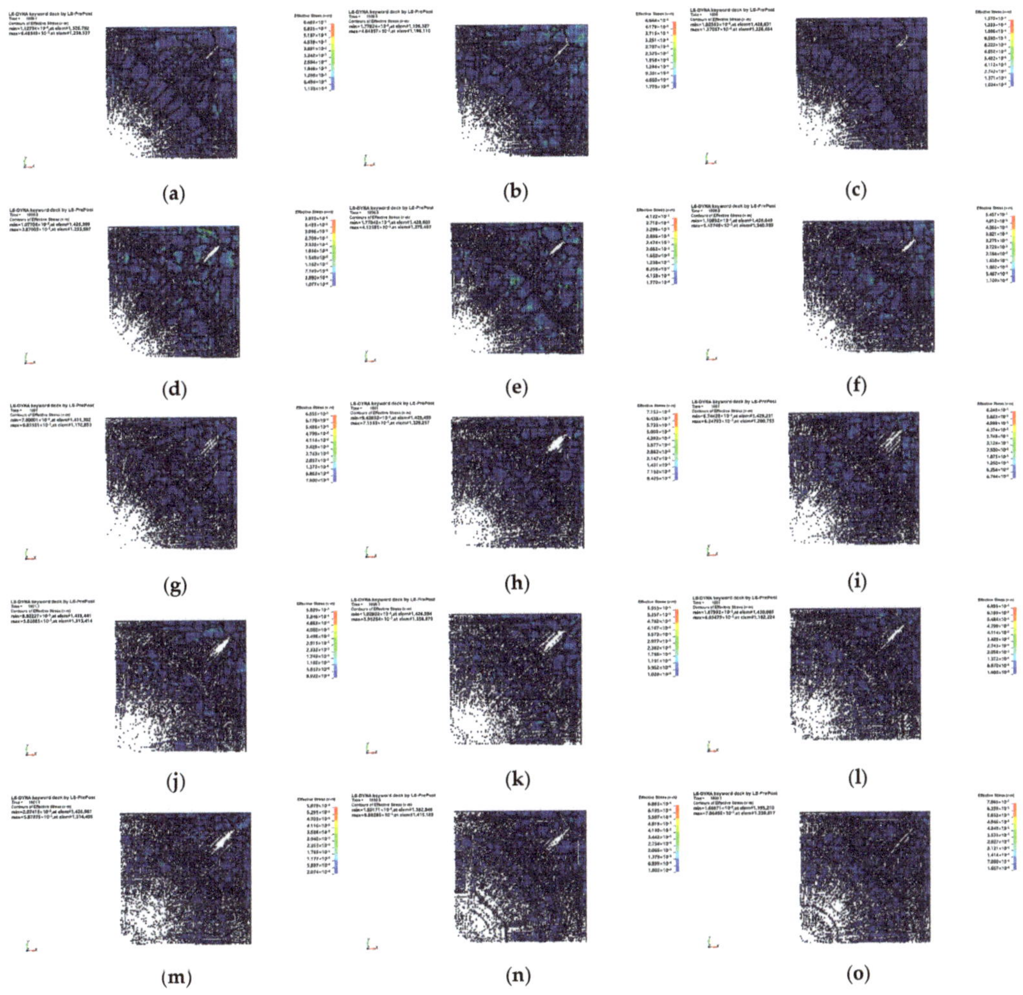

Figure 9. (**a**) the upper end of the explosive is 30 cm from the ice layer; (**b**) the upper end of the explosive is 40 cm from the ice layer; (**c**) the upper end of the explosive is 50 cm from the ice layer; (**d**) the upper end of the explosive is 60 cm from the ice layer; (**e**)the upper end of the explosive is 70 cm from the ice layer; (**f**) the upper end of the explosive is 80 cm from the ice layer; (**g**) the upper end of the explosive is 90 cm from the ice layer; (**h**) the upper end of the explosive is 100 cm from the ice layer; (**i**) the upper end of the explosive is 110 cm from the ice layer; (**j**) the upper end of the explosive is 120 cm from the ice layer; (**k**) the upper end of the explosive is 130 cm from the ice layer; (**l**) the upper end of the explosive is 140 cm from the ice layer; (**m**) the upper end of the explosive is 150 cm from the ice layer; (**n**) the upper end of the explosive is 160 cm from the ice layer; (**o**) the upper end of the explosive is 170 cm from the ice layer.

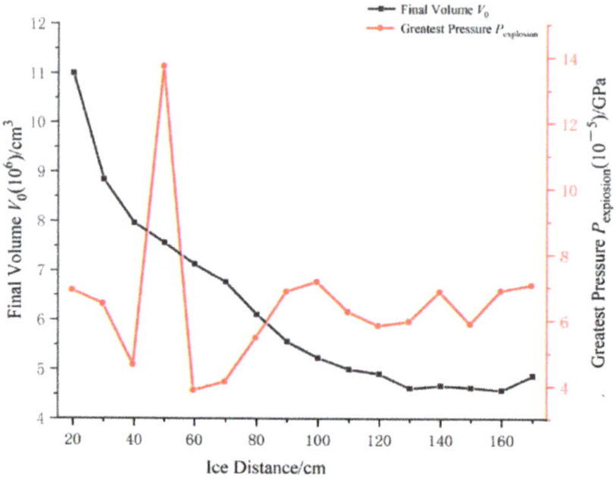

Figure 10. The relationship between ice distance, final volume, and final maximum pressure.

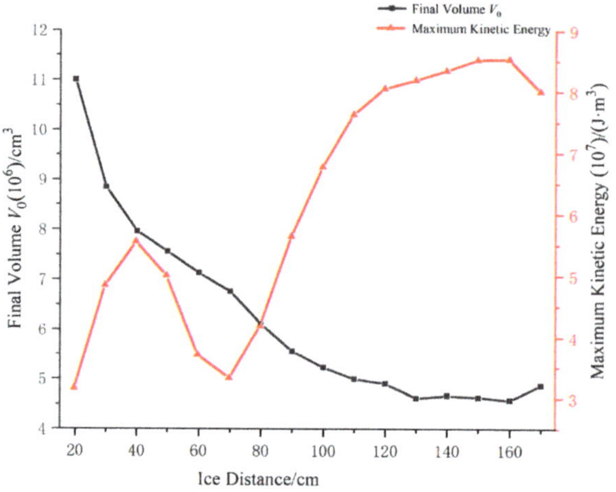

Figure 11. Relationship between ice distance, final volume, and maximum kinetic energy.

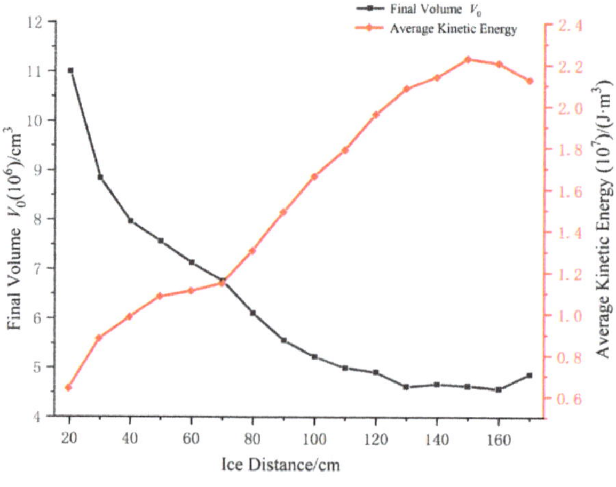

Figure 12. Relationship between ice distance, final volume, and average kinetic energy.

Figure 10 shows that when the distance between the dynamite and ice model was 20 to 120 cm, the larger the distance between the dynamite and ice model, the smaller the residual volume of the ice model after explosion and the more obvious the blasting effect. When the distance between the explosive and the ice model exceeded 160 cm, the residual volume of the ice model after explosion increased with the increase in the distance. As the distance between the dynamite and ice model increased, the maximum pressure at the end of the ice model changed in a wave-like manner, similar to the propagation of underwater explosion waves.

As the distance between the explosive and ice model increased, the maximum kinetic energy of the ice model reached a peak at 40 cm between the explosive and ice model. When the distance between the explosive and ice model was 70 to 160 cm, the maximum kinetic energy of the ice model increased with the increase in distance. When the distance between the explosive and ice model exceeded 160 cm, the maximum kinetic energy decreased with the increase in the distance.

Figure 12 shows that when the distance between the explosive and ice model was 20 to 150 cm, the average kinetic energy of the ice model increased with the increase in the distance; when the distance between the explosive and the ice model exceeded 150 cm, its average kinetic energy decreased with the increase in the distance.

4. Discussion

In the downstream section of Beiji Village, Mohe, between navigation mark 812 and 818, the ice thickness of approximately 100 cm was selected for the blasting experiments with three different ignition methods. The experimental results show that the central ignition blasting method produces the best explosion effect and the largest explosion radius, and the measured explosion radius is 387.5 cm, which is in line with the results obtained in condition 1 (working conditions of three different ignition methods), thus proving the reasonability of the simulation. In addition, after the explosion, it was found by observing the shape of the ice body that the explosion in the center produces a nearly circular blasting pit, but there are circular cracks in its outer ring, indicating that the ice explosion is not finished at one time, but carries on the energy transfer in the form of waves.

With the increase in the explosive mass, the blasting effect decreased gradually, and the impact could be increased by increasing the number of charge bags; however, as the number of individual charge bags increased, the blasting effect gradually approached the limit. Experimental verification indicated that when 1.6 kg TNT explosive explodes in a river section with an ice thickness of 1 m, the effective blasting area is a circular area with a radius of 8 m, thereby proving the reasonability of the simulation results in working condition 2.

In this case, it can be concluded that the explosive bags should be reasonably laid in a staggered arrangement in two rows of ice holes in the adjacent upstream and downstream sections along the direction of water flow, and the distance between the holes in the row and the distance of the adjacent row should be roughly equal. In addition, the effect of the broken ice sheet is the best in the case of an equal charge amount of each hole. The layout of explosive package hole distance is restricted by the water depth, ice thickness, and single hole charge, and the hole distance cannot be too large, or it cannot blast through the ice hole, and the icebreaking effect will be unfavorable as well, even if the single hole charge is large. If the blasting ice holes are arranged in the direction of the water flow along the river and the front and back rows are staggered, the blasting effect will be rather favorable and can effectively destroy the ice sheet and ensure that the scale of the ice is in line with the requirements of the river conditions for transporting the ice sheet.

Under the condition of an ice thickness of approximately 100 cm in the river section and an explosive mass of 1.6 kg, the blasting test was carried out in the ice area where the distance between the explosive and the ice was 20–200 cm (with 20 cm as the difference value). In order to facilitate the analysis, the explosion radius was measured, and it was found that the explosion effect on site is highly similar to the results simulated using the

ANSYS model. The experimental results show that the blasting radius is the largest and the fracture effect of the ice body is the best in the case of a distance of 160 cm, which is in line with the results obtained in condition 3.

At the same time, the breakup of the ice of the main stream of the Heilongjiang River was performed after blasting for 4 days, during which the typical ice dam was observed near the 10 km overland route opening in the upper reach of the Mohe cross section at the blasting location during the period, since the ice blocks from the upper reach were accumulated and blocked. The field observation and theoretical reasoning results indicate a high probability that the ice dam will occur in the blasting river reach, and the resulting force will also be greater if the lower reach cross section of the ice dam is not blasted, without conditions for ice block transport in the blasting area. Since the slope of the river reach is small and has a great degree of curvature, it is considered the place where the water flow is mixed violently in the river. In this case, it is rather obvious that the river ice blasting plays a critical role in preventing an ice disaster in the Heilongjiang River.

5. Conclusions

The main conclusions drawn are as follows:

1. It can be concluded through model simulation and experimental comparison that the blasting effect in the center is the best, and the blasting volume and radius are the largest when the explosive is ignited under the ice;
2. With the increase in the explosive mass, the blasting effect will decrease gradually, and the impact can be increased by increasing the number of charge bags but will gradually approach the limit as the number of individual charge bags increases. Experimental verification indicates that 1.6 kg TNT can blast the river section with an ice thickness of 1 m, and the effective blasting area is a circular area with a radius of about 8 m;
3. The explosive bags should be reasonably laid, in staggered arrangement in two rows of ice holes in the adjacent upstream and downstream section along the direction of water flow, and the distance between the holes in the row and the distance of the adjacent row should be roughly equal;
4. In addition, the effect of the broken ice sheet is the best in the case of an equal charge amount of each hole. An ice blast is not finished at once, but experiences energy transfer in the form of waves. When the distance between the explosive and the ice is about L = 150 cm, the blasting effect is the best.

However, the simulation of the model needs to be improved in some areas. In the simulation, the temperature was not set, but the effect of temperature on the strength of the ice sheet is indeed the main factor. In general, the ice sheet intensity decreases with the decrease in temperature, and the blasting effect of river ice should be different at different temperatures. According to the experimental and field observation data of the blasting observation, ice sheet blasting can be carried out in the case of a stable and positive maximum temperature of the day, when the ice sheet is still hard in texture and high in efficiency. Breaking ice will break too early and will refreeze, and even destroy all previous efforts, resulting in waste. In this case, it is advisable to carry out ice blast 3–5 days before the opening of the river. During the opening period, the existence time of river ice is short, making it difficult to grasp the timing of the blasting simulation and field experiment, and the number of experiments is rather limited. Considering the influence of river hydraulic and thermal factors, the numerical simulation is combined under hydraulic and thermal conditions, and it is more difficult to simulate under complex hydraulic conditions. To this end, the following solutions to close the gap are hereby proposed: (i) to improve the cross application of multiple disciplines; (ii) to fully combine the method with various coupling factors, consider turbulent conditions of hydrothermal conditions, and carry out a deeper theoretical analysis of the model under the boundary treatment and scene setting; and (iii) to combine with practice, perform more field tests, provide data for laboratory tests, and create similarity analyses at the same time.

Author Contributions: Conceptualization, D.L. and C.D.; methodology, D.L.; software, D.L.; validation, C.D., C.Y. and Y.M.; formal analysis, D.L.; investigation, C.D. and Y.M.; resources, C.D.; data curation, D.L.; writing—original draft preparation, D.L.; writing—review and editing, C.D. and C.Y.; visualization, D.L. and Y.M.; funding acquisition, C.D. All authors have read and agreed to the published version of the manuscript.

Funding: This work was funded by the National Natural Science Foundation of China, the State Key Laboratory of Permafrost Engineering Open Fund Grant and Research on optimal matching and regulation of soil and water resources in the Sanjiang Plain, and numbers are as follows: National Natural Science Foundation of China, No. 41202171, The State Key Laboratory of Permafrost Engineering Open Fund Grant, No. SKLFSE201310, Research on optimal matching and regulation of soil and water resources in the Sanjiang Plain, No. 22053.

Institutional Review Board Statement: Not applicable.

Informed Consent Statement: Not applicable.

Data Availability Statement: Not applicable.

Conflicts of Interest: The authors declare no conflict of interest.

References

1. Yang, X.; Pavelsky, T.M.; Allen, G.H. The past and future of global river ice. *J. Nat.* **2020**, *577*, 69–73. [CrossRef] [PubMed]
2. Prowse, T.D. River-Ice ecology. I: Hydrologic, geomorphic, and water-quality aspects. *J. Cold Reg. Eng.* **2001**, *15*, 1–16. [CrossRef]
3. Beltaos, S.; Burrell, B.C. Hydrotechnical advances in Canadian river ice science and engineering during the past 35 years. *J. Can. J. Civ. Eng.* **2015**, *42*, 583–591. [CrossRef]
4. Shen, H.T.; Wang, D.S. Under cover transport and accumulation of frazil granules. *J. Hydraul. Eng.* **1995**, *121*, 184–195. [CrossRef]
5. Chen, J.G.; Zeng, Q.H.; Wang, Z.Y. The transport properties of ice sheet currents. *J. Sediment Res.* **1992**, *1*, 61–69. (In Chinese) [CrossRef]
6. Lu, S.B.; Huang, Q.; Wu, C.G.; Gao, F. Ice disaster in the Ningmeng section of the Yellow River and measures to prevent flooding in reservoirs. *J. Nat. Dis.* **2010**, *19*, 43–47. (In Chinese) [CrossRef]
7. Wang, Y.S.; Yang, K.L.; Zhang, B. *Design Manual for River Ice Management and Control Engineering*; China Water Resources and Hydropower Press: Beijing, China, 2013.
8. Ashton, G.D. *River and Lake Ice Engineering*; Water Resources Publications: Littleton, CO, USA, 1986.
9. Shen, H.T.; Su, J.; Liu, L. SPH simulation of river ice dynamics. *J. Comput. Phys.* **2000**, *165*, 752–770. [CrossRef]
10. Shen, H.T. Mathematical modeling of river ice processes. *J. Cold Reg. Sci. Technol.* **2010**, *62*, 3–13. [CrossRef]
11. Beltaos, S.; Prowse, T.D. Climate impacts on extreme ice-jam events in Canadian rivers. *Hydrol. Sci. J.* **2001**, *46*, 157–181. [CrossRef]
12. Beltaos, S. *River Ice Formation*; Committee on River Ice Processes and the Environment: Whitehorse, YT, Canada, 2013.
13. Boucher, É.; Bégin, Y.; Arseneault, D. Impacts of recurring ice jams on channel geometry and geomorphology in a small high-boreal watershed. *Geomorphology* **2009**, *108*, 273–281. [CrossRef]
14. Song, C.; Zhu, X.; Han, H.; Lin, L.; Yao, Z. The influence of riverway characteristics on the generation and dissipation of ice dam in the upper reaches of Heilongjiang River. *J. Hydraul. Eng.* **2020**, *51*, 1256–1266. [CrossRef]
15. Spyros, B. *River Ice Jams*; Water Resource Publication: New York, NY, USA, 1995.
16. Spyros, B. *River Ice Breakup*; Water Resource Publication: New York, NY, USA, 2008.
17. Shen, H.T. Research on river ice processes: Progress and missing links. *J. Cold Reg. Eng.* **2003**, *12*, 135–142. [CrossRef]
18. Shen, H.T. River Ice Transport Theories: Post, Present, and Future. In Proceedings of the 15th IAHR Symposium on Ice, Gdansk, Poland, 28 August–1 September 2000.
19. Tao, W.; Kailin, Y.; Yongxin, G. Application of artificial neural networks to forecasting ice conditions of the Yellow River in the Inner Mongolia Reach. *J. Hydrol. Eng.* **2008**, *13*, 811–816. [CrossRef]
20. Malcolm, M. *Breaking Ice with Explosives*; Cold Regions Research &Engineering Laboratory: New York, NY, USA, 1982.
21. United States Army Corps of Engineers. *Ice Engineering: Engineering and Design*; Department of the Army: Washington, DC, USA, 2002.
22. He, B.S.; Ding, L.Q.; Yan, X.G. *Research Report on Artificial Intervention in Kaihe River Prevention Measures and Program Development*; China Water Resources and Hydropower Science Research Institute: Beijing, China, 2009. (In Chinese)
23. Liang, X.Q.; He, B.S.; Xie, W.H. Explosive ice breaking and action effect test of the Yellow River ice. *Eng. Blast.* **2012**, *2*, 83–85. (In Chinese) [CrossRef]
24. Liang, X.Q.; Xiong, F.; Lu, X.L. Experimental study on underwater shock wave characteristics of the Yellow River blasting ice. *Blasting* **2014**, *4*, 1–4. (In Chinese) [CrossRef]
25. Tong, Z.; Yang, X.S.; Xue, P.X. Research on the mechanism of ice blasting in the Yellow River. In *China Blasting Technology III*; Metallurgical Industry Press: Beijing, China, 2012. (In Chinese)

26. Duan, Y.S.; Wang, X.H.; Liu, S.B.; Li, Y.S. Discussion on ice blasting technology. *J. Glaciol. Geocryol.* **2003**, *25*, 220–226. (In Chinese) [CrossRef]
27. Yin, H.T.; Yang, X.H.; Jiang, M.; Jin, J.L. Experimental study on blasting ice-breaking with extended charge pack in ice-cold water. *Eng. Blast.* **2010**, *3*, 12–15. (In Chinese) [CrossRef]
28. Mair, H.U. Review: Hydrocodes for structural response to underwater explosions. *Shock Vib.* **1999**, *6*, 81–96. [CrossRef]
29. Hu, H.P. Influence of Tunnel Blasting Vibration on Nearby Bridge and Safety Assessment. Master's Thesis, Chongqing Jiaotong University, Chongqing, China, 2013.
30. Yan, G.B.; Yu, Y.L. Numerical simulation of air and water medium decoupling charge blasting. *Eng. Blast.* **2009**, *15*, 13–19, 65. [CrossRef]
31. Yang, X.M. *Numerical Simulation of the Phenomenon of Explosion Shock*; University of Science and Technology of China Press: Hefei, China, 2010.
32. Wang, H. Numerical Simulation of Underwater Drilling Blasting. Master's Thesis, North University of China, Taiyuan, China, 014.
33. Zhang, X.; Lu, M.W.; Wang, J.J. Research progress of arbitrary Lagrange-Eulerian description method. *Chin. J. Comput. Mech.* **1997**, *1*, 93–104. Available online: http://a.xueshu.baidu.com/usercenter/paper/show?paperid=185f3a1b64261c77a90f6af3e315eadf (accessed on 7 March 2022).
34. Zhang, J.P.; Yu, Y.L. Field experiments and theoretical analysis on blasting craters of ice layers by near charge. *Eng. Blast.* **2006**, *12*, 5–9. [CrossRef]
35. Zufelt, J.E.; Ettenma, R. Fully coupled model of ice-jam dynamics. *J. Cold Reg. Eng.* **2000**, *14*, 24–41. [CrossRef]
36. Li, S.B.; Dong, Z.X.; Qi, Y.J.; Jiao, J.F. Numerical simulation for spread decay of blasting shock wave in different media. *J. Vib. Shock.* **2009**, *28*, 115–117. [CrossRef]
37. Guo, Y.K.; Han, X.Y.; Meng, W.Y. Finite element analysis of the bridge pier safety under ice collisions. *J. North China Univ. Water Res. Electr. Power* **2013**, *34*, 16–18. [CrossRef]
38. David, B. *MSC/DYTRAN User's Manual*; MacNeal-Schwendler Corporation: New York, NY, USA, 1991.
39. Gao, X.N.; Wu, Y.J. Numerical calculation and influence parameters for TNT explosion. *Chin. J. Explos. Propellants* **2015**, *38*, 32–39. [CrossRef]

water

MDPI

Article

Prediction of Ice-Resistance Distribution for R/V Xuelong Using Measured Sea-Ice Parameters

Li Zhou, Yingjie Gu, Shifeng Ding * and Renwei Liu

School of Naval Architecture and Ocean Engineering, Jiangsu University of Science and Technology, Zhenjiang 212100, China; 201600000078@just.edu.cn (L.Z.); guyingjie01@163.com (Y.G.); 202100000140@just.edu.cn (R.L.)
* Correspondence: 201800000060@just.edu.cn

Abstract: Increasing human activity in polar areas is making ice-going ships more indispensable in multiple operation scenarios. An improvement in ice-resistance prediction, which cannot be performed without accurate ice parameters, will promote the development of hull design and operational safety in ice-infested waters. The Nataf transformation is applied to generate correlated pseudo-random numbers which represent ice parameters; then, as a numerical method, the circumferential crack method is introduced to calculate the ice resistance of R/V Xuelong in level ice. The main factors which may have a large influence on simulated ice load are studied. The simulation results show that the Burr distribution is the most suitable model to describe the distribution of ice resistance calculated and ice-force amplitude concentrated at a lower level. The statistical results are also discussed and compared with similar research through empirical formulas and Monte Carlo methods. The present simulation can obtain more detailed information during the icebreaking process compared to similar research: the ice force at each time step is achieved; the key ice-force amplitude can be collected, which can benefit studies on hull structure; and potential stress generated by sea ice can be predicted. The present numerical tools and simulation results can provide a reference for ice-going ships sailing in level ice in most scenarios with regard to ice resistance and operational safety.

Keywords: level ice; ice resistance; probability distribution; polar ship; circumferential crack method

Citation: Zhou, L.; Gu, Y.; Ding, S.; Liu, R. Prediction of Ice-Resistance Distribution for R/V Xuelong Using Measured Sea-Ice Parameters. *Water* **2022**, *14*, 517. https://doi.org/10.3390/w14040517

Academic Editor: Hung Tao Shen

Received: 12 January 2022
Accepted: 7 February 2022
Published: 9 February 2022

1. Introduction

With the increase in human activity in polar areas, ice-going ships play a significant role in commercial transportation, logistics supply, and personnel transportation. These ships cannot avoid colliding with sea ice when navigating in cold regions; therefore, such ships require the capability to break ice or be equipped with an ice-strengthened hull in order to sail in waters covered by ice and snow. An accurate prediction of ice resistance encountered by a ship in various situations is necessary at the design stage or during transportation planning, thereby promoting the development of structural design and operational safety in polar areas.

To date, a large number of studies on the ice-breaking process have been carried out. Lindqvist [1] and Riska et al. [2] proposed some semi-empirical and analytic forecast models, which provided effective methods for the early design of icebreakers. Su et al. [3] simulated ship maneuvers in ice areas involving the surge, sway, and yaw motions; however, only a single ice-failure pattern was considered. Lubbad and Løset [4] applied PhysX to simulate the motions of ice floes in the calculation domain. Zhou et al. [5,6] studied the relationship between the hull slope angle, ice friction coefficient, and sea-ice failure pattern, and analyzed the mathematical model of each component of ice load. In this paper, we apply a numerical simulation method based on a semi-empirical formula to simulate the straight operation of a ship in level ice, combining the advantages of the numerical simulation and empirical formula to make sure that accuracy and efficiency is guaranteed.

This numerical tool can calculate ice resistance based on the interaction between the hull and sea ice. Geometric contact between the hull and ice was carried out at each time step in order to judge the ice-failure pattern and whether the sea ice was broken, and the ice force at each hull node was calculated and collected. Key ice force at certain time steps can be detected and extracted, and such results can be used as a reference resource for ship design and ship operation in polar areas.

Ice parameters such as physical and mechanical properties are required as calculation input in these methods, e.g., ice thickness, ice density, ice flexural strength, ice crushing strength, temperature, and elastic modulus [7]. These parameters show a high degree of randomness in nature [8], which have a great influence on ice–hull interaction, breaking the pattern of sea ice and ice resistance during interactions with different kinds of structures. Sodhi and Chin [9] used Gaussian, Rayleigh, and Weibull distributions for indentation tests in urea-model ice. The ice was then failed by crushing, and the distribution of peak ice forces followed the Weibull distribution. Takeuchi et al. [10] conducted research on the spatial distribution of the uniaxial compressive strength of ice, as well as the correlation with temperature and density. Uniaxial strength and other mechanical parameters fitted well with a normal distribution. Johnston et al. [11] reviewed the measured data of multi-year ice thickness in polar areas and proposed a lognormal distribution for the data. In 2012, Strub-Klein [12] analyzed the spatial and temporal distribution of uniaxial compressive strength, temperature, density, and salinity of first-year ice, then applied a methodology for statistical analysis of uniaxial compressive strength to full-scale ice data in 2017 [13]. Ni [14] applied a one-way computational fluid dynamics–discrete element coupling method to simulate the motion of a ship sailing in level ice. The ice resistance showed strong oscillation and was mainly affected by the ship's velocity and ice thickness. Wang [15] studied the trend of ice resistance under crushed ice. The ice resistance increased with the increase in speed, but when the ship's speed reached a certain value, the ice resistance no longer increased, and even reduced the trend. Zong [16] studied an ice–ship collision. Hydrodynamic influence turned out to be significant for elastic and elastoplastic modes of ice. Chai [17] studied the statistics of ice thickness and the ice strength of first-year sea ice; a scheme based on empirical formulas was proposed to investigate the correlation between different ice parameters. A novel probabilistic model was introduced to describe the distribution of ice flexural strength. However, statistical studies of long-term analysis and estimations of ice resistance are rare. Measured short-term ice-induced loads show that the Weibull probability distribution [18–21] or the exponential or logarithmic normal probability distribution [22,23] match well with measured data after later statistical research. Based on measured data collected by R/V Xuelong, Ji [24] found that ice-induced hull vibration amplitude increases with both ice thickness and ship speed. Wang et al. [25] studied the probability distribution of ice parameters based on measurements collected by R/V Xuelong in scientific voyages, fixed ice sites, or empirical equations. It was found that ice resistance follows the distribution of Gamma. Ice resistance was calculated based on Lindqvist's semi-empirical formulae [1], which provided only an approximate estimate of ice resistance. However, this method ignored geometric contact and weakened the interaction between the hull and the ice. The distribution of ice resistance needs to be analyzed by a more refined method.

On the basis of the probabilistic model proposed by Wang [25], a method which can generate pseudo-random numbers is adopted to produce the distribution and correlation of ice parameters. Ice data are then used as input for the empirical numerical simulation tool mentioned above in order to calculate the ice resistance of R/V Xuelong [26,27]. The ice resistance is obtained after the whole sailing motion is accomplished. Finally, the probability distribution of the ice resistance is analyzed and compared with related research.

2. Numerical Method

The ice-load model was first introduced, which refined ice-failure patterns into common bending-failure patterns, and crushing-failure patterns occurred at a vertical frame angle. Dynamic global ice force was calculated based on hull–ice interaction and semi-empirical formulae; then, we used Nataf transformation to process statistical model and generate corresponding ice parameters as input into the simulation [28].

2.1. Circumferential Crack Method

Vertical compression and horizontal tension of ships act continuously on sea ice during voyages in polar areas; then, ice fails at bending and crushing on most occasions. Circumferential cracks parallel to the contact surface will take the lead under compression and tension. The sea-ice-failure mechanism adopted is the occurrence of circumferential crack based on this physical phenomenon observed from model tests and ice trials, which assumes that ice blocks break from ice sheets in a vertical direction, and that their contact surface is flat and their contact area can be determined by its length and depth [4].

The geometric shape of the ice floes broken from level ice can be assumed to be wedge-shaped, as shown in Figure 1. The shape of the crack is represented by a red curve, and the original intact ice sheet is represented in blue. The angle of ice wedge is θ, and the ice wedge is shaped based on the icebreaking radius R, the expression of which is given in reference [29]:

$$R = C_l l \left(1.0 + C_v v_n^{rel}\right) \tag{1}$$

where C_l and C_v are empirical parameters, v_n^{rel} is the relative normal velocity between ice and hull, and l is the characteristic length of ice;

$$l = \left\{E_i h_i^3 / \left[12\left(1 - v^2\right)\rho_w g\right]\right\}^{1/4} \tag{2}$$

where E_i is the young's modulus of ice, h_i is the thickness of ice, v is the poisson ratio of ice, ρ_w is the density of water, and g is the acceleration of gravity.

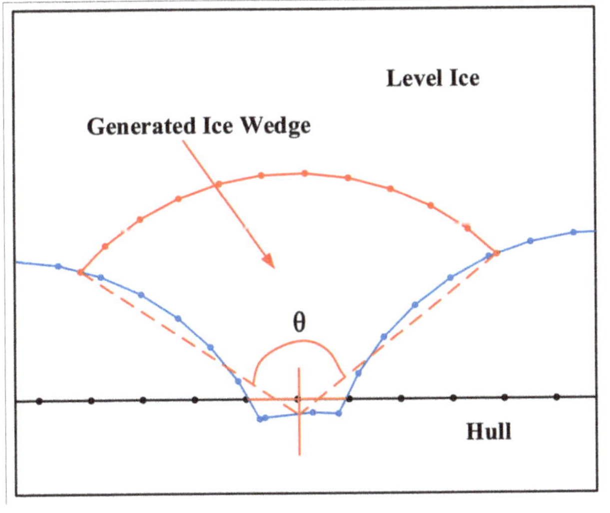

Figure 1. Ice wedge in the horizontal plane.

2.1.1. Mechanical Model of Interaction between Ice and Hull

The localized crushing force F_{cr} generated on the contact surface of hull and sea ice can be expressed as:

$$F_{cr} = \sigma_c \cdot A_c \tag{3}$$

where A_c is the contact area between ice and hull, σ_c is the crushing strength of ice.

Figure 2 shows velocities and ice forces during the icebreaking process. The parameters v_n^{rel} and v_τ^{rel} are the normal and tangential components of the relative velocity between hull and ice along the hull contact nodes. The parameters $v_{n,1}^{rel}$ and $v_{n,2}^{rel}$ are the components of v_n^{rel}, along and perpendicular to the contact surface in the vertical plane, respectively. f_H and f_V are the horizontal and vertical components of the frictional force, respectively. The combined crushing force and the component of the frictional force can be divided into the horizontal component F_H and vertical component F_V. Forces mentioned above can be expressed in Su et al.'s study, where, μ_i is the frictional coefficient between hull and ice, φ is the slope angle of hull, α is the water entrance angle.

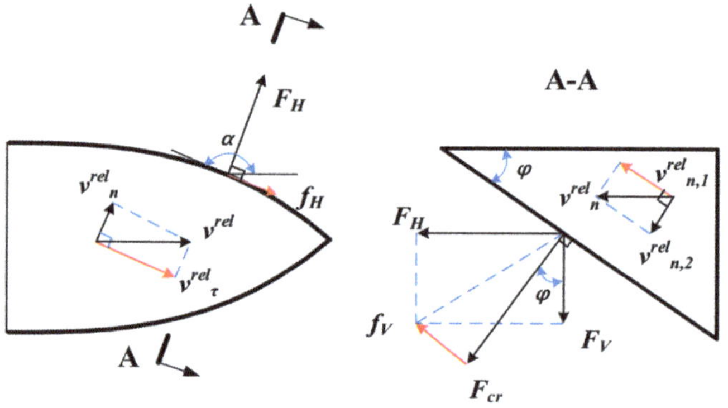

Figure 2. Forces and velocities during interaction.

According to Zhou et al. [26], the failure patterns of ice vary under the influence of different hull slope angles. Friction coefficient will affect the slope angle to distinguish the failure patterns. The difference in failure patterns will lead to different sizes of ice wedge and different ice forces, especially at the shoulder parts of the ship where crushing failure of ice usually occurs in turning or transverse towing operations, which should be distinguished in the simulation. Considering the ice-structure friction coefficient as 0.0394, a limit angle of 87° is chosen to separate the different ice-failure patterns.

2.1.2. Bending-Failure Pattern

When slope angle φ of hull is less than the limit angle, and the vertical component F_V of crushing force and friction force is large enough to exceed the ice-bearing capacity, bending failure occurs. Ice will break from the ice sheet as an intact ice wedge. In the simulation, the waterline of ship is represented with a bunch of discrete nodes which form a closed polygon as the hull; level-ice boundary is also distinguished into discrete nodes. The interaction between ship and ice is transformed into the interaction between a closed polygon and discrete nodes. Whether the ice boundary consisting of discrete nodes enters the closed polygon at each time step in the simulation is detected; contact area and other parameters are collected through the program.

The vertical component F_V increases with the increase in contact area. When F_V meets the ice-bearing capacity, initial ice boundary around the contact area will be updated to a new ice wedge. Due to the distribution of low slope angle, bending failure always occurs around the bow area. Ice force at this time step is then calculated and stored in the program for calculation in the next step. According to the circumferential crack method mentioned above, when bending failure happens to level ice under the action of icebreaker hull, the opening angle and the radius of an ice wedge formed due to bending failure can use parameters θ and R in expression (1) to present. The top part of the ice wedge is under the load in vertical direction. The bearing capacity of ice P_f can then be introduced based on reference [30]:

$$P_f = C_f(\theta/\pi)^2\sigma_f h_i^2 \quad (4)$$

where σ_f is the flexural strength of ice, h_i is the ice thickness, C_f is the empirical parameter. According to the previous work conducted by Su et al. [31], the value of C_f is 2.9.

2.1.3. Crushing-Failure Pattern

When the slope angle φ exceeds limit angle and the icebreaking condition is achieved, according to ISO/FDIS 19906-2019 [32] and without the limit mechanism, the localized crushing force acting on the hull can be expressed as:

$$F_{cr} = P_G \cdot A_c \quad (5)$$

$$P_G = C_R\left[(h/h_1)^n(L/h)^m + f_{AR}\right] \quad (6)$$

$$C_R = 2.8/2.6\sigma_c \quad (7)$$

$$f_{AR} = \exp[-L/(3h)]\sqrt{1+5(h/L)} \quad (8)$$

where P_G is the global average ice pressure; L is the projected width of the hull; h is the ice thickness; h_1 is the reference thickness of 1 m; m and n are empirical parameters, the value of m is -0.16 and n is equal to $-0.5 + h/5$ for $h < 1.0$ m and to -0.3 for $h > 1.0$ m; and C_R is the ice-strength coefficient, which changes under different regions.

As shown in Figure 3, the length of the contact area projected on the hull structure indicated as a red line is calculated by program at each time step, and increases with the advance of icebreaker; when the crushing length L reaches the predefined limitation of crushing length, ice will break from the sheet. In the program, the initial ice boundary will be replaced by the updated ice boundary represented as red curve; ice force at this time step will be calculated and stored. According to the model test, the predefined limitation of crushing length we choose as the crushing failure limit condition is ch, where c is a constant. The limitation of crushing length increases with the increase in ice thickness, which indicates that ships need to reach wider crushing lengths in order to break thicker ice.

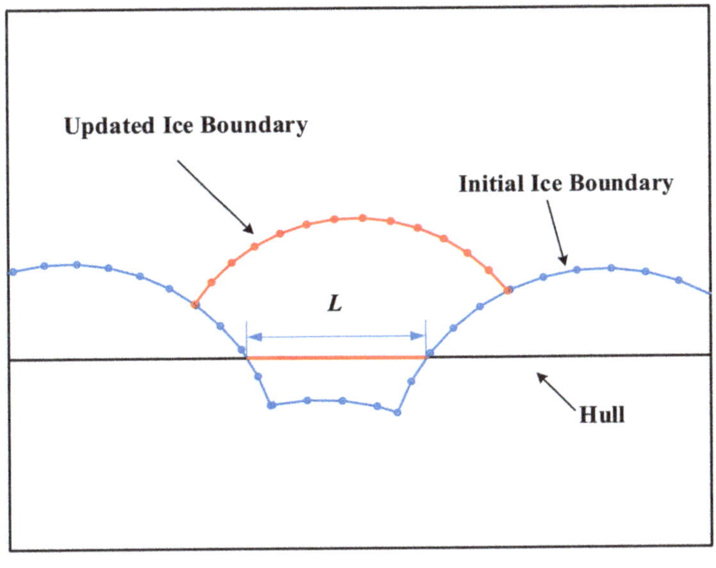

Figure 3. Ice-crushing failure.

2.2. Excitation Forces and Moments

The excitation forces and moments involved in simulation mainly include the ice force $F^{ice}(t)$, generated due to the interaction between hull and level ice. The detailed calculation formula of each part of ice force is explained below.

Ice force includes the continuous ice-breaking forces $F^{brk}(t)$ and ice-submerging forces. The total continuous icebreaking forces are calculated based on the local icebreaking forces, which change under the different parts of the hull at each time step. Local icebreaking forces are calculated and collected with ice-failure mechanism summarized above. The ice-submerging forces are calculated according to ice-force formula [1]; then the ice force in general can be expressed as:

$$F_1^{ice} = F_1^{brk} + R_S\left[1 + 9.4\left(v^{rel}/\sqrt{gL}\right)\right] * \left(v_1^{rel}/v^{rel}\right) \tag{9}$$

$$F_2^{ice} = F_2^{brk} + R_S\left[1 + 9.4\left(v^{rel}/\sqrt{gL}\right)\right] * \left(v_2^{rel}/v^{rel}\right) \tag{10}$$

$$F_6^{ice} = F_6^{brk} \tag{11}$$

where v^{rel} is the relative velocity between ice and hull; v_1^{rel} is the forward component of v^{rel}; v_2^{rel} is the transverse components of v^{rel}; L is the waterline length of the ship; and R_S is the submersion component of ice force [1], which can be written as:

$$R_S = (\rho_w - \rho_i)gh_iB\left(T\frac{B+T}{B+2T}\right) + \mu\left[\left(0.7L - \frac{T}{tan\varphi}\right) - \frac{B}{4tan\alpha}\right] + Tcos\varphi \cdot cos\psi\sqrt{(1/sin^2\varphi) + (1/tan^2\alpha)} \tag{12}$$

where ρ_w is the sea density; ρ_i is the density of the level ice; B is the beam; T is the draught; μ is the frictional coefficient between ice and hull; φ is the stem angle of the ship; and α is the water entrance angle, $\psi = \arctan\left(\frac{tan\varphi}{sin\alpha}\right)$.

After obtaining each component of ice force, the contact area and position between hull and ice is processed by computer at each time step in the simulation. Ice force and updated level-ice boundary can be used as input for the next time step; then ice channels created by ship and ice resistance in level ice can be achieved, as well as the time series curves of ice force.

2.3. Nataf Transformation

According to ice-property measurements along R/V Xuelong's scientific sailing and ice sites in polar areas, it can be found that ice thickness follows distribution of Burr XII, while ice density follows Beta distribution. Gamma distribution and GEV distribution can be applied to describe the distribution of ice flexural strength and ice effective modulus, respectively [25]. Based on the distribution of ice parameters and correlation coefficient matrix, we can use Nataf transformation to generate pseudo-random numbers as corresponding ice parameters. These ice parameters can be used in the numerical simulation to calculate the ice resistance exposed to R/V Xuelong in polar areas.

The Nataf transformation is a tool to rebuild joint distribution once the distribution of input parameters is obtained. We can obtain ice properties by following specific distribution through inverse Nataf transformation. In Nataf transformation, $X = [x_1, x_2, \ldots, x_n]$ is the n-dimensional random variable. The probability density function and cumulative distribution function of random variable x_i are $f_i(x_i)$ and $F_i(x_i)$, respectively. According to literatures [33,34], the relationship between random variable $X = [x_1, x_2, \ldots, x_n]$ and standard normal random variable $Y = [y_1, y_2, \ldots, y_n]$ can be described as:

$$\begin{cases} \phi(y_i) = F_i(x_i), \\ y_i = \phi^{-1}(F_i(x_i)), \end{cases} i = 1, 2, \ldots n, \tag{13}$$

where $\phi(\cdot)$ is the cumulative distribution function of standard normal distribution, assuming that correlation coefficient matrix of X and Y are ρ and ρ_0, respectively. According to Nataf transformation, elements ρ_{ij} and ρ_{0ij} (i ≠ j) at corresponding positions in matrix follow the below relationship:

$$\begin{aligned}\rho_{ij} &= \int_{-\infty}^{+\infty}\int_{-\infty}^{+\infty} \frac{x_i-\mu_i}{\sigma_i}\cdot\frac{x_j-\mu_j}{\sigma_j} f_{ij}(x_i,x_j)dx_i dx_j \\ &= \int_{-\infty}^{+\infty}\int_{-\infty}^{+\infty} \frac{F_i^{-1}(\phi(y_i))-\mu_i}{\sigma_i}\cdot\frac{F_j^{-1}(\phi(y_j))-\mu_j}{\sigma_j}\phi_2(y_i,y_j,\rho_{0ij})dy_i dy_j\end{aligned} \quad (14)$$

where, μ_i and μ_j are the expectation of x_i and x_j, respectively; σ_i and σ_j are the standard deviation of x_i and x_j, respectively; and $\phi_2(y_i,y_j,\rho_{0ij})$ is the joint distribution function of two-dimensional standard normal random variables y_i and y_j with correlation coefficient of ρ_{0ij}.

Correlation coefficient ρ_{ij} of ice properties is shown in Table 1. Correlation coefficient ρ_{0ij} of standard normal random variable is calculated by solving the root of nonlinear function in program; the results are shown in Table 2.

Table 1. Correlation coefficient ρ_{ij} between ice parameters.

Ice Parameters	Thickness	Density	Flexural Strength	Effective Modulus
Thickness	/	0.2	0	0
Density		/	−0.3	−0.3
Flexural strength			/	0.9
Effective modulus				/

Table 2. Correlation coefficient ρ_{0ij} between standard normal random variable.

Correlation Coefficient	ρ_{011}	ρ_{012}	ρ_{013}	ρ_{014}
ρ_{011}	/	0.2072	0	0
ρ_{012}		/	−0.3173	−0.3281
ρ_{013}			/	0.9794
ρ_{014}				/

The correlation coefficient matrix ρ_0 can be expressed using Cholesky factorization:

$$\rho_0 = L_0 L_0^T \quad (15)$$

where L_0 is lower triangular matrix of ρ_0 after Cholesky factorization, and T represents matrix transposition. Correlation standard normal random variables Y can be transformed into independent standard normal random variables U through L_0:

$$U = L_0^{-1} Y \quad (16)$$

Nataf transformation is then established. Inverse Nataf transformation of independent standard normal random variables U transformed into correlation standard normal random variable Y can be expressed as:

$$Y = L_0 U \quad (17)$$

$$x_i = F_i^{-1}(\phi(y_i)) \; i = 1,2,\ldots n. \quad (18)$$

N-dimensional random variable $X = [x_1, x_2, \ldots, x_n]$ that presents different ice parameters can be generated and obtained; 1000 samples are generated for each ice parameter.

The distribution of each generated ice parameter follows the same distribution as revealed in [25], with a small margin of error. Ice thickness follows the distribution of Burr XII, with parameters of a = 145.1, c = 3.5, and k = 3.1; compared to a = 156.5, c = 3.4, and k = 4.4 in the previous study. Ice density follows the distribution of Beta with parameters of a = 15.4 and b = 3.6, while a = 15.2 and b = 3.6 in the previous study. Ice flexural strength follows the distribution of Gamma with the parameters of a = 19.8 and b = 13.1, while a = 19.3 and b = 13.4 in the previous study. Ice effective modulus follows the distribution of GEV with the parameters of k = −0.02, c = 0.3, and μ = 1.0; while k = −0.01, c = 0.3, and μ = 1.0 in the previous study.

Due to sea-ice-failure mechanism considering the crushing failure of sea ice in ice model, the crushing strength of sea ice is required; however, many previous studies revealed only the distribution of crushing strength. Quantifying the relationship of crushing strength with other ice parameters is a difficult task and has not been clearly studied. We used other methods to generate this key parameter. According to Timco and Weeks's research [7], the ratio of crushing strength and flexural strength of sea ice is 0.5–5, while model scale measurements show a ratio of 1.5–2.7 in model test by Zhou et al. [5], which narrows this range. The ratio of crushing strength and flexural strength we choose is 2.12, which is the average ratio of 10 model test cases. Crushing strength follows the distribution of Gamma, with the parameter of a = 19.8, b = 27.8.

3. Sensitivity Analysis of Ice Parameters

3.1. Numerical Simulation Setup

The hull used in the simulation is the icebreaker R/V Xuelong, which undertook the scientific expedition and the field measurements. This vessel was built in 1993 and has the capability to break level ice with a thickness of 1.2 m and snow of 0.2 m at 0.8 m/s. The main parameters of the ship and ice parameters are shown in Table 3.

Table 3. The main parameters of the simulation.

Parameter	Value
Length (m)	147
Breadth (m)	23
Draught (m)	8
Waterline entrance angle (°)	20
Stem angle (°)	24
Ice–hull kinetic friction coefficient	0.0394
Poisson ratio	0.3
Sea-water density (kg/m^3)	1025

The ship was set to sail in level ice with a fixed forward velocity of 0.8 m/s. The ship sailed for 200 s in total, and the time interval was 0.1 s in the simulation. The ice thickness in one case was constant. At the initial stage, the ship kept a certain distance away from the ice and was going to collide at the boundary of level ice. The random ice boundary settled around the hull; however, there was no contact, as shown in Figure 4.

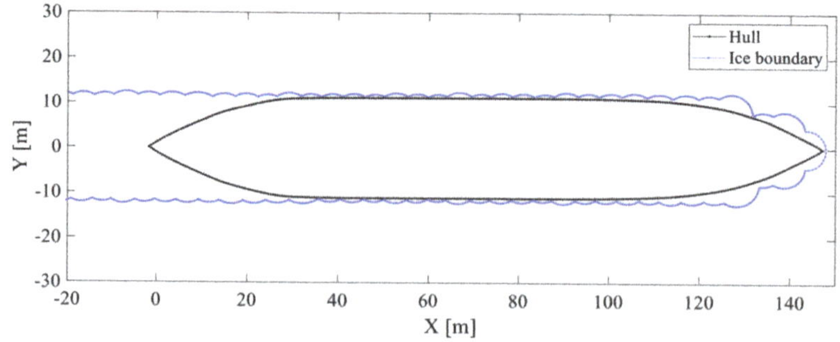

Figure 4. Ship in level ice at initial stage.

The ice boundary is represented by the blue dotted line. It distinguishes the open water from the level ice, and between the two parallel blue curves is the ice channel created for the icebreaker. The waterline of the ship is represented by the black dotted line. With ship parameters and generated pseudo-random numbers, a simulation can be performed to predict ice-resistant ship encounters.

3.2. *Sensitivity Analysis*

In order to deeply reveal the relationship of each ice parameter (ice thickness, flexural strength, crushing strength, effective modulus, and ice density) with ice resistance, sensitivity analysis on parameters was carried out. Based on the initial case, we changed the value of one parameter at a time; meanwhile, other parameters remained the same. It should be noted that the ice parameters selected in the initial case are summarized by the ship's trial in the Arctic in order to achieve appropriate representativeness, while such a goal is hard to reach via pseudo-random numbers. A total of eight cases were conducted for each parameter. The parameters of the initial case and the corresponding changing intervals are listed in Table 4.

Table 4. Parameters of 001 and responding standard deviation.

	Ice Thickness (m)	Flexural Strength (kPa)	Crushing Strength (kPa)	Effective Modulus (GPa)	Ice Density (g/cm^3)
Initial case	1	500	2300	5.4	0.88
Changing interval	0.1	50	200	0.4	0.02

3.2.1. Ice-Thickness Sensitivity Analysis

We conducted sensitivity analysis on ice thickness first. Ice thickness is one of the most intuitive parameters affecting the icebreaking process. The changing range of ice thickness is from 0.6 m to 1.3 m, with intervals of 0.1 m. The trend of ice resistance under rising ice thickness is shown in Figure 5.

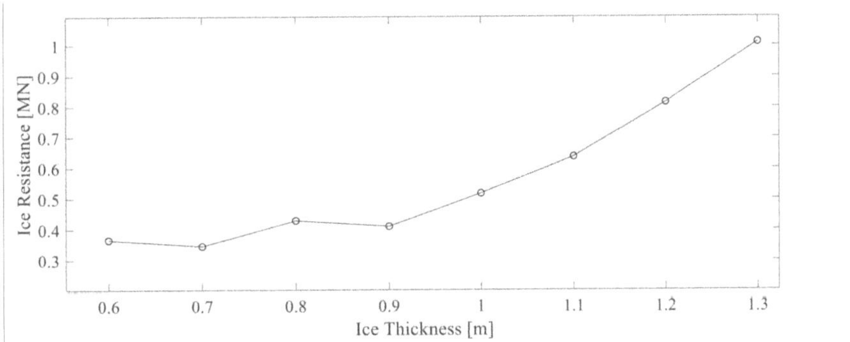

Figure 5. Ice resistance under different ice thickness.

As shown in Figure 5, the trend of ice resistance roughly increases with the increase in ice thickness; the trend fluctuation is small. Ice resistance increases rapidly when ice thickness exceeds 0.9 m. Ice thickness has a great influence on ice resistance; ice resistance ranges from 0.34 MN to 1.01 MN—a great span of 0.67 MN—while ice thickness ranges from 0.6 m to 1.3 m.

The reason why a strict upward trend was not presented may be that the change in ice thickness will lead to a change in failure size and bearing capacity of sea ice; this will make interactions between the hull and the ice more complex, so certain unclear trends in simulation are acceptable.

3.2.2. Flexural Strength Sensitivity Analysis

The sensitivity analysis of flexural strength was then carried out, the parameters of which are involved in the bearing capacity of ice in the ice model. The changing range of flexural strength is from 300 kPa to 650 kPa, with intervals of 50 kPa. The trend of ice resistance under rising flexural strength is shown in Figure 6.

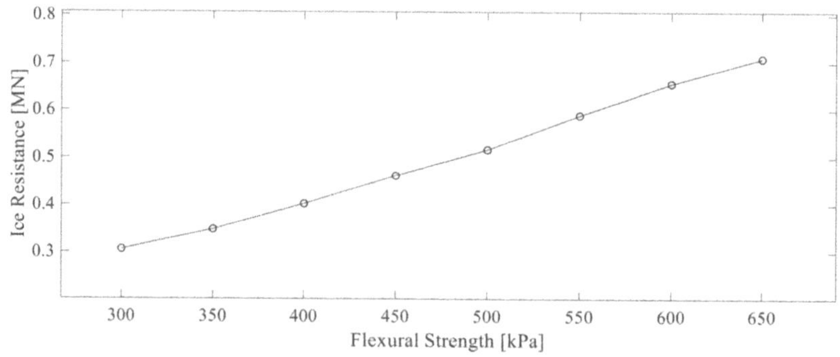

Figure 6. Ice resistance under different flexural strength.

Ice resistance increases strictly with the rise in flexural strength, as shown in Figure 6. Ice resistance ranges from 0.30 MN to 0.70 MN. The span of ice resistance is 0.40 MN. This parameter is related to the bearing capacity of ice in the ice model; the increase in flexural strength will make sea ice harder to break, which matches the total trend of resistance.

3.2.3. Crushing Strength Sensitivity Analysis

The changing range of crushing strength is from 1500 kPa to 2900 kPa, with intervals of 200 kPa. The trend of ice resistance under rising crushing strength is presented, shown in Figure 7.

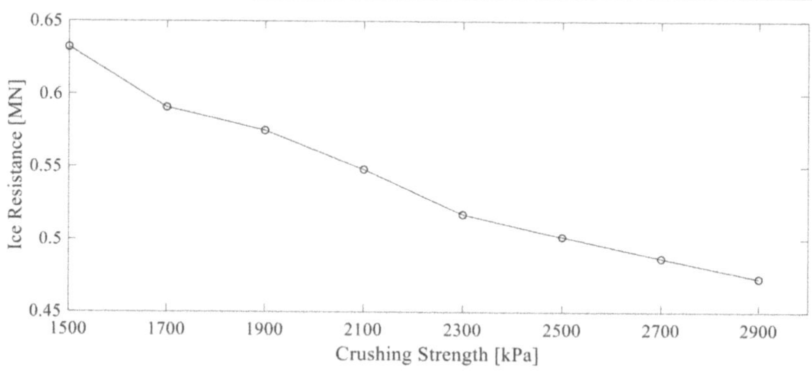

Figure 7. Ice resistance under different crushing strength.

Ice resistance decreases as crushing strength increases. Crushing strength has little impact on ice resistance compared to other parameters; the presented curve is flat. The changing range of ice resistance is from 0.632 MN to 0.473 MN; the span of ice resistance is 0.159 MN.

3.2.4. Effective Modulus Sensitivity Analysis

The changing range of effective modulus is from 3.8 GPa to 6.6 GPa, with intervals of 0.4 GPa. The trend of ice resistance under rising effective modulus is shown in Figure 8.

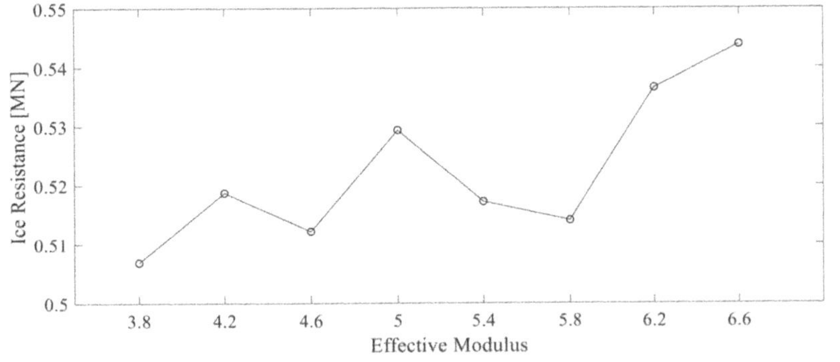

Figure 8. Ice resistance under different effective modulus.

The trend of ice resistance under changing effective modulus is hard to describe and an overall fluctuating trend is presented, as shown in Figure 8. The influence of a different effective modulus is quite small. Ice resistance oscillates at around 0.52 MN, ranging from 0.506 MN to 0.543 MN. The span of ice resistance is 0.037 MN.

3.2.5. Ice-Density Sensitivity Analysis

Finally, the sensitivity analysis of ice density was studied. The changing range of ice density is from 0.80 g/cm^3 to 0.94 g/cm^3, with intervals of 0.02 g/cm^3. The trend of ice resistance under rising ice density is shown in Figure 9.

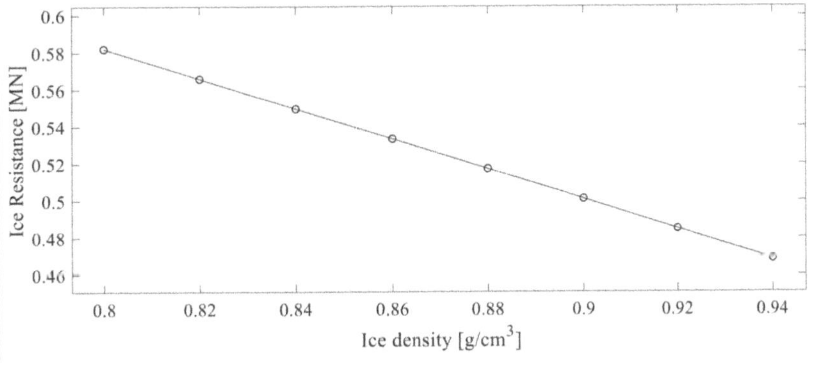

Figure 9. Ice resistance under different ice density.

Ice resistance decreases as ice density increases, as shown in Figure 9. The changing rate of this parameter is stable. Ice resistance ranges from 0. 582 MN to 0.468 MN, with a span of 0.114 MN; the impact of ice density on ice resistance is stable, and this parameter only affects the ice-submerging force; such a trend can be explained in simulation.

3.2.6. Influence of Different Ice Parameters on Resistance

The influence of different ice parameters on resistance is studied based on the initial case. The dimensionless number is introduced to study the extent of influence, shown in Figure 10.

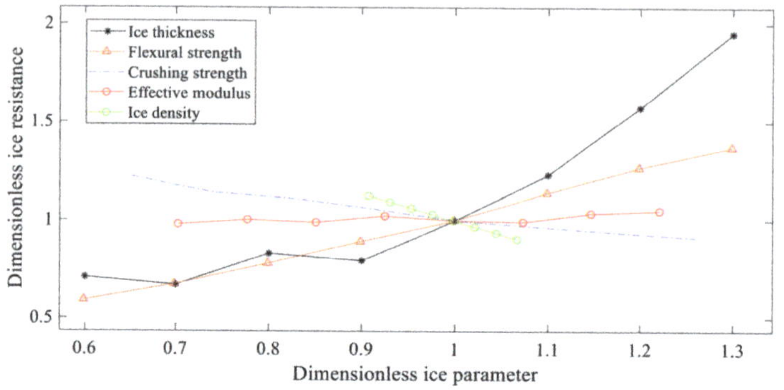

Figure 10. Comparison of dimensionless ice resistance.

As shown in Figure 10, different ice parameters are represented by different colors and shapes; their relationships with ice resistance are revealed by dimensionless numbers. The extent of influence of ice parameters can be reflected directly by the changing rate of curves. Ice thickness, represented by the black line, has the greatest positive influence on resistance among all parameters. Ice density, represented by the green line, has a great effect on resistance, second only to ice thickness; however, such effect is negative. Flexural strength, represented by orange triangles, shows the third-greatest impact on ice resistance. According to the figure, effective modulus, represented by the red line, has the least effect on the result; the curve is almost flat. Finally, crushing strength, represented by the blue dotted line, has the second-smallest impact on the results.

4. Analysis of Numerical Simulation Results

Considering the calculation time required for simulation and efficiency, 400 sets of cases using generated ice parameters were carried out in order to study the distribution pattern of ice resistance. After completing all 400 cases, the ice parameters and corresponding ice resistance of five random cases were collected and are listed in Table 5 to study individual cases.

Table 5. Ice parameters and corresponding ice resistance.

Case	Ice Thickness (m)	Flexural Strength (kPa)	Crushing Strength (kPa)	Effective Modulus (GPa)	Ice Density (g/cm^3)	Ice Resistance (MN)
001	1.12	257.78	546.48	1.18	0.88	0.62
002	1.63	341.37	723.71	1.73	0.79	2.00
003	0.27	327.97	695.30	1.45	0.73	0.15
004	1.23	247.16	523.97	1.18	0.77	0.85
005	1.04	236.55	501.48	0.97	0.87	1.22

According to the sea-ice-failure mechanism, a difference in ice properties will not only affect ice resistance directly, but also lead to different ice-failure shapes created by the ship or the breaking frequency of sea ice. The detailed icebreaking process is studied to reveal such a phenomenon. The ice channel created during the icebreaking process and the corresponding ice force encountered by the ship in the surge direction are shown in Figures 11 and 12.

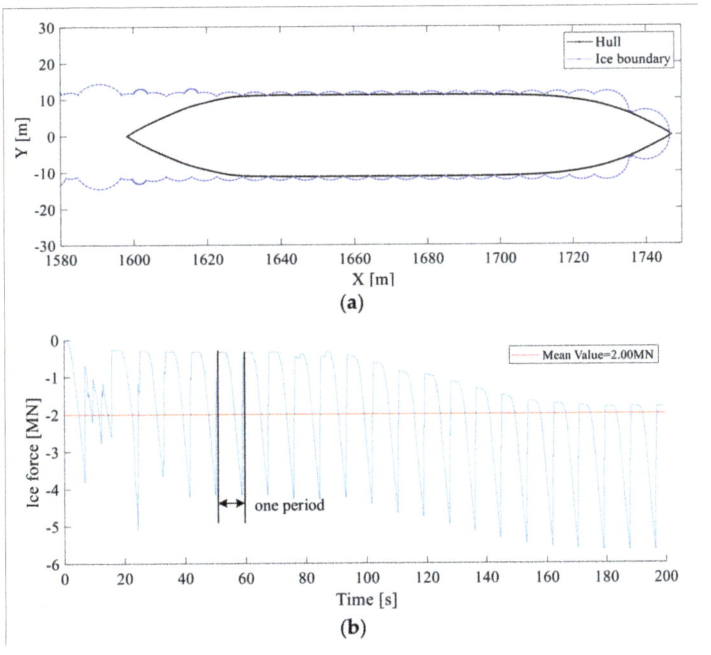

Figure 11. Simulation results of Case 002. (**a**) Ice channel created in the simulation; (**b**) Ice force in surge direction.

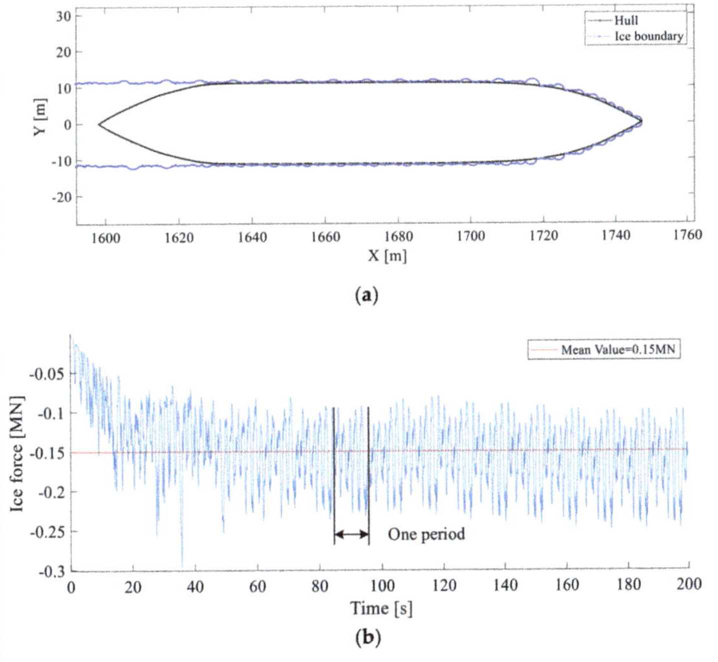

Figure 12. Simulation results of Case 003. (**a**) Ice channel created in the simulation; (**b**) Ice force in surge direction.

As shown in Figures 11 and 12, two typical cases are studied. The simulation results of two cases are significantly different, with a span of 1.85 MN. The radius of the ice wedge induced by ice–hull interaction in case 2 is larger than that of case 3, according to Figures 11a and 12a. The number of ice wedges created during the contact in case 3 is more than that of case 2. One period in simulation is identified by the corresponding maximum value of ice force; then, ice-force time history curves show that ice force duration in the period of case 3 is similar to that of case 2, which is about 8.6 s in case 2 and 9 s in case 3. However, one period of duration in case 3 can be separated into several load events due to the increasing number of ice wedges and the weakening ice-bearing capacity. Both cases indicate that the peak value of ice force will flatten over time, and the icebreaking process will follow a regular pattern.

The ice-resistance distribution of 400 sets of cases is shown in Figure 13. Ice resistance ranged from 0.07 to 3.42 MN, and the average ice resistance was 0.74 ± 0.40 MN with a median value of 0.70 MN. The Burr distribution turned out to be the most suitable model to describe the distribution of ice resistance obtained, which was also accepted by K–S test. The parameters of Burr distribution are estimated as a = 0.90, c = 2.92, and k = 1.91, with standard errors of 0.09, 0.18, and 0.38, respectively.

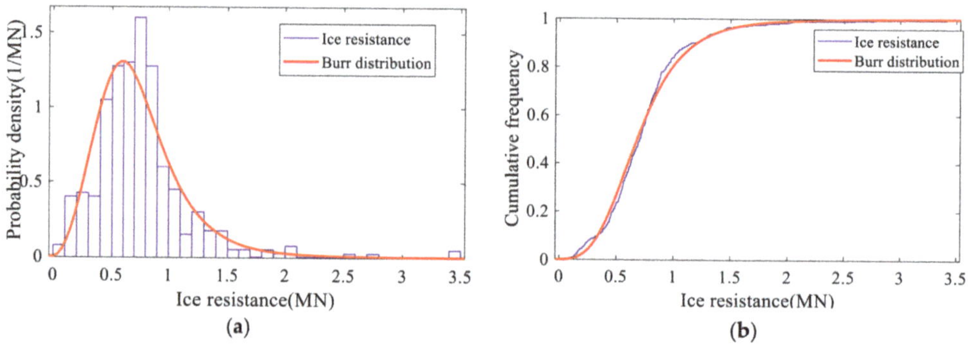

Figure 13. Ice-resistance distribution based on present method. (**a**) The probability density curve; (**b**) The cumulative frequency curve.

In addition, based on the rain flow counting method [35], we processed all of the recorded ice force in the surge direction and obtained the amplitude of ice force. A total of 66,599 ice force amplitudes were extracted from 400 cases. These ice-load amplitudes are more convincing than ice resistance in regard to hull safety, since ice resistance only reflects the average value of ice force in the surge direction during one sail, whereas a large amount of ice force exceeds that value in the icebreaking process. If we focused only on the average value and ignored the high value of ice force, the ice force could be underestimated and cause damage to the ship during sailing in polar areas. The ice-force amplitudes and probability density curves of all ice-force amplitudes are shown in Figures 14 and 15.

As shown in the figure, 91.3% of ice-force amplitudes are concentrated at a lower level of 0–1 MN. The number of ice-force amplitudes decreases rapidly when the amplitude exceeds 1 MN, and the number of ice-force amplitudes exceeding 2 MN is quite rare, close to 2.3%. The maximum ice-force amplitude is 8.845 MN in the study; extreme ice force amplitude may be caused by extreme ice parameters. Such situations cannot be ignored, because these situations still can happen in real life, albeit with a low probability. It was concluded that the ice force encountered by ships seldom exceeds 2 MN in most navigation scenarios; however, some extreme cases are still worth a deep study in the future.

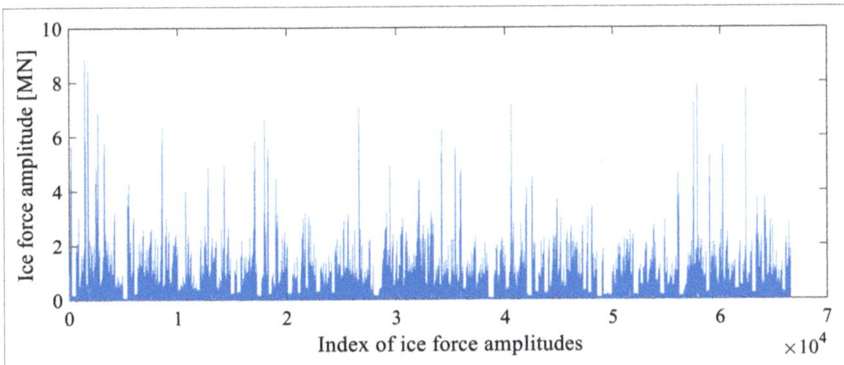

Figure 14. Extracted ice-force amplitudes.

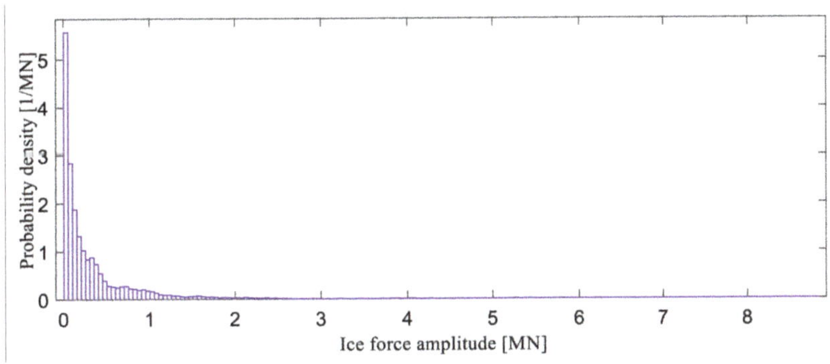

Figure 15. The probability density curve of extracted ice-force amplitudes.

5. Comparison and Discussion

Figure 16 shows a comparison of the probability density curve and cumulative frequency curve, presented by Wang et al. [25] and simulated with the present method.

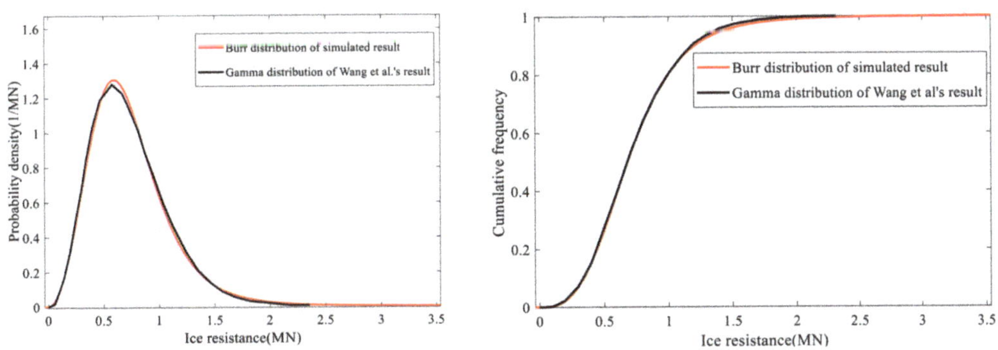

(**a**) (**b**)

Figure 16. Comparison of ice-resistance distribution (**a**) The probability density curve; (**b**) The cumulative frequency curve.

The ice resistance calculated with the present numerical tools, represented by the red line, ranges from 0.07 to 3.42 MN, and the average ice resistance is 0.74 ± 0.40 MN with a median value of 0.70 MN, while ice resistance ranges from 0.06 to 2.69 MN, and the average ice resistance is 0.73 ± 0.34 MN with a median value of 0.68 MN by the Monte Carlo method, represented as black line. The differences in average value and median value between two results are quite small, and two probability density curves show the same positively skewed bell shape. Both distributions reveal that ice resistance tends to be concentrated in the area of low resistance, around 0.4–0.8 MN. The probability decreases significantly when the resistance exceeds about 1.5 MN. In addition, the upper limit of ice resistance is higher than that of the Monte Carlo method, which is also reflected in the figures; the value of the upper limit is 3.42 MN in the present numerical simulation. Ice resistance obtained through numerical simulation has high probability in the range of 0.7–0.8 MN, while ice resistance calculated by Wang has high probability in the range of 0.5–0.6 MN.

Although the probability density curves of the two results are quite similar, results follow different distributions due to the diversity in results. Gamma distribution is the most fitting model for the Monte Carlo method's results, with parameters of a = 4.6 and b = 0.2, and corresponding standard errors of 0.06 and 0.002, respectively. The Burr distribution turned out to be the most suitable model to describe the distribution in the present research. Reasons for such differences can be summarized as follows: crushing failure occurrences on the steep hull surface and geometric contact between the ice and hull were considered in the numerical simulation, while the Monte Carlo method was based on semi-empirical formulae taken before. In addition, 400 cases were performed due to the computational efficiency of the program, which may have led to errors caused by a small number of samples. Although there are no full-scale measured results to compare with the simulated results, the ice resistance of M/V Tianen based on measurements was estimated to verify the results [25]. The probability density curve shows a positively skewed bell shape, which matches with the probability density curves of both results. With the ice parameters of one area, the ice resistance of the area may be obtained and the ship safety operation could be achieved.

Sensitivity analysis of ice parameters conducted indicates that ice resistance shows a rising trend with the increase in ice thickness and flexural strength in general. Ice resistance decreases with the decrease in crushing strength and ice density, while ice resistance shows an unclear and low-changing range trend under the influence of effective modulus. Changing in ice thickness and flexural strength will affect ice resistance significantly. Crushing strength and effective modulus shows a small impact on ice resistance. Moreover, the rain flow counting method is adopted to obtain the amplitude of ice force, which provides more detailed information to study the actions of ice force on structures than the Monte Carlo method.

6. Conclusions

Ice parameters following certain distributions are generated. A numerical simulation method based on circumferential crack assumption is then applied to calculate the ice load. The ice-failure pattern is refined into bending and crushing effects; then, the trigger conditions of failure patterns are introduced. Simulated results show more detailed information than the Monte Carlo method; ice force at each time step is obtained, key ice force amplitude is extracted, and significant differences in sea-ice break cracks are also observed.

The main findings based on numerical simulations and comparisons with another researcher are summarized as follows:

(1) The change in ice thickness and flexural strength will affect ice resistance significantly, while effective modulus has the least effect on ice resistance among all the parameters;
(2) Ice resistance obtained by pseudo-random numbers tends to be concentrated in the area of low resistance, around 0.4–0.8 MN. The probability decreases significantly when the resistance exceeds about 1.5 MN;
(3) Ice-force amplitude is concentrated at 0–1 MN. Only 2.3% of ice-force amplitudes exceed 2 MN. The maximum ice-force amplitude is 8.845 MN. The results of ice

amplitude could be a useful reference for ship design and structural analysis, and potential stress generated by sea ice could be predicted at an early stage;

(4) The distribution of ice resistance follows the Burr distribution in the simulation. The probability density curve shows the same positively skewed bell shape with results obtained by the Monte Carlo method, a trend matched well with measurements of another ship.

In the future, the combination of this study in ice resistance and real-time image processing may provide more detailed ice data for simulation in field operation, continuous accurate ice-force predictions could be achieved, and serious damage to ships may be avoided.

It should be noted that some assumptions made in the empirical formula, or errors occurred in the generation of ice parameters, will affect the simulation results. An insufficient number of samples also may lead to certain deviations.

Author Contributions: Conceptualization, L.Z. and S.D.; methodology, L.Z.; software, L.Z. and Y.G.; validation, Y.G.; formal analysis, L.Z.; investigation, Y.G.; resources, L.Z.; data curation, L.Z.; writing—original draft preparation, Y.G.; writing—review and editing, R.L.; visualization, Y.G.; supervision, L.Z. and S.D.; project administration, L.Z. and S.D.; funding acquisition, L.Z. All authors have read and agreed to the published version of the manuscript.

Funding: This research was funded by the National Natural Science Foundation of China, grant number 52171259, 51809124, 51911530156.

Data Availability Statement: All analyzed data in this study have been included in the manuscript.

Acknowledgments: The authors would like to thank the Jiangsu University of Science and Technology (JUST).

Conflicts of Interest: The authors declare no conflict of interest.

References

1. Lindqvist, G. A Straightforward Method for Calculation of Ice Resistance of Ships. In Proceedings of the International Conference on Port and Ocean Engineering under Arctic Conditions (POAC), Luleaa, Sweden, 12–16 June 1989; pp. 722–735.
2. Riska, K.; Wilhelmson, M.; Englund, K.; Leiviskä, T. *Performance of Merchant Vessels in the Baltic*; Research Report No 52; Winter Navigation Research Board: Espoo, Finland, 1997.
3. Su, B.; Riska, K.; Moan, T. A Numerical Method for the Prediction of Ship Performance in Level Ice. *Cold Reg. Sci. Technol.* **2010**, *60*, 177–188. [CrossRef]
4. Lubbad, R.; Løset, S. A numerical model for real-time simulation of ship–ice interaction. *Cold Reg. Sci. Technol.* **2011**, *65*, 111–127. [CrossRef]
5. Zhou, L.; Riska, K.; Moan, T.; Su, B. Numerical modelling of ice load on an icebreaking tanker: Comparing simulations with model tests. *Cold Reg. Sci. Technol.* **2013**, *87*, 33–46. [CrossRef]
6. Zhou, L.; Gao, J.L.; Li, D.Q. An engineering method for simulating dynamic interaction of moored ship with first-year ice ridge. *Ocean Eng.* **2019**, *171*, 417–428. [CrossRef]
7. Timco, G.W.; Weeks, W.F. A review of the engineering properties of sea ice. *Cold Reg. Sci. Technol.* **2010**, *60*, 107–129. [CrossRef]
8. Sinsabvarodom, C.; Chai, W.; Leira, B.J.; Høyland, K.V.; Naess, A. Uncertainty assessments of structural loading due to first year ice based on the ISO standard by using Monte-Carlo simulation. *Ocean Eng.* **2020**, *198*, 106935. [CrossRef]
9. Sodhi, D.S.; Chin, S.N. Indentation Tests Using Urea Ice and Segmented Indentors. In Proceedings of the 11th International Offshore Mechanics and Arctic Engineering Conference (OMAE), Calgary, AB, Canada, 7–12 June 1992; Volume 4, pp. 223–230.
10. Takeuchi, T.; Enoki, K.; Okamoto, S.; Saeki, H. On the Characteristics of Uniaxial Compressive Strength and Elastic Modulus of Natural Sea Ice. In Proceedings of the 3rd International Offshore and Polar Engineering Conference, Singapore, 6–11 June 1993.
11. Johnston, M.; Masterson, D.; Wright, B. Multi-year ice thickness: Knowns and unknowns. In Proceedings of the 20th International Conference on Port and Ocean Engineering under Arctic Conditions, Lulea, Sweden, 9–12 June 2009.
12. Strub-Klein, L.; Høyland, K.V. Spatial and Temporal Distributions of Level Ice Properties: Experiments and Thermo-Mechanical Analysis. *Cold Reg. Sci. Technol.* **2012**, *71*, 11–22. [CrossRef]
13. Strub-Klein, L. A statistical analysis of first-year level ice uniaxial compressive strength in the Svalbard area. *J. Offshore Mech. Arct. Eng.* **2017**, *139*, 011503. [CrossRef]
14. Ni, B.Y.; Chen, Z.W.; Zhong, K.; Li, X.A.; Xue, Y.Z. Numerical Simulation of a Polar Ship Moving in Level Ice Based on a One-Way Coupling Method. *J. Mar. Sci. Eng.* **2020**, *9*, 692. [CrossRef]

15. Wang, C.; Feng, Z.; Li, X.; Li, P. Analysis on ice resistance and ice response of ships sailing in brash ice. *Chin. J. Ship Res.* **2018**, *13*, 73–78. (In Chinese)
16. Zong, Z.; Zhou, L. A numerical study of hydrodynamic influence in structure-broken ice interaction. In Proceedings of the 16th National Congress on Hydrodynamics and 32nd National Conference on Hydrodynamics, Wuxi, China, 30 October 2021; pp. 526–531. (In Chinese)
17. Chai, W.; Leira, B.J.; Høyland, K.V.; Sinsabvarodom, C.; Yu, Z. Statistics of thickness and strength of first-year ice along the Northern Sea Route. *J. Mar. Sci. Technol.* **2021**, *26*, 331–343. [CrossRef]
18. Kujala, P.; Suominen, M.; Riska, K. Statistics of Ice Loads Measured on MT Uikku in the Baltic. In Proceedings of the 20th International Conference on Port and Ocean Engineering under Arctic Conditions (POAC 09), Luleå, Sweden, 9–12 June 2009.
19. Suominen, M.; Kujala, P. *Analysis of Short-Term Ice Load Measurements on Board MS Kemira during Winters 1987 and 1988. Series AM. Report AALTO-AM-22*; Aalto University, School of Science and Technology: Espoo, Finland, 2010.
20. Suyuthi, A.; Leira, B.J.; Riska, K. Short term extreme statistics of ice loads on ship hulls. *Cold Reg. Sci. Technol.* **2012**, *82*, 130–143. [CrossRef]
21. Suyuthi, A.; Leira, B.J.; Riska, K. Statistics of local peaks on ship hulls. *Struct. Saf.* **2012**, *40*, 1–10. [CrossRef]
22. Kujala, P.; Vuorio, J. On the statistical nature of the ice-induced pressures measured on board I.B. In Proceedings of the POAC 1985, Narssarssuaq, Greenland, 8–13 September 1985; pp. 823–837.
23. Suominen, M. Statistical Modeling of Ice Induced Loads on a Ship Hull in the Baltic Sea. Master's Thesis, Aalto University, Espoo, Finland, 2011.
24. Ji, S.Y.; Lei, R.B.; Li, C.H.; Cai, K. Measurement of ice-induced local vibration of R/V Xuelong icebreaker during its navigation in ice-covered fields. *Chin. J. Polar Res.* **2017**, *29*, 427–435. (In Chinese)
25. Wang, Q.K.; Zong, Z.; Lu, P.; Zhang, G.Y.; Li, Z.J. Probabilistic estimation of level ice resistance on ships based on sea ice properties measured along summer arctic cruise paths. *Cold Reg. Sci. Technol.* **2021**, *189*, 103336. [CrossRef]
26. Zhou, L.; Riska, K.; Ji, C. Simulating transverse icebreaking process considering both crushing and bending failures. *Mar. Struct.* **2017**, *54*, 167–187. [CrossRef]
27. Zhou, L.; Diao, F.; Song, M.; Han, Y.; Ding, S.F. Calculation Methods of Icebreaking Capability for A Double Acting Polar Ship. *J. Mar. Sci. Eng.* **2020**, *8*, 179. [CrossRef]
28. Nataf, A. Determination des distribution dont les marges sont donnees. *C. R. Acad. Sci.* **1962**, *225*, 42–43. (in French).
29. Wang, S. *A Dynamic Model for Breaking Pattern of Level Ice by Conical Structures*; Department of Mechanical Engineering, Helsinki University of Technology: Helsinki, Finland, 2001.
30. Kerr, A.D. The bearing capacity of floating ice plates subjected to static or quasi-static loads. *J. Glaciol.* **1976**, *17*, 229–268.
31. Su, B.; Riska, K.; Moan, T. Numerical Simulation of Ship Turning in Level Ice. In Proceedings of the ASME 2010 29th International Conference on Ocean, Offshore and Arctic Engineering, Shanghai, China, 6–11 June 2010; pp. 751–758.
32. ISO/FDIS 19906. *Petroleum and Natural Gas Industries—Arctic Offshore Structures*; International Standard Organisation: Geneva, Switzerland, 2019; pp. 199–200.
33. Der Kiureghian, A.; Liu, P.L. Structural reliability under incomplete probability information. *J. Eng. Mech.* **1986**, *112*, 85–104. [CrossRef]
34. Liu, P.L.; Der Kiureghian, A. Multivariate distribution models with prescribed marginals and covariances. *Probab. Eng. Mech.* **1986**, *1*, 105–112. [CrossRef]
35. Matsuishi, M.; Endo, T. Fatigue of metals subjected to varying stress. *Jpn. Soc. Mech. Eng.* **1968**, *68*, 37–40.

water

Article

Experimental and Numerical Study on Ice Blockage Performance of Propeller in Cavitation Flow

Li Zhou [1], Sijie Zheng [1], Feng Diao [2], Shifeng Ding [1] and Junliang Gao [1,*]

1 School of Naval Architecture and Ocean Engineering, Jiangsu University of Science and Technology, Zhenjiang 212100, China; 201600000078@just.edu.cn (L.Z.); 209010123@stu.just.edu.cn (S.Z.); 15001945469l@163.com (S.D.)
2 China Ship Scientific Research Center, Wuxi 214000, China; diaofeng@702sh.com
* Correspondence: gaojunliang@just.edu.cn

Abstract: Cavitation greatly affects the ice blockage performance of propellers in polar areas, while the combined effect of cavitation and ice blockage on propellers has rarely been considered. In this work, the propeller model test in the cavitation tunnel and the viscous flow CFD numerical simulation based on RANS were conducted. In the cavitation tunnel test, the ice blockage model was simulated by a water-insoluble rectangular solid block, and the ice blockage was measured by the distance between the solid block and the propeller. The thrust and torque in tests and simulations were discussed under the uniform flow and ice blockage scenarios, as well as the variation of cavitation excitation force, pressure distribution of blades, cavitation characteristics and vortex intensity with advance coefficient when σ_n = 1.5, L/D = 0.15 in an ice blockage environment. The research shows that the numerical simulation results based on overlapping grids are in good agreement with the model test results, and the mean hydrodynamic errors are within 5%. In the uniform flow test, when the advance coefficient is small, the thrust and torque of the propeller will experience a sharp drop due to the influence of heavy cavitation. In the ice blockage test, the thrust and torque increase with the decrease of ice-propeller spacing, and the ice blockage becomes more serious as the cavitation grows. The propeller oscillates violently due to the cavitation excitation force, and the oscillation frequency increases with the increase of the advance coefficient. The cavitation is generated in the low-pressure area of the suction surface, and the cavitation shape captured in the present numerical simulation is consistent with the experimental phenomenon. Since the cavitation reduces the contact area between the water and the blade, the vortex strength will be reduced for the attachment of cavitation, and the vortex strength increases with the increase of the advance coefficient. This study will explore more hydrodynamic regularities with ice-class propellers in an ice blockage environment when cavitation occurs, and provide technical support for the design of propellers of polar ships.

Keywords: polar ships; propeller; ice blockage; cavitation effect; EFD-CFD

Citation: Zhou, L.; Zheng, S.; Diao, F.; Ding, S.; Gao, J. Experimental and Numerical Study on Ice Blockage Performance of Propeller in Cavitation Flow. *Water* **2022**, *14*, 1060. https://doi.org/10.3390/w14071060

Academic Editors: Francesco Gallerano and Giuseppe Pezzinga

Received: 27 February 2022
Accepted: 25 March 2022
Published: 28 March 2022

1. Introduction

With global warming and the accelerated melting of sea ice, polar oil and gas and waterway resources will gradually become valuable for development, and are becoming the focus of countries around the world [1]. As a key equipment of polar engineering, the polar ship has an important role in the exploration and exploitation of polar resources. When polar ships sail in an ice channel where ice and water are mixed, the ice may interact with the propeller, which will have a great impact on the safety of the structure and hydrodynamic performance of the propeller. The reason for the affected hydrodynamic performance is the ice blockage effect of the propeller; when the propeller rotates at high speed, the ice block will be sucked into the flow field before the propeller, making the propeller blocked by the sucked ice block, so the actual rotating speed of the propeller will be lower [2]. However the polar ship navigates in an ice field, the ice resistance is much greater than open-water resistance, so more power is needed to keep navigating speed. The power of the polar ship is obtained from the propeller, and the propeller needs to maintain a high rotating speed, which puts the propeller in a heavy load state [3]. When the propeller rotates at a high speed, the pressure of the field around the propeller will be decreased, which very easily causes the surface pressure of the blades to drop sharply. At the same time, cavitation will occur when the surface pressure of the blades is lower than the saturated vapor pressure. The appearance of cavitation will adversely affect the propeller propulsion performance, blade erosion, hull vibration and other aspects [4].

In order to study the hydrodynamic performance of the propeller at low cavitation numbers, scholars have performed relevant research on the cavitation tunnel test and numerical simulation of the propeller. As for experiments, Sampson et al. [5] conducted ice-propeller model tests with different cavitation numbers in the Emerson cavitation tunnel (ETC) and proved that the cavitation effect is an important factor affecting the hydrodynamic performance of the propeller, and the hydrodynamic performance of the propeller is also related to the propeller and ice shapes, working environment, motion, and so forth. Wu et al. [6–8] conducted a series of ice propeller cavitation performance tests in the cavitation water cylinder of the China Ship Scientific Research Center, and tested the blockage parameters and different cavitation numbers in the uniform flow and ice blocking environments.

With respect to numerical simulations, Rhee et al. [9] used the computational fluid dynamics (CFD) method combined with unstructured grids to study the cavitation performance of a marine propeller. Based on an incompressible Reynolds-averaged Navier–Stokes (RANS) equation, Liu et al. [10] utilized a single complete cavitation model to study the cavitation performance of an open-water propeller. Zhu et al. [11] adopted a RANS model based on unstructured grids to predict different cavitation phenomena of the propeller under the condition of uniform flow. Shin et al. [12] carried out numerical simulations on the cavitation performances for a conventional propeller and a skewed propeller located behind the ship. Ju [13] used a RANS solver to calculate the cavitation performances of an open-water propeller when located behind the ship, and the calculation results were good. Zheng [14] utilized Open FOAM to establish a numerical prediction method for the cavitation of a propeller behind the ship, and the numerical results were compared with the experimental values. Wu et al. [6] studied the influence of ice-propeller spacing on the thrust, torque and cavitation of a propeller by numerical calculations, and their numerical results were compared with the experimental data as well. However, the combined effect of cavitation and ice blockage on propellers has rarely been considered and its regularities (e.g., cavitation excitation force, pressure distribution of blades, cavitation characteristics and vortex intensity) are not investigated systematically in these works.

In this paper, the propeller tunnel test technique and CFD numerical simulation based on RANS were used to investigate the propeller cavitation, and the propeller model test data and numerical simulation results were compared and analyzed. The effect of cavitation on the open-water performance of the propeller in the uniform flow and ice blockage scenarios were studied, as well as the cavitation excitation force, the pressure distribution of the blades, the cavitation characteristics and the vortex strength. This study will explore more hydrodynamic regularities with the cavitation performance of the ice-class propeller in polar, provide technical support for the design of propellers, and accumulate experience for polar development and utilization.

2. Model Test

2.1. Test Model

Experiments for a propeller model under uniform flow and ice blockage conditions were carried out in a cavitation tunnel of the China Ship Research Center. The main parameters of the propeller model are shown in Table 1. The propeller material is aluminum alloy, and the rotation direction is right. According to the requirements of the marine industry standard for the roughness of the propeller model, the height of blade surface roughness is less than 1.6 μm [15].

Table 1. Main parameters of the propeller.

Parameters	Full Scale	Model Scale
Scale ratio λ	1	28
Diameter D (m)	7.0	0.25
Number of Blades Z	4	4
Pitch/Diameter Ratio $(P/D)_{0.7R}$	0.84	0.84
Blade area ratio A_E/A_0	0.75	0.75
Hub Diameter Ratio d_h/D	0.21	0.21

In order to simulate the ice blockage environment in the cavitation tunnel, an ice blockage model in front of the propeller model was designed and manufactured. The ice blockage was simulated by a water-insoluble rectangular solid block with a length of 430 mm, a width of 250 mm and a height of 125 mm. The axial relative position L between the ice blockage and the propeller disk was determined by the axial drive system. The installation diagram of the ice blockage and the propeller model in the cavitation tunnel is shown in Figure 1.

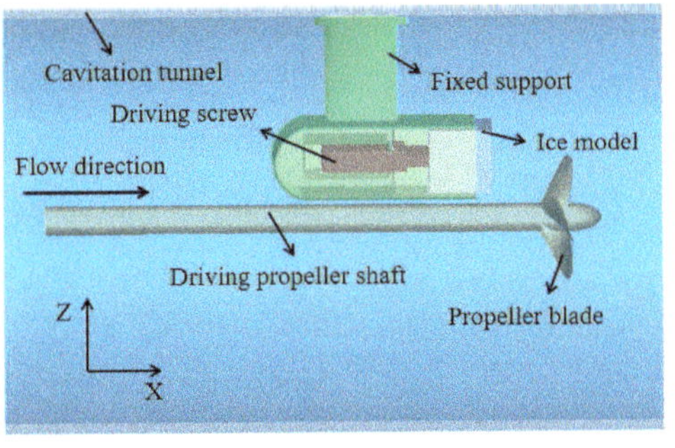

Figure 1. Schematic installation of the ice and propeller models in the cavitation tunnel.

2.2. Test Device

The diameter and the length of the tunnel in the working section are 0.8 m and 3.2 m, respectively. The flow speed range is 3~20 m/s, the pressure adjustment range is 8~400 kPa, and the minimum cavitation number is 0.15. Measuring instruments adopted in the test and their measuring range are listed in Table 2.

Table 2. Measurement instruments and their range.

Instruments	Model Number	Parameters	Range
Pressure transmitter	3151	Pressure measurement Velocity measurement	0–0.3 MPa 3–20 m/s
Rotating speed measuring instrument	FC4-490F	Rotating speed measurement	0–4500 rpm
Long Axis dynamo meter	1415	Thrust Force Torque Rotating speed	0–3000 N 0–150 N·m 0–3000 rpm

2.3. Similarity Criteria

From the model test, the numerical simulation of the propeller must meet certain similarity criteria. In the test, the similarity relationships are considered in terms of the aspects of propeller geometry, motion, viscous force and cavitation.

a. Geometric similarity

The propeller model made strictly according to the scale ratio of 1:28 ensures the geometric similarity between the model and the actual one.

b. Motion similarity

The similarity of motion is ensured by the definitions of propeller advance coefficient J, thrust coefficient K_T, torque coefficient K_Q, and open-water efficiency η_0, which are defined as follows:

$$J = \frac{V}{nD} \tag{1}$$

$$K_T = \frac{T}{\rho n^2 D^4} \tag{2}$$

$$K_Q = \frac{Q}{\rho n^2 D^5} \tag{3}$$

$$\eta_0 = \frac{J K_T}{2\pi K_Q}, \tag{4}$$

where V is the flow speed, n is the rotating speed of the propeller, D is the diameter of the propeller, T is the thrust of the propeller, Q is the torque of the propeller, and ρ is the density of water.

c. Viscous force similarity

To ensure that the Reynolds number $Rn_{(0.75R)}$ of the chord length of the blade section at 0.75 R of the propeller exceeds the critical Reynolds number [16], that is:

$$Rn_{(0.75R)} = \frac{L_{0.75R}\sqrt{V^2 + (0.75\pi nD)^2}}{\nu} > 3.0 \times 10^5, \tag{5}$$

where $L_{0.75R}$ is the chord length of blade section at 0.75 R, and ν is the kinematic viscosity coefficient of water.

d. Cavitation number similarity

The pressure similarity of the test is met by confirming the cavitation number. The cavitation number of flow σ_v and the cavitation number of rotation σ_n are defined as follows:

$$\sigma_v = \frac{P - P_v}{\frac{1}{2}\rho V^2} \tag{6}$$

$$\sigma_n = \frac{P - P_v}{\frac{1}{2}\rho V^2}, \tag{7}$$

where P and P_v denote the pressure of the shaft center and the saturated vapor pressure, respectively.

2.4. Test Scheme

To study the hydrodynamic performance of the propeller at a low cavitation number, the experiments were carried out in a cavitation tunnel with two different environmental conditions, that is, the uniform flow condition and the simulated ice blockage condition.

For the uniform flow, the flow speed V was 4.0 m/s, the cavitation number of flow σ_v was 2.5, and five advance coefficients J were tested. The range of five advance coefficients was 0.4–0.8, and the ΔJ is 0.1.

For the ice blockage, the ice blockage performance of the propeller was studied according to the cavitation number of rotation σ_n, the advance coefficient J and the ice-propeller spacing L/D. The method of constant rotating speed with variable flow speed was used to adjust the advance coefficient J. The rotating speed n was 35 rps, and the pressure of the cavitation tunnel was adjusted according to the cavitation number of rotation σ_n. Because the propeller is under heavy load due to ice blockage, this paper will study the hydrodynamic performance of the propeller under low cavitation numbers and low advance coefficients. The test operating conditions are shown in Table 3.

Table 3. Test operating conditions of the propeller for ice blockage.

Operating Conditions	σ_n	J	L/D
1	1.5	0.35	0.15, 0.20, 0.25, 0.30, 0.40, 0.50
2		0.45	
3		0.55	
4	4.0	0.35	
5		0.45	
6		0.55	

3. Test Results and Analyses

3.1. Uniform Flow Environment

Figure 2 shows the variation curves of the thrust coefficient K_T, the torque coefficient K_Q and the open-water efficiency η_0 of the propeller with advance coefficient under uniform flow conditions at $\sigma_v = 2.5$.

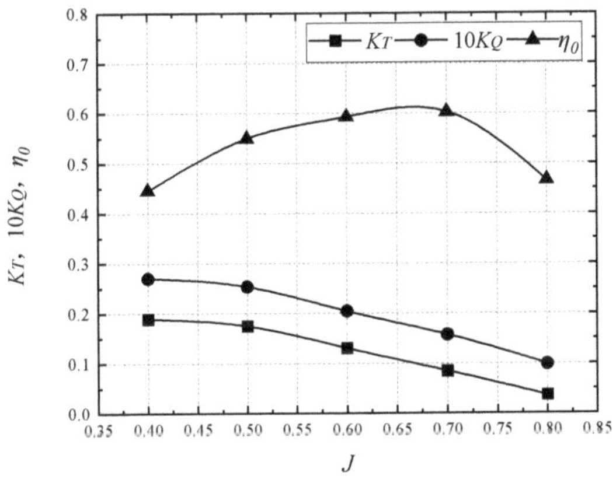

Figure 2. Open-water performance curves of propeller in the uniform flow environment.

As shown in Figure 2, when the advance coefficient J increases from 0.4 to 0.8, the thrust coefficient K_T decreases from 0.189 to 0.036. The torque coefficient 10 K_Q also decreases with the increase of advance coefficient J, from 0.027 to 0.099, and the two curves of K_T and K_Q are almost parallel to each other. As the advance coefficient J increases from 0.4 to 0.7, the open-water efficiency η_0 increases with the increase of the advance coefficient J, as J further increases to 0.8, η_0 decreases rapidly. The decline rates of both the thrust coefficient and the torque coefficient are almost constant as the advance coefficient increases from 0.5 to 0.8, which are slightly larger than their corresponding decline rates as the advance coefficient increases J from 0.4 to 0.5; therefore, there is a sharp drop, and the same conclusion can be seen from Reference [17].

In the uniform flow environment, the flow speed is kept constant. When the advance coefficient increases, both the rotating speed and the flow speed decrease. They would cause a decrease in the pressure difference on the blade surface. Under low cavitation number conditions, as the rotating speed of the propeller increases and the pressure of flow field decreases, the pressure on the blade surface becomes lower than the saturated the vapor pressure, resulting in cavitation [18]. The cavitation coverage area increases with the decrease of the advance coefficient, and the most serious cavitation phenomenon occurs at $J = 0.4$, which makes both the thrust coefficient and the torque coefficient drop sharply.

3.2. Ice Blockage Environment

Figure 3 shows the variation of the thrust coefficient K_T, the torque coefficient K_Q and the open-water efficiency η_0 with the ice-propeller spacing L/D at different rotating speeds and advance coefficients in the ice blockage environment.

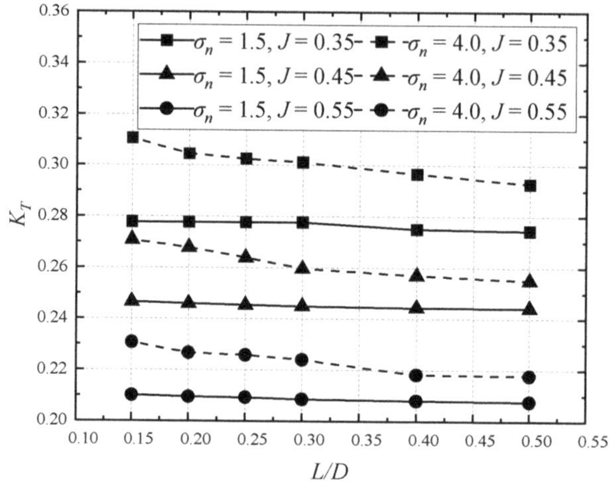

(a)

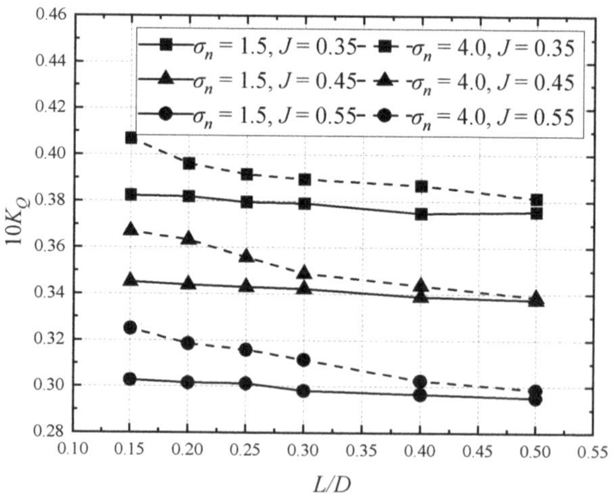

(b)

Figure 3. *Cont.*

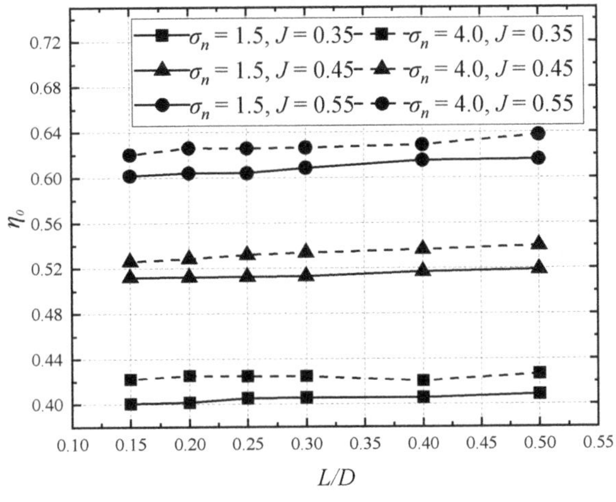

(c)

Figure 3. Open-water performance curves of the propeller in ice blockage environment. (**a**) The influence of ice blockage on the thrust coefficient. (**b**) The influence of ice blockage on the torque coefficient. (**c**) The influence of ice blockage on the open-water efficiency.

As shown in Figure 3, when the cavitation number σ_n is 1.5 and 4.0, both the thrust and the torque coefficients are insensitive to the ice-propeller spacing. When σ_n = 1.5 and J = 0.35, the thrust coefficient K_T fluctuates between 0.275 and 0.278, the torque coefficient $10K_Q$ increases from 0.375 to 0.383, and the open-water efficiency η_0 decreases from 0.409 to 0.401. Both the thrust and the torque coefficients show a downward trend with the increase of the ice-propeller spacing, while the open-water efficiency increases with the latter. Similarly, with the increase of the advance coefficient, both the thrust and the torque coefficients decrease, and the open-water efficiency increases [19]. All three parameters (i.e., the thrust coefficient, torque coefficient and open-water efficiency) are shown to increase with the rise of the cavitation number.

In the propeller test under the condition of ice blockage, the rotating speed of the propeller is constant. The inflow speed increases with the increase of the advance coefficient, and the pressure of the flow field increases with the increase of rotating speed and the cavitation number of rotation. When the cavitation number of rotation decreases, the pressure of the flow field decreases accordingly; the more serious the cavitation, the more it leads to lower values of the thrust coefficient, the torque coefficient and the open-water efficiency. When the advance coefficient increases, the inflow speed increases, causing the decrease of pressure in the flow field, and both the thrust and the torque of the propeller decrease.

The existence of ice blockage reduces the inflow speed and increases both the thrust and the torque of the propeller [20]. With the increase of ice-propeller spacing, the blockage effect decreases, resulting in the decrease of the thrust and torque, and the increase of open-water efficiency. When the cavitation number of rotation decreases, the pressure in the flow field also decreases, and the cavitation phenomenon on the blade surface becomes more serious. The cavitation reduces the influence of the ice blockage effect on both the thrust and the torque so that both the thrust and torque coefficients change slightly with the ice-propeller spacing.

Figure 4 presents some field pictures for the propeller test in an ice blockage experiment. We find that the various degrees of the cavitation occur under different conditions and that the cavitation shape directly determines the changing trend of the hydrodynamic performance of the propeller. The cavitation phenomenon becomes more serious with the decrease of cavitation and ice-propeller spacing, and the increase of the advance coefficient [21]. In the left of Figure 4c, there is cavitation on the back of the blade, continuous tip vortex cavitation appears at the tip, and hub vortex cavitation appears in the center of the hub. With the increase of the advance coefficient, the inflow speed increases accordingly, the cavitation coverage area decreases and its shape becomes narrow and long. With the impact of the high-speed inflow, the sheet-like cavitation breaks into cloud-like cavitation, and the tip vortex cavitation increases significantly (see Figure 3c,d). Meanwhile, the hydrodynamic performance of the propeller decreases, accompanied by strong erosion and noise, and the exciting force of the propeller is obvious. With the decrease of the ice-propeller spacing, the cavitation develops more fully; both the cavitation coverage area and the tip vortex cavitation increase significantly (see Figure 3c). With the decrease of the cavitation number of rotation, the pressure of the flow field decreases, and the cavitation phenomenon becomes serious (see Figure 3a,c). For the same ice-propeller spacing and advance coefficient, the cavitation phenomenon at a lower cavitation number becomes more obvious, and both the number and the coverage area of the cavitation increase sharply.

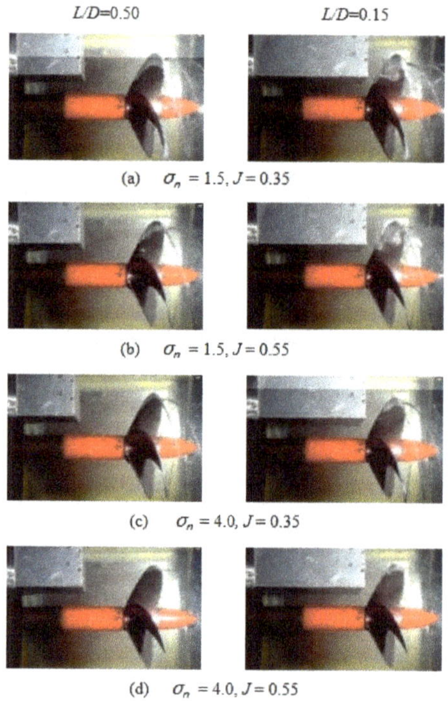

Figure 4. Photos for the propeller test in ice blockage experiment.

4. Numerical Simulation Calculation

The CFD method based on RANS was also adopted to investigate the hydrodynamic performance of the propeller in uniform flow and ice blockage environments. The open-water performance, pressure distribution, cavitation characteristics and the vortex strength of the propeller were further discussed in typical conditions.

4.1. Numerical Model Configuration

4.1.1. Governing Equations

The governing equations of the cavitation flow are the continuity equation and the momentum equation, which are given as:

$$\frac{\partial \rho}{\partial t} + div(\rho U) = 0 \tag{8}$$

$$\frac{\partial(\rho U)}{\partial t} + \nabla\cdot(\rho UU) = \nabla p + \frac{\partial \rho}{\partial t} + \nabla((\mu + \mu_t)\nabla \cdot U) + \rho g + F, \tag{9}$$

where U, ρ and μ respectively denote the velocity, the density and the viscosity coefficient of the vapor-liquid two-phase flow.

$$\rho_m = \rho_l \alpha_l + \rho_v(1 - \alpha_l) \tag{10}$$

$$\mu_m = \mu_l \alpha_l + \mu_v(1 - \alpha_l), \tag{11}$$

where the subscript v and l represent the gas phase and liquid phase respectively, and the subscript m represents the mixed medium; α_l is the volume fraction of liquid phase.

4.1.2. Turbulence Model

μ_t within Equation (9) is the eddy viscosity coefficient caused by turbulence. In this work, the *SST k-ω* turbulence model was utilized to close the governing equations. The relationship between the eddy viscosity and the turbulent energy dissipation rate is expressed as:

$$\mu_t = \rho k/\omega. \tag{12}$$

The transport equations of turbulent kinetic energy k and the turbulent dissipation rate ω are formulated as:

$$\rho\frac{\partial k}{\partial t} + \rho U_j \frac{\partial k}{\partial x_j} = \tau_{ij}\frac{\partial U_i}{\partial x_j} - \beta^* \rho k \omega + \frac{\partial}{\partial x_j}\left[(\mu + \sigma^* \mu_t)\frac{\partial k}{\partial x_j}\right] \tag{13}$$

$$\rho\frac{\partial \omega}{\partial t} + \rho U_j \frac{\partial \omega}{\partial x_j} = \partial\frac{\omega}{k}\tau_{ij}\frac{\partial U_i}{\partial x_j} - \beta \rho \omega^2 + \frac{\partial}{\partial x_j}\left[(\mu + \sigma \mu_T)\frac{\partial \omega}{\partial x_j}\right], \tag{14}$$

where U_j is the average velocity component in the direction of the coordinate axis x_j, τ_{ij} is the Reynolds stress tensor, and β, β^*, σ and σ^* are four turbulence model constants.

4.1.3. Cavitation Model

The volume of fluid (VOF) method based on the Eulerian grids was used to trace the interface between the vapor and the liquid via the transport equation of the volume fraction γ, which is expressed as:

$$\frac{D\gamma}{Dt} = \frac{\partial \gamma}{\partial t} + \nabla \cdot (\gamma U) = \frac{\dot{m}}{\rho_1}. \tag{15}$$

For $\gamma = 1$ refers to the liquid phase, $0 < \gamma < 1$ refers to the vapor–liquid interface, and $\gamma = 0$ refers to the vapor phase. $\dot{m}$ is the mass transport rate, and ρ_1 is the density of water.

The Schnerr–Sauer model [22] is used to simulate the cavitation, and the mass transport rate of the phase transition process is expressed as:

$$\dot{m} = sign(p_v - p)\frac{n_0}{1 + n_0\frac{4}{3}R\pi^3}4\pi R^2\sqrt{\frac{2}{3}\frac{|p_v - p|}{\rho_1}}, \tag{16}$$

where *sign* is the sign function, n_0 is the density of vapor nuclei per unit volume and R is the radius of the vapor nuclei.

4.2. Numerical Scheme

Consistent with the parameters used in the model test for the uniform flow and ice blockage field, the same scale calculation model was established. Figure 5 presents the geometric model of the propeller calculation. The size of the ice blockage and its relative position with the propeller are shown in Figure 6.

Figure 5. The geometric model of the propeller.

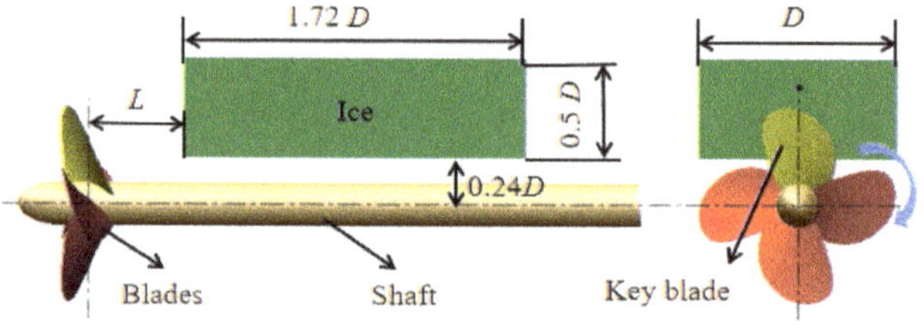

Figure 6. The size of ice blockage and its relative position with the propeller.

To accurately simulate the hydrodynamic performance of the propeller in the cavitation tunnel, the calculation domain was divided into the stationary domain in the far field and the rotation domain in the near field, as shown in Figure 7. The multi-reference frame (MRF) method was used to simulate the motion of the propeller.

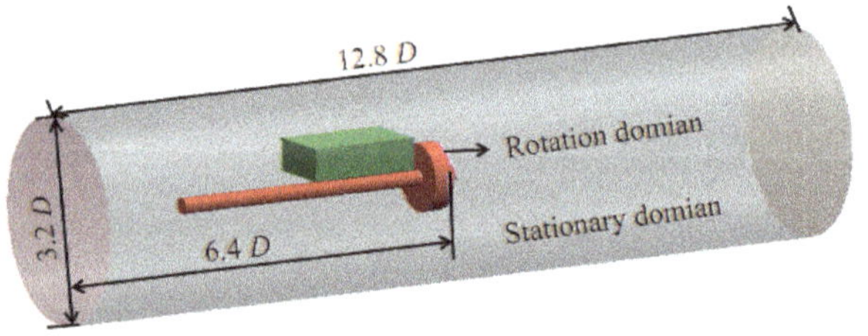

Figure 7. Schematic diagram of calculation domain.

Figure 8 illustrates the schematic diagram of boundary conditions. A fixed velocity was used as the boundary of flow inlet, the outlet boundary of fluid used the pressure control, and the side boundary utilized the wall to simulate the real wall of the cavitation tunnel. The boundary conditions of the ice blockage and the propeller were set to the wall as well. The interface between the rotation domain and the stationary domain adopted the overlapping grids, and the data transmission was realized by linear interpolation.

The trimmer volume grids generator was used to discretize the calculation domain. To ensure the calculation accuracy, the area around the propeller was refined. The sliding grids and overlapping grids were compared in this work. The total number of grids is 3.2 million, as shown in Figure 9.

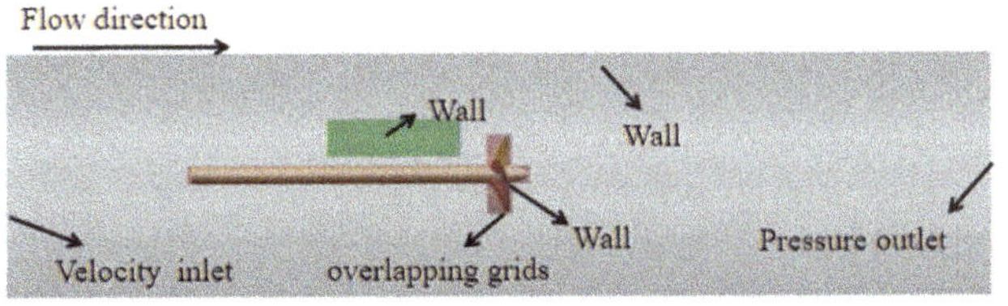

Figure 8. Schematic diagram of boundary conditions.

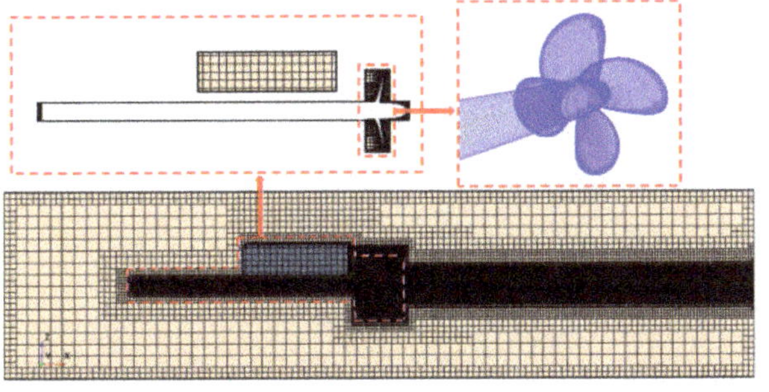

Figure 9. Grids in the computational domain.

The numerical simulations include the following parts. Firstly, the steady hydrodynamic calculation for the propeller was performed to obtain the approximately stable flow field. Secondly, the unsteady hydrodynamic force of the propeller was calculated, and a relatively stable flow field was obtained. Finally, the cavitation model was introduced to calculate the hydrodynamic performance in unsteady cavitation.

5. Simulation Results and Analyses

5.1. Open-Water Performance

To simulate the hydrodynamic performance of the propeller in the cavitation tunnel accurately, the mean thrust and torque in the uniform flow and the ice blockage environments were simulated, and the error from the experimental results was investigated in this subsection.

5.1.1. Uniform Flow Environment

Table 4 presents the numerical results based on the sliding grids and the overlapping grids, and their percentage errors with the experimental data in the uniform flow environment.

From Table 4, the numerical simulation results are in good agreement with the experimental data in the uniform flow. The deviation between the results of overlapping grids and the experimental data is small, and the errors are less than 3%. The results of overlapping grids are better than the sliding grids. The errors between the simulation results and the experimental results decrease with the decrease of advance coefficient, indicating that the numerical simulations based on overlapping grids can reproduce the experiments well at lower cavitation numbers and lower advance coefficients.

5.1.2. Ice Blockage Environment

With the experimental results in the ice blockage simulation, both the cavitation number of rotation and advance coefficient have great influences on the open-water performance of the propeller. This subsection further discusses the hydrodynamic performance of the propeller in the ice blockage environment when $\sigma_n = 1.5$. Figure 10 shows the comparison between the simulation results and the experimental data for the mean thrust and torque coefficient when $\sigma_n = 1.5$, and $J = 0.35$, 0.45 and 0.55.

Table 4. Numerical results and the errors from experimental data.

Grids	J	K_T	$10K_Q$	Error (%)	
				K_T	$10K_Q$
Sliding grids	0.4	0.1854	0.2678	−1.827	−0.750
	0.5	0.1701	0.2538	−2.702	0.191
	0.6	0.1237	0.2070	−1.667	1.209
	0.7	0.0833	0.1603	−1.828	1.989
	0.8	0.0348	0.1030	−3.875	3.993
overlapping grids	0.4	0.1886	0.2721	−0.125	0.851
	0.5	0.1740	0.2572	−0.464	1.531
	0.6	0.1246	0.2067	−1.734	1.058
	0.7	0.0826	0.1592	−2.725	1.289
	0.8	0.0354	0.1006	−2.291	1.613

As shown in Figure 10, the numerical results for the mean hydrodynamic variation of the propeller are almost identical to the results obtained by experiment. The errors between the simulation results and the experimental data for the thrust coefficient and the torque coefficient are less than 5%.

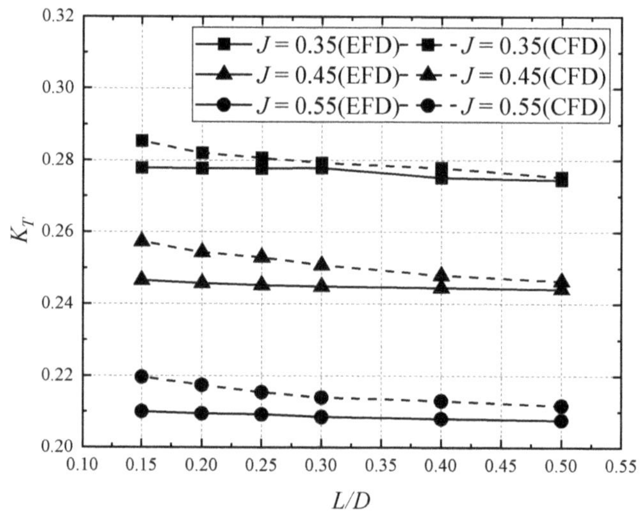

(a)

Figure 10. *Cont.*

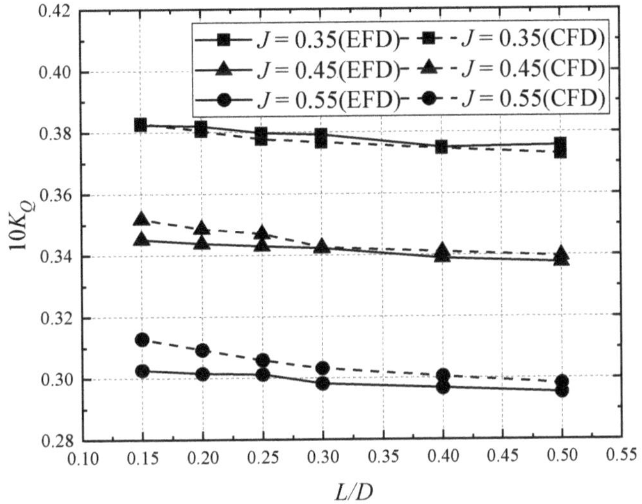

(b)

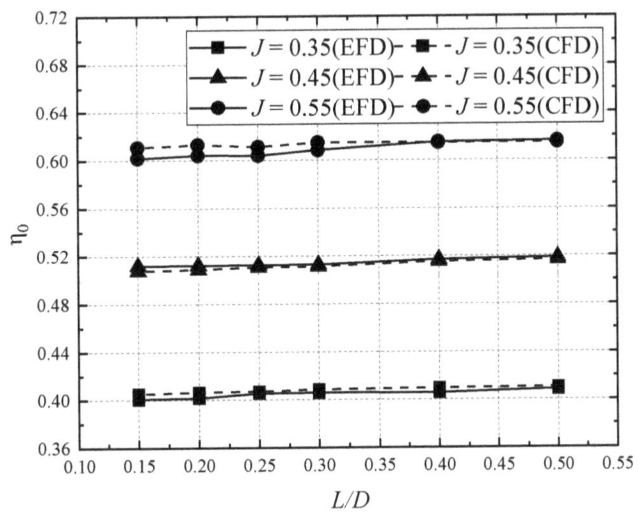

(c)

Figure 10. Comparison of mean open-water performance between the numerical results (marked with CFD) and the experimental data (marked with EFD) in ice blockage environment. (**a**) The influence of ice blockage on the thrust coefficient. (**b**) The influence of ice blockage on the torque coefficient. (**c**) The influence of ice blockage on the open-water efficiency.

5.2. Cavitation Excitation Force

In the process of experiments and numerical simulations, it was found that with the decrease of the ice-propeller spacing, both the blockage effect and the oscillation of thrust and torque become more obvious. When $\sigma_n = 1.5$, $L/D = 0.15$, the hydrodynamic performance of the propeller is the worst. In this subsection, the oscillation law of thrust coefficient and torque coefficient with the excitation force of cavitation are studied when $J = 0.35$, 0.45 and 0.55.

As shown in Figure 11, when $\sigma_n = 1.5$ and $L/D = 0.15$, both the thrust coefficient and the torque coefficient oscillate violently within two cycles of propeller rotation, with 57 cycles at $J = 0.35$, 57 cycles at =0.45 and 59 cycles at $J = 0.55$. Both the mean values and the oscillation amplitudes for both coefficients decrease with the increase of the advance coefficient. When $J = 0.35$, the thrust coefficient oscillates in the range of 0.266~0.308, and the torque coefficient oscillates in the range of 0.036~0.043. With the increase of the advance coefficient, both the thrust coefficient and the torque coefficient decrease, and their amplitudes also decrease.

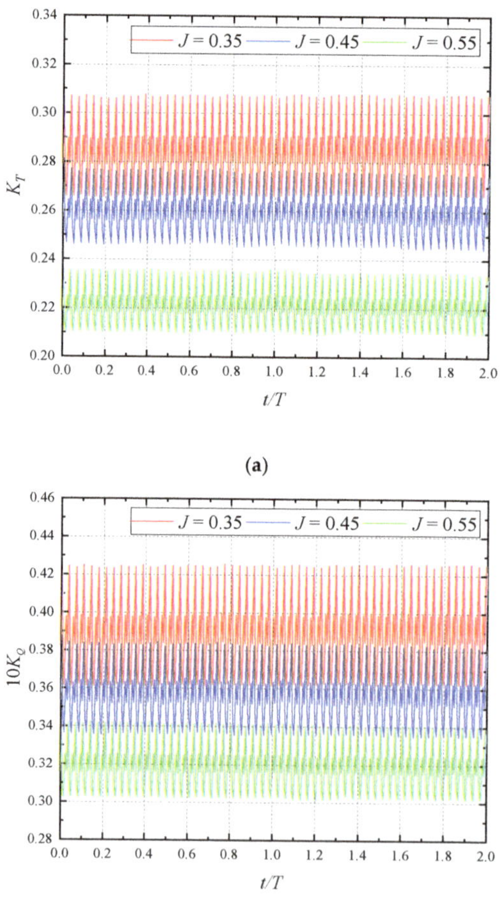

Figure 11. Open-water performance of propeller with cavitation excitation force. (**a**) The influence of ice blockage on the thrust coefficient. (**b**) The influence of ice blockage on the torque coefficient.

The interference among the four blades during the rotation of the propeller, the oscillation frequency of the propeller is much higher than that of the single blade. Due to the stability of the flow field, the hydrodynamic performance of the propeller in different oscillation periods is relatively stable. With the increase of the advance coefficient, the inlet velocity increases, and the pressure, the thrust coefficient and the torque coefficient decrease. With the flow velocity increasing, the cavitation tends to be dispersed by the inflow when it adheres to the blade, and the influence of cavitation on the hydrodynamic performance is reduced, but the oscillation frequency of the propeller increases.

5.3. Pressure Distribution

To further reveal the hydrodynamic performance of the propeller, the pressure distribution of the propeller blade at $\sigma_n = 1.5$, $L/D = 1.5$, and $J = 0.35$, 0.45 and 0.55 is studied, as shown in Figure 12, and the pressure coefficient C_P is defined as follows:

$$C_p = \frac{P - P_0}{\rho (nD)^2}, \tag{17}$$

where P is the pressure of the propeller, P_0 is the pressure of domain.

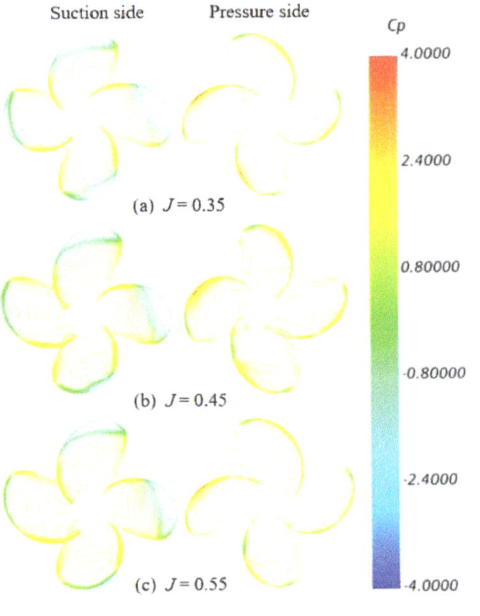

Figure 12. Distribution of pressure coefficient on the blade surface.

As shown in Figure 12, the pressure on the blade surface changes slightly with the increase of the advance coefficient. The pressure coefficient on the leading edge of the blade surface (roughly between 0.6 R and 1.0 R) is the largest, and the pressure coefficient is about 3.5, while the pressure coefficient at the blade tip is negative. The pressure on the suction surface is the lowest on the leading edge of the surface, and the pressure coefficient over most of the area is about—2.0, which easily generates cavitation. With the increase of advance coefficient, the cavitation is difficult to adhere to the blade surface, and the low-pressure area decreases.

5.4. Cavitation Characteristics

To further explore the hydrodynamic performance of the propeller, the cavitation shape and thickness are studied when $\sigma_n = 1.5$, $L/D = 1.5$, and $J = 0.35$, 0.45 and 0.55, as shown in Figure 13.

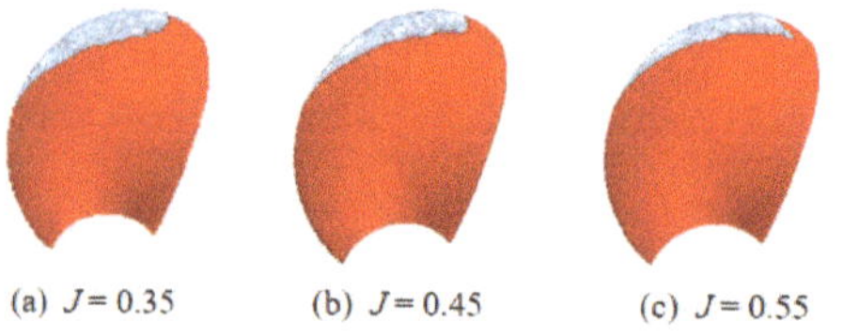

Figure 13. Cavitation shapes of suction surface for the key blade.

Figure 13 shows that the cavitation shape on the surface of the key blade is consistent with that obtained by the physical experiments. The cavitation coverage area is the same as its low-pressure area, and the cavitation occurs about between 0.6 *R* and 1.0 *R*. With the increase of the advance coefficient, the shape of the cavitation region becomes narrow and long, and the coverage area also decreases, and the cavitation leaks into the wake soon. The most serious cavitation effect occurs at around 0.90 *R*.

Figure 14 shows the distribution of the cavitation thickness along the chord length of the blade section at 0.85 *R*, 0.90 *R* and 0.95 *R*. It is observed that the thickness of the cavitation varies greatly at different radii. The thickness of the cavitation at 0.85 *R* is the smallest, and the coverage of the cavitation is also the smallest. With the increase of the advance coefficient, the thickness of the cavitation at the leading-edge decreases, and its peak value moves forward. When the blade section is 0.90 *R*, both the thickness and coverage area change the most with the increase of the advance coefficient. Specifically speaking, as the advance coefficient increases, the thickness of the cavitation decreases and the peak value moves back. When the blade section is 0.95 *R*, the cavitation almost covers the whole chord length. The maximum value of the cavitation thickness is about 6.4 mm, and the peak value of the cavitation thickness moves backwards with the increase of the advance coefficient. The cavitation thickness at the leading edge of 0.85 *R* and 0.90 *R* blade sections is greater than 0, and decreases with the advance coefficient. At 0.95 *R*, the thickness of the blade section is distributed behind the leading edge, and the cavitation is separated from the blade surface.

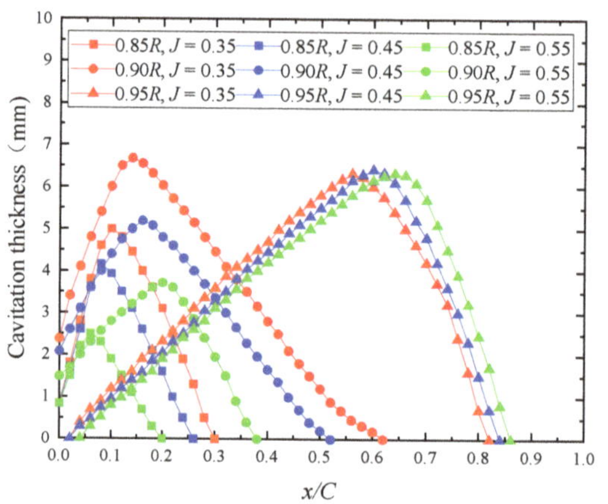

Figure 14. Distribution of cavitation thickness along chord length of blade section.

5.5. Vortex Intensity

To investigate the influence of the cavitation on the vortex intensity on the surface of the key blade in the ice blockage effect, the vortex intensity distributions on the key blade when $\sigma_n = 1.5$, $L/D = 0.15$ and $J = 0.35$, 0.45 and 0.55 are compared in Figure 15. The vortex intensity distribution on the pressure surface is relatively uniform. The largest and the smallest vortex intensities occur at the blade edge and at the blade root, respectively. As the radial distance decreases, the vortex intensity gradually decreases as well. With the increase of the advance coefficient, the vortex intensity and its distribution on the pressure surface are almost unchanged. Compared with the pressure surface, the vortex intensity on the suction surface is disorderly distributed near the blade tip. Because the cavitation is attached to the blade tip, the contact area between the viscous water and the blade is reduced, and hence the vortex intensity is reduced. With the increase of the advance coefficient, the low-pressure area and the area covered by the cavitation become narrower and longer, and the vortex intensity also shows a similar trend.

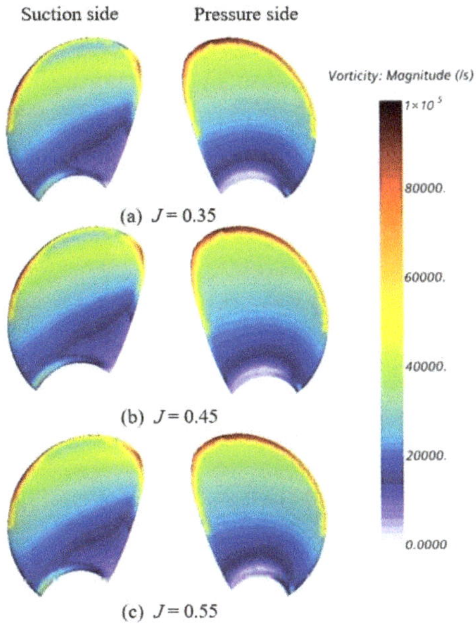

Figure 15. Distribution of blade vortex intensity.

6. Conclusions

In this work, the propeller model test and CFD numerical simulations based on RANS were performed to study the hydrodynamic performance of the propeller in the uniform flow and the ice blockage environments.

Based on both the experimental and the numerical results, the following main conclusions are drawn:

(1) The numerical simulation results are in good agreement with the model test data, and the error is within 5%.
(2) In the uniform environment, the smaller advance coefficient leads to the higher rotating speed of the propeller, and with the cavitation more heavy, the thrust and torque of the propeller will drop sharply due to the influence of the cavitation.
(3) In the ice blockage environment, the thrust and torque increase with the decrease of the ice-propeller spacing. The cavitation becomes more obvious when the ice blockage is more serious. When $\sigma_n = 1.5$ and $L/D = 0.15$, the hydrodynamic performance of the propeller is the worst.

(4) The propeller oscillates violently under the action of the cavitation excitation force, and the oscillation frequency of the propeller increases with the increase of the advance coefficient.
(5) The cavitation is generated in the low-pressure area of the suction surface, the coverage area of the cavitation decreases with the increase of the advance coefficient, the sheet-like cavitation is broken into cloud-like cavitation, the tip vortex cavitation increases significantly, and the cavitation shapes of the numerical simulation are consistent with the experimental phenomenon.
(6) Since the cavitation reduces the contact area between the water and the blade, the vortex strength at the attachment of the cavitation decreases, and the vortex strength increases with the increase of the advance coefficient.

Author Contributions: Experiment, L.Z. and F.D.; methodology, L.Z. and S.Z.; simulation, S.Z.; validation, S.D. and J.G.; analysis, L.Z. and S.Z.; investigation, S.Z.; resources, L.Z. and F.D.; data curation, S.Z.; writing—original draft preparation, S.Z.; writing—review and editing, J.G.; visualization, L.Z. and S.Z.; supervision, L.Z. and S.D.; project administration, L.Z. and S.D.; funding acquisition, L.Z. All authors have read and agreed to the published version of the manuscript.

Funding: This research was funded by the National key research and development program, grant number 2022YFE0107000 and National Natural Science Foundation of China, grant number 52171259, 51809124, 51911530156.

Data Availability Statement: All analyzed data in this study have been included in the manuscript.

Acknowledgments: The authors would like to thank the Jiangsu University of Science and Technology (JUST).

Conflicts of Interest: The authors declare no conflict of interest.

References

1. Zhang, Z.; Huisingh, D.; Song, M. Exploitation of trans-Arctic maritime transportation. *J. Clean. Prod.* **2019**, *212*, 960–973. [CrossRef]
2. Tsarau, A.; Lubbad, R.; Løset, S. A numerical model for simulating the effect of propeller flow in ice management. *Cold Reg. Sci. Technol.* **2017**, *142*, 139–152. [CrossRef]
3. Ryan, C.; Huang, L.; Li, Z.; Ringsberg, J.W.; Thomas, G. An Arctic ship performance model for sea routes in ice-infested waters. *Appl. Ocean Res.* **2021**, *117*, 102950. [CrossRef]
4. Aktas, B. A Systematic Experimental Approach to Cavitation Noise Prediction of Marine Propellers. Ph.D. Thesis, Newcastle University, Newcastle, UK, 2007.
5. Sampson, R.; Atalr, M.; Sasaki, N. Cavitation analysis of a double acting podded drive during ice milling. In Proceedings of the 7th International Symposium on Cavitation, Ann Arbor, MI, USA, 16–20 August 2009.
6. Wu, S.; Liu, Y.; Zeng, Z.B. Influence of ice block on hydrodynamic performance and cavitation of propeller. *Shipbuild. China* **2018**, *59*, 110–121.
7. Wu, S.; Rui, W.; Zeng, Z.B. Investigation on the characteristics of cavitation excited force of propeller in ice block condition. *Ship Sci. Technol.* **2019**, *41*, 21–26.
8. Wu, S.; Zeng, Z.B.; Zhang, G.P. Experimental research on the influence of ice block parameters on propeller hydrodynamic performance. *J. Ship Mech.* **2018**, *22*, 156–164.
9. Rhee, S.H. A study of propeller cavitation using a RANS CFD method. In Proceedings of the 8th International Conference on Numerical Ship Hydrodynamics, Busan, Korea, 22–25 September 2003; pp. A292–A303.
10. Liu, D.C.; Hong, F.; Zhao, F. The CFD analysis of propeller sheet cavitation. In Proceedings of the 8th International Conference on Hydrodynamics, Nantes, France, 30 September–3 October 2008; pp. 171–176.
11. Zhu, Z.F.; Wang, X.Y.; Fang, S.L. Propeller cavitation research by an unstructured grid based RANS solver. *Ocean. Eng.* **2009**, *27*, 103–107.
12. Shin, K.W.; Andersen, P.; Mikkelsen, R. Cavitation simulation on conventional and highly-skewed propellers in the behind-hull condition. In Proceedings of the Second International Symposium on Marine Propulsors, SMP 11, Hamburg, Germany, 15–17 June 2011.
13. Ju, L. Unsteady Cavitation Numerical Simulation of Propeller behind Ship Hull. Master's Thesis, Harbin Engineering University, Harbin, China, 2012.
14. Zheng, C.S.; Zhang, Z.R. Prediction Method for Open-Water Performance of Propeller Based on OpenFOAM. *Chin. J. Ship Res.* **2012**, *7*, 30–35.

15. CB/T 346–1997. Open Water Test Method of Propeller Model. S. Available online: http://std.samr.gov.cn/hb/search/stdHBDetailed?id=8B1827F17E3BBB19E05397BE0A0AB44A (accessed on 26 February 2022).
16. Misra, S.C. *Design Principles of Ships and Marine Structures*; CRC Press: Boca Raton, FL, USA, 2015.
17. Xu, P.; Wang, C.; Ye, L.; Guo, C.; Xiong, W.; Wu, S. Cavitation and Induced Excitation Force of Ice-Class Propeller Blocked by Ice. *J. Mar. Sci. Eng.* **2021**, *9*, 674. [CrossRef]
18. Yusvika, M.; Prabowo, A.R.; Tjahjana, D.D.D.P.; Sohn, J.M. Cavitation Prediction of Ship Propeller Based on Temperature and Fluid Properties of Water. *J. Mar. Sci. Eng.* **2020**, *8*, 465. [CrossRef]
19. Guo, C.Y.; Xu, P.; Wang, C.; Xiong, W.P. Experimental Investigation of the Effect of Ice Blockage on Propeller Hydrodynamic Performance. *Math. Probl. Eng.* **2019**, *2019*, 3712012. [CrossRef]
20. Chao, W.; Sheng-Xia, S.; Xin, C.; Li-Yu, Y. Numerical simulation of hydrodynamic performance of ice class propeller in blocked flow–using overlapping grids method. *Ocean Eng.* **2017**, *141*, 418–426. [CrossRef]
21. Gu, L.; Wang, C.; Hu, J. Application Study of Cartesian Grid in Numerical Prediction of a Marine Propeller. In Proceedings of the 35th International Conference on Offshore Mechanics and Arctic Engineering, American Society of Mechanical Engineers, Busan, Korea, 19–24 June 2016.
22. Sauer, J.; Schnerr, G.H. Development of a New Cavitation Model based on Bubble Dynamics. *ZAMM-J. Appl. Math. Mech./Z. Für Angew. Math. Und Mech.* **2001**, *81*, 561–562. [CrossRef]

water

MDPI

Article

Research on Sea Spray Distribution of Marine Vessels Based on SPH-FEM Coupling Numerical Simulation Method

Jiajing Chen , Xu Bai *, Jialu Wang, Guanyu Chen and Tao Zhang

School of Naval Architecture & Ocean Engineering, Jiangsu University of Science and Technology, Zhenjiang 212003, China
* Correspondence: baixu@just.edu.cn

Abstract: Due to the effect of the maritime environment and low temperature factor, ice phenomena are easily produced while a ship is sailing in a polar location. Types of ice accumulation include sea spray icing, which accounts for 90% of all ice accumulation, and, therefore, sea spray generation is a crucial step in ice accumulation prediction research. In order to investigate the phenomenon of ice formation on ships, this paper uses the SPH-FEM (Smooth Particle Hydrodynamics-Finite Element) coupling approach in this paper, and tracks the data pertaining to the wave current particles by simulating the impact of a single wave on the ship hull under the wave height standard of various sea conditions. Following the numerical simulation, it was discovered that when the sea state reaches five levels, the waves will produce marine sea spray on the simulated hull and distribute them in an arc on both sides of the hull; the phenomenon is influenced by the sea state level, meaning that the arc becomes more obvious as the sea state level rises. Furthermore, the number of sea spray particles formed in different sea state levels accounts for about 9–13% of the range of individual waves, and the rest of the sea spray particles will flow back to the surface quickly after passing through the hull.

Keywords: ship ice accumulation; sea spray; marine structures; SPH-FEM coupling method

Citation: Chen, J.; Bai, X.; Wang, J.; Chen, G.; Zhang, T. Research on Sea Spray Distribution of Marine Vessels Based on SPH-FEM Coupling Numerical Simulation Method. *Water* **2022**, *14*, 3834. https://doi.org/10.3390/w14233834

Academic Editor: Mariano Buccino

Received: 28 October 2022
Accepted: 18 November 2022
Published: 24 November 2022

1. Introduction

The impact of low temperature elements and the marine environment causes ice formation in hull structures easily when ships are cruising in polar waters. High levels of ice buildup on the hull structure can result in the obstruction of different exposed systems for ventilation and equipment, as well as safety risks including ship capsizes and fatalities of crew members [1]. Therefore, the primary focus and direction of this field's research has been on the issues linked to hull freezing and ice accumulation, in order to assure the safety of ship navigation and operation. The subject of marine icing forecasting is more complex and depends on many variables, such as ice accumulation will depend on meteorological conditions (wind speed, air temperature, water temperature, and seawater freezing point), the wave height generated, and the vessel speed and motion relative to the wave. The amount of spray, its density, duration, and frequency depend on the ship design features, speed and heading with respect to the waves. In typical polar locations, there are two types of ship hull icing, those being atmospheric icing and sea spray icing [2]. The most influential factor of atmospheric icing depends on meteorological conditions, and the most influential factor of sea spray icing depends on the state of motion of the waves and the ship. The findings of similar investigations indicate that wave-induced marine sea spray account for 90% of the water sources that cause the icing of marine hull structures [3]. The process of oceanic sea spray icing includes sea spray generation, air flow, and heat transfer [4], among which sea spray generation is crucial to the study of pre-ice prediction. There is a dearth of physical modeling and flux measurement of sea spray sources, and the majority of current research on sea spray formation is based on the empirical equation of droplet flux [5,6]. Therefore, further research on the process of sea spray generation is of great significance for icing forecasting.

In recent years, scientists have created a variety of physical models that anticipate the sea spray icing which results from "wave-ship" interactions (Lozowski et al., 2000; Anton Kulyakhtin, 2012; Samuelsen et al., 2017 [7–9]). Most of these models do not include the process of ocean sea spray generation, which includes the procedure of wave impact on the hull and the distribution of sea spray on the hull after breaking up to form sea spray, instead directly combine the droplet flux of the empirical equation with the heat transfer icing process. Despite Lozowski's [7] more recent research findings suggesting that the generation of sea spray can be predicted by using linear wave theory and potential flow theory in a constant uniform wind field and suggesting that the ice accretion rate is determined by the sea spray flux and heat conduction on the surface of the structure, this ignores the actual process of sea spray generation, so the model does not describe the real hull ice accumulation behavior. Anton Kulyakhtin [8] proposed the ice growth model-MARICE after ignoring the effect of thermodynamics in the interaction between sea spray and simulating the flow of sea spray in relation to the surrounding environment of the ship, combined with the icing process of heat transfer, but the droplet flux of this model is obtained from empirical equations, so there is still some room for the research of this icing model on the generation of sea spray and the problem of heat transfer between ice layers. The MINICOG model proposed by Samuelsen [9] is modified by two different empirical formulas of droplet flux to propose an empirical formula based on the relationship between local wind speed and effective wave height and wave period, but the model is affected by the actual sea spray generation process and flux problems, and has certain conditions and limitations.

In general, most ocean icing models have historically relied on empirical equations for ocean droplet fluxes due to the limitations of theoretical development and observational tools [10–12]. This lack of physical modeling of sea spray sources, measurements of sea spray generation and actual fluxes, eventually resulted in discrepancies between icing predictions and actual conditions [13]. The droplet trajectory in the empirical equation depends on gravity, wind speed, initial droplet size and velocity, and many other parameters. In the past studies, droplet concentration was usually ignored because droplets were assumed to be diluted and moved individually over the container without interacting with each other. Individual droplets experience more drag during their flight and therefore require high initial velocities to overcome gravity and drag. The air flow field influenced by the surrounding droplets has less drag on a particular droplet and, therefore, the droplet travels at an increased speed and maximum height. This suggests that a cloud of droplets traveling as droplets or filaments can reach higher heights with lower jet velocities [3]. Therefore, the theoretical model is currently limited to predicting mono-droplet trajectories, but CFD can simulate numerous droplets in a single domain. Therefore, one of the key components of current research on icing phenomena in maritime structures and a key factor in improving icing forecasts is the investigation of physical modeling and flux measurements of the sea spray formation process [14,15]. Regarding sea spray sources, sea spray generation can be divided into a variety of stages, such as free surface modeling, wave thumping on the bow, air entrainment during impact, formation of water film and jets on the wall after wave thumping, wind-induced water film and sea spray fragmentation, and sea spray trajectory in contact with the ship's deck surface, and others [16–18]. Modern mainstream research on icing of marine structures has mainly focused on the formation of sea spray after wave impact on marine structures, and the research has achieved certain results, but there are also some controversies and shortcomings.

Horjen and others [19–21] optimized the procedure used to simulate sea spray generated by wave impact based on the RIGICE and MARICE model and derived a new equation for liquid water content, but their observations only matched their own data. In recent years, many scholars have begun to observe sea spray through experimental modeling, and many related models based on computational fluid dynamics (CFD) have been developed. Amir Banari et al. [22,23] combined the VOF (Volume of fluid) method to propose the trajectory of the sea spray form after wave thumping when it comes into contact with the hull, which can visualize the changing form of the sea spray, but cannot predict its droplet flux. Kees et al. [24,25] introduced a new Level-Set method to solve the robustness of large-scale complex free-surface deformation. However, after all, the simulation of airflow entrainment is very important in the study of wave-hull impact problems [26], so the method also has limitations and do not apply to morphologies where the air is intercepted, or where the liquid contains gas; Greco and Lugni et al. [27] developed a numerical and experimental method to estimate various statistical analyses of the performance of wave lapping on the deck, and the relevant parameters such as pressure and collision position on the deck were obtained from the sea spray morphology. The shortcoming of this method is the lack of microscopic application of the rupture mechanism; After this, Dehghani et al. [28] developed a new three-dimensional model to simulate wave impact on MFV ships using droplet size and velocity characteristics in an attempt to derive the droplet size and velocity distribution of sea spray after wave impact on the hull from a microscopic perspective. In addition, Shipilova et al. [29] used the LWC model of Zakrzewski [30] to generate sea spray in front of two types of ships and assumed that the sea spray was generated in the form of squares and used the Eulerian-Lagrangian method into sea spray simulation, but this method requires setting the sea spray duration by itself and cannot simulate the periodic motion of the sea spray. Finally, due to the computational cost, most Lagrangian methods cannot be used for sharp deformations of the free boundary at large scales.

The present study is based on the idea proposed by Anton Kulyakhtin [31] and other scholars that the phenomenon of ice accretion on marine structures is caused by the periodic superposition of marine sea spray formed by waves impacting on the hull of a ship after icing on the hull. Considerable progress has been made in recent years in simulating ocean sea spray and icing phenomena using numerical methods, but none of them have studied the periodicity of sea spray superposition [32]. The droplet flux includes droplet collection and collision coefficient, droplet velocity, liquid water content, frequency and duration of droplet occurrence, of which the last three basic parameters are the key ones to be estimated. In this study, the effect of the wind speed, heave and pitch of the ship is neglected to simplify the problem. In order to address this issue, this paper analysis first makes the assumption that the ship moves in a regular wave and that the outcomes of the two periodic periods are unrelated, reducing the ship motion to the degree of freedom of forward translation only. Second, the periodic superposition problem is divided into smaller, more independent parts for the research based on the multi-scale character and intricacy of the sea spray problem. This study examines the distribution of marine sea spray on a ship's hull caused by a single cycle of regular waves striking the hull in a variety of sea states. The process of wave impact on ship hull is numerically simulated using SPH-FEM (Smooth Particle Hydrodynamics-Finite Element Method), taking into account the viscous effect and wave-hull interaction effect. A new method of ocean sea spray improvement and the distribution probability of sea spray on the hull are established based on the probability distribution law of single period sea spray on the hull. This establishes the groundwork for later research on the periodicity of ice accretion brought on by ocean sea sprays.

2. Theoretical Methods

2.1. Governing Equations

This paper studies the phenomenon of sea spray caused by the interaction of wave and ship structure phase, which does not involve the subsequent steps of icing proliferation. Therefore, the influencing factors of wind speed and heat transfer are temporarily ignored. In order to detect the proportion of a single period of sea spray particles on the hull, the SPH-FEM coupling method is used in this paper. SPH particles are used as fluid particles of waves, and the hull structure of the phase action is divided into FEM grids [33]. The coupling contact algorithm is added to achieve the effect of phase action. This method effectively improves the efficiency of operation [34]. In the following, wave particles are regarded as water fluid, and the N-S equations describing the mass conservation and momentum conservation of fluid motion are used:

$$\frac{D\rho}{Dt} = -\rho\nabla\cdot\mathbf{u} \tag{1}$$

$$\frac{D\mathbf{u}_w}{Dt} = -\frac{\nabla p}{\rho} + v\nabla^2\mathbf{u} + \mathbf{g} \tag{2}$$

In the formula: $\mathbf{u}$ is the velocity of fluid particles, p is inter-fluid pressure, ρ is the density of water fluid and v is the kinematic viscosity coefficient of water. After the correction of the Riemann solver, particle-to-particle contact algorithm and the CSPM method [35], the SPH particle approximate expression of Equations (1) and (2) is:

$$\frac{d\rho_i}{dt} = \frac{2\rho_i \sum\limits_{j=1}^{N} \frac{m_j}{p_j}(u_i^R - u_{ij}^R)\cdot\nabla_i W_{ij}}{\sum\limits_{j=i} N(r_{j-r_i}) \otimes \nabla_i W_{ij}\frac{m_{ij}}{\rho_j}} \tag{3}$$

$$\frac{D\mathbf{u}_i}{Dt} = \frac{-\sum\limits_{j=1}^{N} m_j \frac{p_{ij}^*}{\rho_i\rho_j}\cdot\frac{\partial W_{ij}}{\partial x_i}}{\sum\limits_{j=i} N(r_{j-r_i}) \otimes \nabla_i W_{ij}\frac{m_{ij}}{\rho_j}} + \sum_{j=1}^{N} m_j\left(\frac{4v\mathbf{r}_{ij}\mathbf{u}_{ij}}{(\rho_i+\rho_j)\left|\mathbf{r}_{ij}\right|^2}\right)\frac{\partial W_{ij}}{\partial x_i} + \mathbf{g} \tag{4}$$

where: $W_{ij} = \alpha_D(1-\frac{q}{2})^4(2q+1), 0 \leq q \leq 2; \alpha_D = \frac{7}{4\pi h^2}, q = \frac{|r_i - r_j|}{h}$

Of which, the normal velocity U_{ij}^R and pressure P_{ij}^* on the particle discontinuity surface are:

$$P_{ij}^* = \frac{p_j\rho_i c_i + p_i\rho_j c_j + \rho_i c_i \rho_j c_j (u_i^R - u_j^R)}{\rho_i c + \rho_j c_j} \tag{5}$$

$$U_{ij}^R = \frac{u_i^R\rho_i c_i + u_j^R\rho_j c_j - (p_j + p_i)}{\rho_i c + \rho_j c_j} \tag{6}$$

The parameters of the left and right sides of the discontinuity are u_i^R, p_i, ρ_i, and U_j^R, p_j, ρ_j, respectively. This discontinuity propagates at sound speed c_i and c_j to the left and right sides. The pressure p in formula (5) selects the equation of state used by Monahan and Kos to calculate the free surface flow problem [36]:

$$p = B\left[\left(\frac{\rho}{\rho_0}\right)^\gamma - 1\right] \tag{7}$$

In the formula: γ is a constant, the general value in the liquid phase is 7; ρ_0 is the initial density, where the density of water is 998 kg/m^3. B is used to limit the maximum change in density, taking $B = \frac{c_0^2\rho_0}{\gamma}$ in the fluid, c_0 is the sound speed.

In the simulation optimization step, a very important point is to determine the appropriate particle spacing and particle resolution (dp). Mintu considered various particle spaces from D/2 to D/6, and the results showed that the droplet trajectory tends to converge when the particle resolution gradually increases, but considering the cost of time and computational memory, the particle spacing of D/3 was found to be the best choice. Since variable particle spacing cannot be set in this algorithm, the validated D/3 optimal particle spacing is chosen for the simulation [33].

2.2. Coupling Wall Boundary Treatment

In this paper, the SPH-FEM coupling technique can be obtained in LS-DYNA to simulate the formation of sea spray, and is based on weakly compressible smooth particle hydrodynamics. The interaction between the FEM and the SPH particles was defined in the numerical simulation via the *CONTACT_NODES_TO_SURFACE contact card, with the master-slave penalty algorithm, in which the shell elements were assigned the role of the master part and the SPH particles the slave part. In the numerical model applied in this paper, the surface of the hull structure is simplified as an immovable solid wall. Since the particles on or near the boundary are cut off by the boundary when integrating, because the particles on or near the boundary are only affected by the particles within the boundary, the particles must not penetrate the solid element during the interaction. Although the particle velocity on the boundary is zero, other variables are not necessarily zero, so this unilateral action may lead to failure of the solution, so the SPH particle algorithm is not suitable for the whole calculation area [37]. In order to solve this problem, the current mainstream methods are virtual particle method and penalty function method. The virtual particle method is unstable to generate virtual particles for a complex boundary, and it is easy to penetrate. The penalty function method is easy to implement, and it can deal with complex geometric boundary conditions [38].

In order to prevent the wave fluid particles from penetrating the hull structural unit, the following conditions need to be satisfied at the boundary:

$$\left(\frac{\partial}{\partial t}\mathbf{u} - \boldsymbol{\upsilon}\right)\cdot\mathbf{n} = 0 \tag{8}$$

In the formula, $\mathbf{u}$, $\boldsymbol{\upsilon}$ and $\mathbf{n}$ denote the displacements of element nodes at the boundary, the velocities of particles at the boundary and the normal vectors of elements at the boundary are represented, respectively.

In this paper, the coupling of SPH particles and FEM elements is solved by the contact algorithm based on the penalty function. The contact model is shown in Figure 1. For each fluid particle in the fluid domain, the adjacent finite element nodes are searched within a given smooth radius H, and all finite element surfaces belonging to FEM nodes are searched. When the SPH particle and the finite element surface satisfy the penetration depth L of the particle to the finite element, it is considered that the particle collides. At this time, a penalty function is introduced, and a boundary contact force $\mathbf{f}_c$ is added to limit the penetration of fluid particles into the hull structure unit.

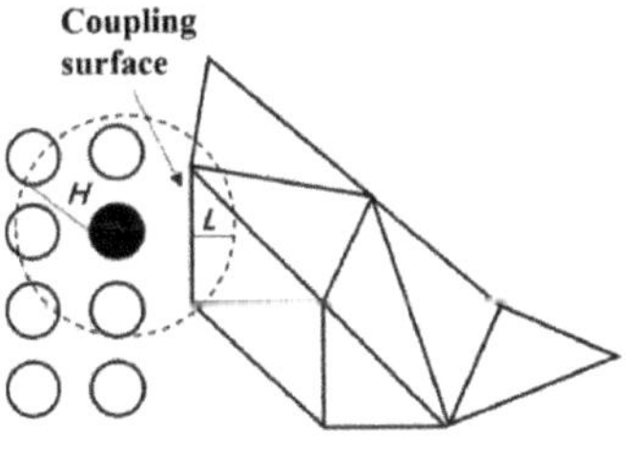

Figure 1. SPH-FEM Coupling surface.

When the hull structure is meshed, the finite element size and the smooth radius of the particles should not differ greatly, otherwise the finding omission phenomenon will occur. However, in order to reduce the memory of the mesh, and at the same time, the number of particles increases with the increase of the sea state level, etc., the missing phenomenon needs to be supplemented by the algorithm correction content. In other words, when the SPH particles at the boundary find the neighboring cells, the search is performed directly on the surface of the FEM cell [39]. Through the surface normal vector of each FEM cell and the position of the fluid particle at the boundary, the distance r_{s-f} between the particle and the face of the cell is calculated, and then the contact force is calculated by the penalty function, and the contact force is applied to the fluid particle at the boundary as an external force. Since the finite element discretization of the solid is carried out by using tetrahedral cell in this paper, and each face of the tetrahedral cell is angular. Therefore, the process of detecting the contact between the particle and the cell surface can be converted into the process of detecting the triangular shape of the particle and the solid surface. The boundary particle search process is shown in the following Figure 2. When particles cannot be retrieved around the mesh, which are defined as boundary particles.

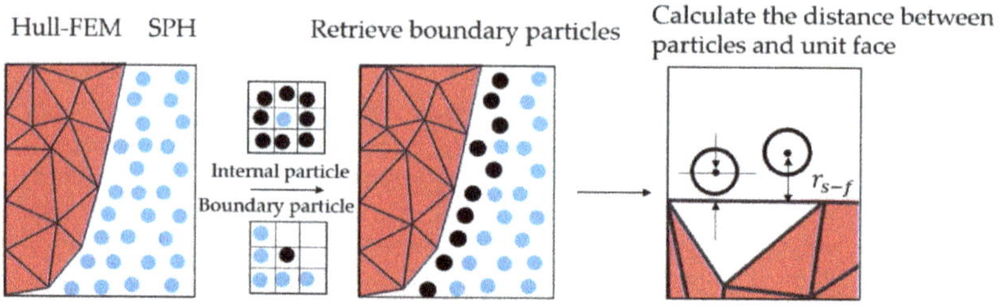

Figure 2. The relationship between boundary particles and cell surface.

The contact force at the boundary can be divided into normal contact force $\mathbf{f}_n$ and tangential contact force $\mathbf{f}_c$. The specific formula is as follows:

$$\mathbf{f}_n = -kl\mathbf{n} \tag{9}$$

$$l = h - r_{s-f} \tag{10}$$

$$\mathbf{f}_t = \eta\left(\frac{\partial u}{\partial t} - \mathbf{v}\right) \tag{11}$$

$$\mathbf{f}_c = \mathbf{f}_n + \mathbf{f}_t \tag{12}$$

where k is the elastic coefficient, η is the elastic potential energy density, h is the smooth radius of the particle, and n is the normal vector of the finite element surface.

Similarly, for the finite element, the contact force is also distributed to the FEM node as an external force for dynamic calculation, as shown in Equation (13):

$$\int_k^t = \frac{2w_k f_{con}}{1 + w_3^2 + w_3^2 + w_3^2}, \ k = \{1,2,3\} \tag{13}$$

where w_k is the barycentric coordinate of the contact point between the fluid particle and the element surface relative to the triangular element surface. After the force on each node element is calculated, it can be summed to obtain the contact force on the whole solid node, which is applied to the solid together with gravity as an external force.

3. Method Validation Analysis

3.1. Methods Analysis

Researchers have previously investigated the generation of sea spray from a single wave striking the hull of a ship. One of the more in-depth studies was conducted by Dehghani (2018) [40] for the examination of the trajectory study of sea spray. In order to determine the position, velocity, and acceleration of the sea spray, his paper makes the assumption that the sea spray is spherical and substitutes Newton's law based on the drag, body force, and added mass force on the sea spray. This study establishes the validation model in Dehghani's publication in order to determine whether the SPH algorithm suggested in this paper can be used as a method for sea spray particle calculation.

The arithmetic model in this paper is shown in Figure 3b, derived from the model of Dehghani et al. in Figure 3a. Since this paper only considers the distribution of droplets rising to descending to the hull as a percentage, the subsequent droplets moving with the wind on the hull are less (Dehghani, 2018), so the wind speed and the size of the droplets can be ignored. The total width of the hull is 7.3 m and the height of the foremast is 8 m, in order to measure the highest humidity of the foremast for comparison with the model calculated by Dehghani and Zakrzewski. The height of the superstructure is 4.5 m, which is the same as the model for comparison, in order to obtain information on the height and position of the sea spray particles reaching the superstructure, and whether any sea spray particles can reach the top of the superstructure or even go beyond the superstructure to cover the rear part of the hull. The above three points are the relevant information points as the comparison. The calculated parameters are shown in Table 1.

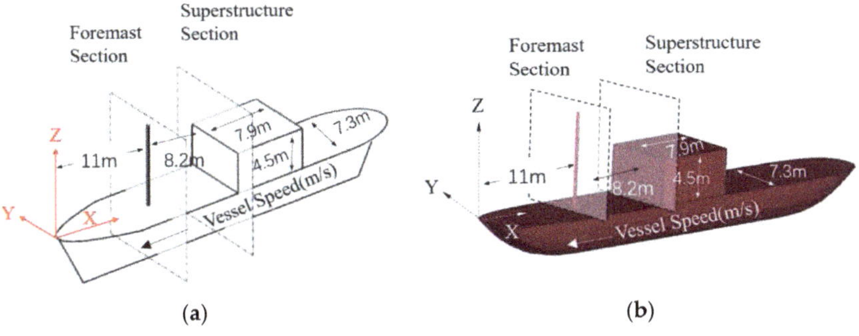

Figure 3. This figure shows the model built imitating the Dehghani model: (**a**) schematic of the MFV used as the ship model for the model of 3-D spray cloud by Dehghani (2018) [40], and (**b**) schematic of the MFV used as the ship model for the model of 3-D spray cloud by using SPH method.

Table 1. The calculation parameters of wave impacting hull forming ocean droplet model calculations.

FEM (Hull) Parameters	Value	SPH Parameters	Value
Material density/(kg/m^3)	7850	Fluid particle density(kg/m^3)	998
Grid spacing/m	0	Initial spacing of fluid particles/m	0.01
Number of grids	1067	Number of fluid particles	26,570
Modulus of elasticity/MPa	2.06×10^5	Smooth kernel functions	Wendland
Poisson's ratio	0.25	Smooth length/m	0.015
Global damping factor	0.00	Pressure correction algorithm	CSPM algorithm
FEM Time step/s	1.0×10^{-4}	Viscosity coefficient	0.3
		Sound velocity coefficient	10
		Fluid dynamic viscosity/(Pa·s)	1.0×10^{-3}
		SPH Time step/s	1.0×10^{-4}

By using the coupled SPH-FEM algorithm for the numerical simulation of wave-induced sea spray, the process of wave impacting the hull and starting to break up to form sea spray is shown in Figure 4. Firstly, 26,570 water particles are generated as shown in the blue part of Figure 4a; through the motion, a wave of 3.45 m wave height is generated, at this time, the front part of the ship starts to contact with the wave current, which makes the wave current particles start to break up and initially form smaller sea spray as shown in Figure 4b. After the wave continues to impact the ship, the sea spray particles gradually increase and start to bifurcate, part of them first land on the ship, and the other part of the separated sea spray particles start to rise as shown in Figure 4c. A part of the separated sea spray particles formed a preliminary rising trend, this part of the particles were continuously impacted and began to rise as shown in Figure 4d. Sea spray particles with the impact force continued to rise to the highest point and began to land on the hull, the highest humidity in the front mast part of the hull, that is, the maximum height of the sea spray particles to reach the front mast is 6.84 m as shown in Figure 4e.

As can be seen from the simulation of flying particles in Figure 4a–e, wave particles begin to form flying particles after impacting the hull, part of the particles land on the hull, and part of the particles rise to the highest point. The SPH particles do not penetrate the FEM grid in this process, demonstrating the effectiveness of the penalty function contact algorithm used in this example. Secondly, to determine whether the coupling algorithm used in this example is effective, it is possible to compare the stress time course at the point where SPH particles and FEM meet. Figure 5 fits the adjacent SPH and FEM pressure comparison graph; however, due to the number of SPH and the calculation method, the pressure rising and falling trend is slightly different from the value of the FEM grid by 0.1 s. As can be seen from the figure, there is no abrupt change point formed at the boundary, the pressure peaks of both systems are comparable, and the pressure values trend in the same direction. As a result, the coupling setting in this calculation can guarantee that the stress data from the near-zone SPH can be transmitted to the FEM unit.

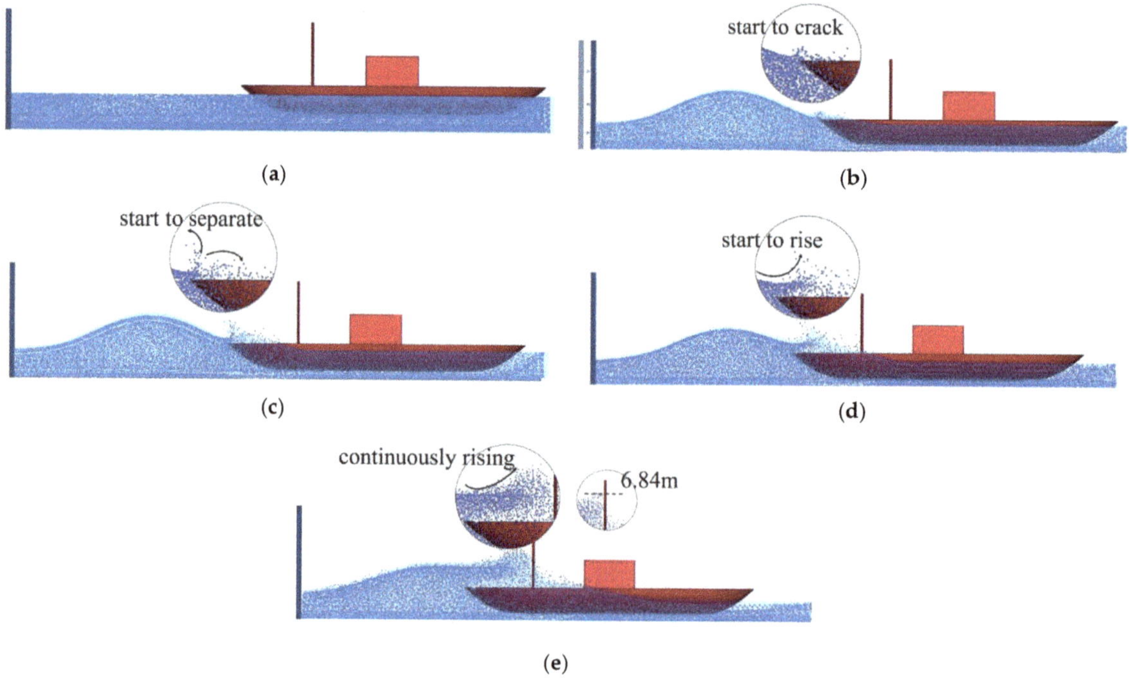

Figure 4. This is the sea spray generation process: (**a**) the ship is placed in the current particles, (**b**) waves begin to form sea spray, (**c**) sea spray particles start to disperse, (**d**) sea spray particles start to rise, and (**e**) sea spray particles continue to rise to the highest point.

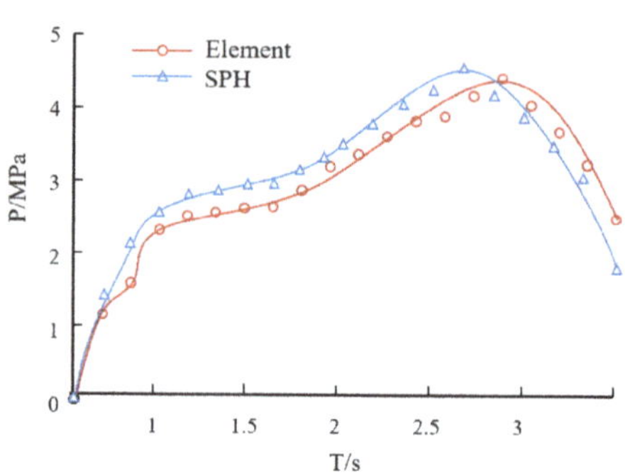

Figure 5. The pressure comparison curve between adjacent SPH particles and FEM grid.

3.2. Method Validation

The highest position of the sea spray particles in this mathematical example SPH model is depicted in Figure 6, according to a comparison of the maximum position of the sea spray trajectory in Dehghani's (2018) study. The greatest value of the sea spray from their model calculation is lower than the highest value from the SPH model because the data provided by Dehghani et al. include the LWC value and the wave height operating on the hull is 3.09 m. The graphs also show that the greatest values for the SPH model and for the Dehghani et al. model are both approximately 5–6 m on the ship's foredeck, and that the rising and decreasing trends of the sea spray particles are identical. In addition, at 11 m at the front mast, the highest humidity of Dehghani's 2-D model is at 6.28 m, the 3-D model is at 6.25 m, and the model of this calculation is at 6.84 m, which is within the range of the decreasing trend. The results of this computation are acceptable because the goal of this publication is to measure the number of occupied particles in a circumstance, which is distinct from the goal of Dehghani et al. study.

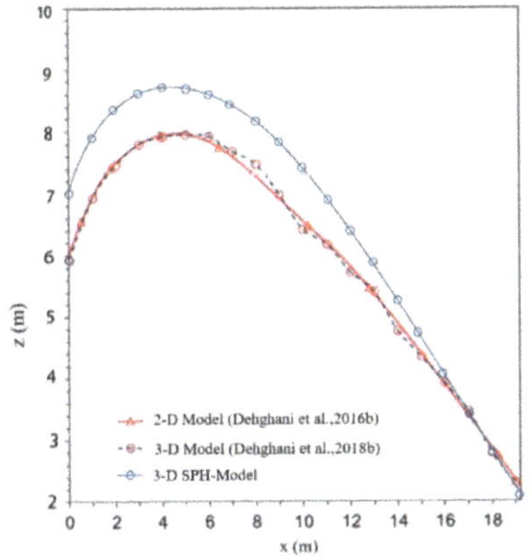

Figure 6. The maximum height of the model in this example is compared with the two-dimensional and three-dimensional models of Dehghani et al. (2016,2018) [40,41].

The calculation range can be broadly defined as 0–12 m and 12–18.5 m by the deck sea spray distribution map of Figure 7a. In the case of Dehghani et al., the water sea spray particles are split into particles of different sizes for normalized statistics. According to the literature model Figure 7b, the smaller water droplets are concentrated between 12 and 18.5 m from the bow, while the larger water droplets are concentrated in great numbers near the tip of the bow. SPH particles are denser at the bow's tip and less dense between 12 and 18.5 m away.

Finally, the comparison between the maximum humidity of the front mast and the maximum humidity of the front side of the superstructure is reported by the field observations of Zakrzewski (1988) et al. and supplemented by the analysis of Dehghani et al. Table 2 shows the reported data from previous studies and the numerical simulation results of the arithmetic example in this paper. According to the results of Figure 6, the front mast's maximum humidity is within the trend of droplet descent and the maximum humidity of the superstructure's front side is within a tolerable range when compared to the findings of earlier research. In contrast to the three earlier experiments, there is no droplet concentration on the top of the superstructure.

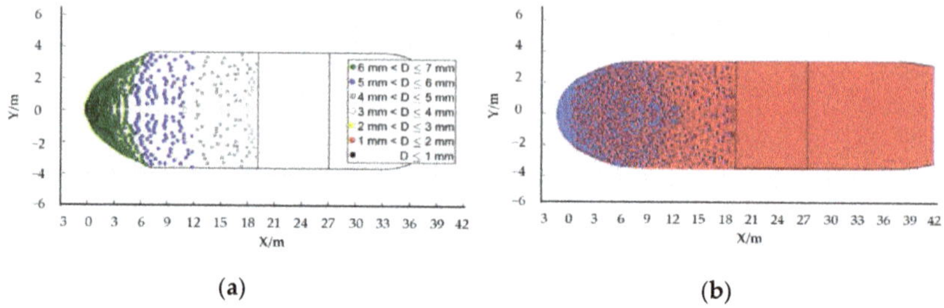

Figure 7. The deck droplet distribution for both models: (a) Dehghani et al. (2018) droplet distribution in the impact deck [40] and (b) SPH model arithmetic example of droplet distribution on impact deck.

Table 2. A comparison between the results of 2-D and 3-D models, field observations, and previous data.

Positions on the Vessel	Result			
	Zakrzewski et al. (1988) [30]	Dehghani et al. 2-D (2016b) [41]	Dehghani et al. 3-D (2018b) [40]	SPH-Model
Wet height of the foremast	5.85 m	6.28 m	6.25 m	6.84 m
Front side of the superstructure	2.07 m	2.33 m	2.6–3.5 m	3.1 m
Roof of the superstructure	No spray	No spray	No spray	No spray

4. Case and Analysis of Calculation Results

There is no report on the droplet distribution of oceanic droplets in the existing research on medium-sized container ships. This paper adopts the SR108 container ship [42], which already has the relevant ship motion performance study, as an example to do the model study. Since this paper only considers the wave droplet conditions of a single cycle, the stationary results at the end of a single cycle need to be obtained, ignoring the phenomenon of smaller droplets in the ship's continuous motion, so the model can ignore the influence of wind speed. Considering that the wave is the most direct factor causing the ocean droplet, and the wave height is small in the 1–3 level sea state, which is not close to the height of the ship, four different standard sea state schemes of level 4, level 5, level 6 and level 7 are selected to do the analysis of the wave impact hull model. Among them, Scheme 1 uses the fourth level sea state wave height, Scheme 2 uses the fifth level sea state wave height, Scheme 3 uses the sixth level sea state wave height, and Scheme 4 uses the seventh level sea state wave height. The length of the ship model is 175 m, the draft is 4.72 m, the width is 8.5 m, the speed of the ship is 24 m/s, and the incident wavelength is 175 m. Figure 8.

The results from the numerical simulation of wave impact on marine hull structures can be used as a reference for ice accumulation model predictions for ships sailing in polar regions. The flux percentage of ocean droplets on the hull is one of the primary reference factors for ice accretion, so the main objective of this research is to determine how many droplet particles remain on the hull following a single wave cycle that impacts it. The number of particles is calculated within one-third of the hull position from the bow, and the following scenario model is simulated using the algorithm settings already verified in Section 3.

One of the scenarios simulates the working condition of wave impacting the hull under the fourth level of sea state Figure 9. Under the realistic sea state operation, when the sea state enters the fourth level, the waves on the sea surface start to have a more obvious shape and will form waves of 1.25 to 2.5 m in height on the sea surface. The outcomes of the simulations determined by deriving the wave values demonstrate that the waves currently lack the capacity to produce sea spray for the ship hull used in the simulation.

There will be no impact on the ice buildup on the ship deck and superstructure because the water sea spray produced by the wave impacting the hull fall back to the surface before they can rise to the hull deck.

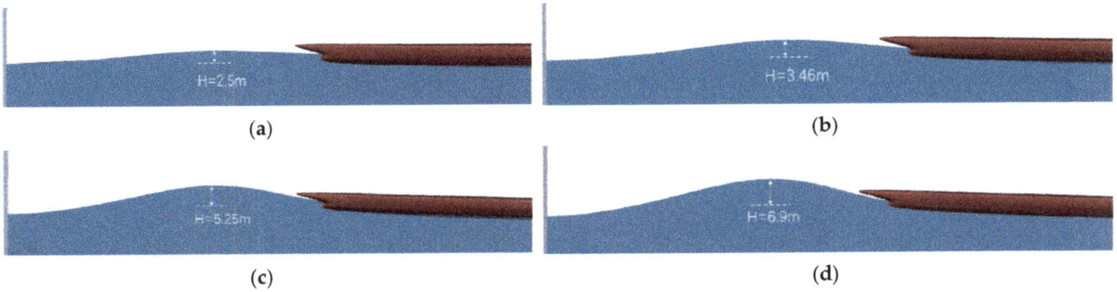

Figure 8. Four models with different levels of wave height: (**a**) four-level sea state simulation scheme, (**b**) five-level sea state simulation scheme, (**c**) six-level sea state simulation scheme, and (**d**) seven-level sea state simulation program.

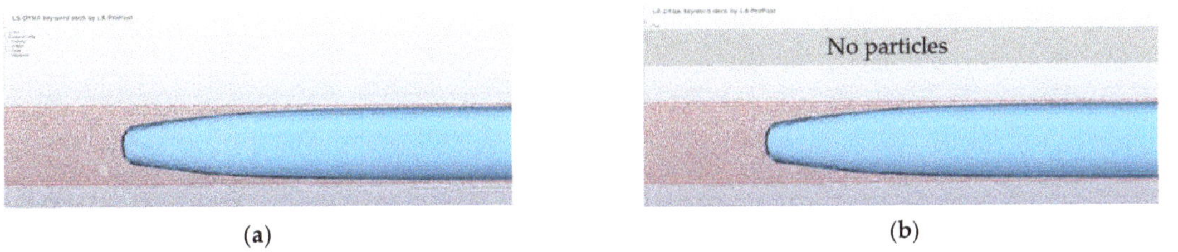

Figure 9. These are the two working conditions of Scenario 1: (**a**) conditions in the fourth-level sea state when waves hit the hull and (**b**) conditions when the fourth-level sea state's sea spray particles settle.

In scenario 2, waves impacting the hull under fifth-level sea conditions are simulated Figure 10. Since the impact on the hull structures at sea begins to increase when the sea state reaches the fifth level, it is often impossible to ignore the wave factor at this time. Larger wave crests start to appear on the sea surface at this level, and the waves take up a significant portion of the ocean and form waves that range in height from 2.5 to 4.0 m. The results of numerical simulation show that when the wave impacts the hull in class V sea state, a single wave has a number of 10,719 SPH particles, and sea spray start to form at the bow of the ship, and the droplets fall onto the deck after a period of operation. The number of ocean droplet particles produced when the wave hit the hull was 992, and the number of droplet particles after hitting the hull and falling into the hull was reduced by 412, showing that 41% of the droplets will return to the ocean surface quickly, and the remaining droplets will stay on the deck and are primarily concentrated on the sides of the deck, accounting for 9.25% of the individual waves.

Scenario 3 simulates the corresponding operating conditions in the sixth-level sea state Figure 11. Under the realistic sea state operation, when entering the sixth-level sea state, the wave crests on the sea surface sometimes take the shape of long waves of a storm, with tall crests everywhere, and the edges of the crests start to break and form huge waves. According to the simulation's findings, which were derived from wave height measurements, the waves in the class VI sea state contain 17,608 SPH particles. When the hull is struck by waves, a wider area of marine droplets has begun to form on the bow as a result of the wave contacting the hull, producing 1905 droplet particles. After the impact, there were still 1368 droplet particles on the hull's surface, which accounted for 10.81% of the individual waves, and their layout range gradually shifted from the hull's sides to the center, eventually creating an arc.

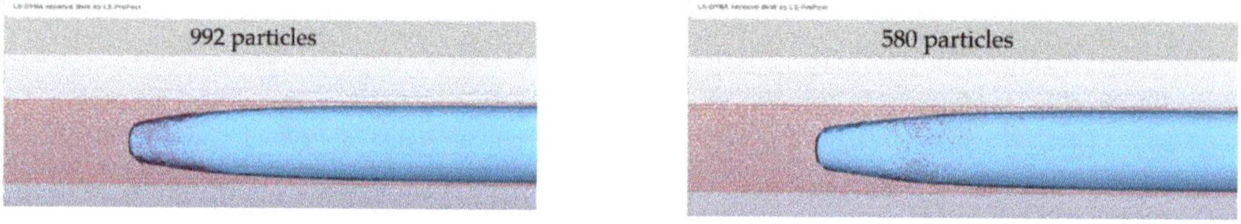

(a) (b)

Figure 10. These are the two working conditions of Scenario 2: (**a**) conditions in the fifth-level sea state when waves hit the hull, and (**b**) conditions when the fifth-level sea state's sea spray particles settle.

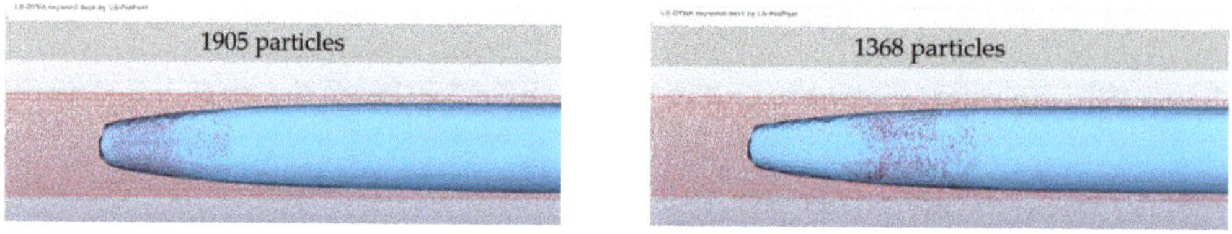

(a) (b)

Figure 11. These are the two working conditions of Scenario 3: (**a**) conditions in the sixth-level sea state when waves hit the hull, and (**b**) conditions when the sixth-level sea state's sea spray particles settle.

The corresponding operating conditions about wave height under the seventh-stage sea state are simulated in Scenario 4 Figure 12. Under the realistic sea state operation, the sea surface begins to lurch, wave crests begin to roll, and flying foam can decrease visibility, which is a bad weather condition and makes navigation difficult for the steamer. According to the numerical simulation method, there are 21,238 SPH particles in the wave of the seventh -level sea state. The number of marine droplet particles created by the wave striking the hull is 2868, a significant increase from the above scheme. Additionally, there are 2153 droplet particles that fall on the hull, accounting for 13.51% of each wave, and the hull is completely covered in numerous droplets from both sides of the hull to the hull, with a large number of water particles and a clear arc-shaped distribution.

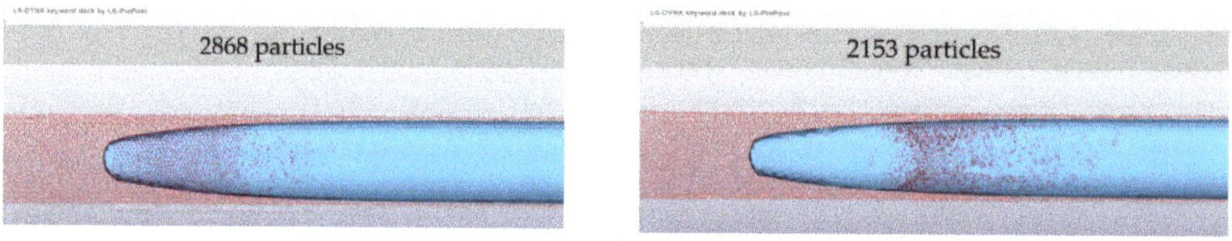

(a) (b)

Figure 12. These are the two working conditions of Scheme 4: (**a**) conditions in the seventh-level sea state when waves hit the hull, and (**b**) conditions when the seventh-level sea state's sea spray particles settle.

5. Conclusions

From the numerical simulation and statistical results, we can analyze the relationship between the percentage of marine droplet particles on the hull according to the SPH particles of individual waves as a reference, and can derive the percentage of marine sea spray particles on the hull under different sea conditions, respectively. In accordance with Figure 13a, the class V sea state will deposit 9.25% of the droplet particles in the hull, the class VI sea state will deposit 10.81% of the sea spray particles and the class VII sea state will deposit 13.51% of the sea spray particles. The ratio will rise as the wave value rises, and the relative particle number follows a similar pattern. At the same time, Figure 12b shows that when the droplet particles fall back to the hull, some particles will return to the sea through their own motion, and the trend of the number of particles on the hull is decreasing, however, the number of returning particles is more stable under different sea conditions, and as sea conditions increase, the change in the number of returning particles is smaller.

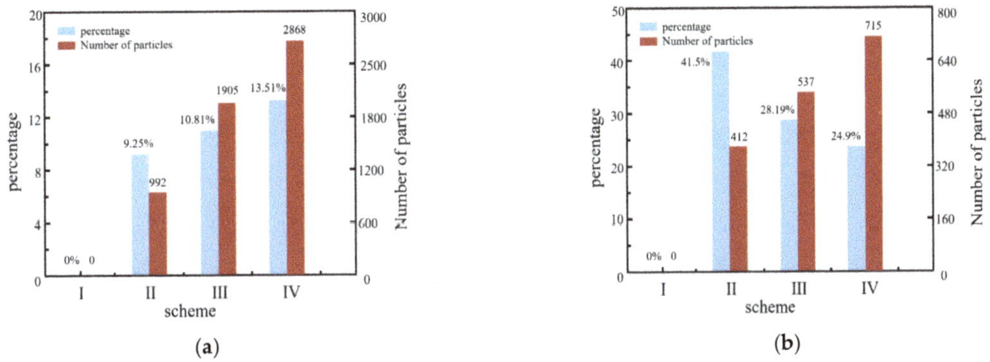

Figure 13. (**a**) The number of particles on the hull and their share in the number of wave particles and (**b**) number of returning particles and their share in the number of particles in the hull.

In this paper, the SR108 container ship model, which already has the corresponding motion state research conclusions, is selected as the numerical simulation research model, and four different sea state simulation scenarios are set up to study the marine droplet working conditions formed by different wave heights impacting the hull in a single cycle, and further statistics are obtained on the effects of different sea state scenarios on the formation of marine droplets. The following conclusions can be drawn from the above study:

(1) After fifth-level sea state, a single cycle of waves will generate approximately 9.25% to 13.5% of the total number of particles on the hull of a single wave, and the higher the sea state level, the higher the number of particles and the higher the percentage of particles—where fifth-level sea state particles account for 9.25% of the total number of particles, six-level sea state accounts for 10.81%, and 13.51% of the seventh sea state.

(2) A small amount of sea spray particles returning to the sea surface soon after generation, and the particles that fall back to the sea surface are less related to the sea condition level and mainly influenced by the shape of the hull structure, which can be studied in more detail in the bow design for ice prevention later.

(3) The sea spray particles left on the hull are mainly distributed at both sides of the hull and gradually spread to the center as the sea state rises. When the sea state level is low, the attention to both sides of the hull needs to be increased when preventing icing, and when the sea state level is high, the attention to the ice accumulation on the superstructure in the center of the hull needs to be increased on the basis of the preliminary work.

(4) The coupled SPH-FEM method can simulate the distribution working conditions of marine droplets after interacting with the ship hull under different sea conditions, provide the physical modeling of droplet sources and effectively track the information of droplet particles, which can provide technical support for later prediction of droplet flux information related to the work of polar ship ice accumulation.

Author Contributions: J.C.; calculation method determined, put together the data, performed the data analysis and took part in the manuscript write-up. X.B.; contributed to the study conception and design, J.W.; collected the conference data and modeling, G.C. and T.Z.; simulation, and writing. All authors have read and agreed to the published version of the manuscript.

Funding: This research was supported by the National Natural Science Foundation of China (No. 51879125), the Postgraduate Research & Practice Innovation Program of Jiangsu Province (No. KYCX22_3777), the Natural Science Foundation of Jiangsu Province (No. BK20211342) and the Key R&D Projects in Guangdong Province (No. 2020B1111500001).

Data Availability Statement: The data presented in this study are available upon request from the corresponding author or the first author.

Acknowledgments: We thank the editor and anonymous reviewers for their comments, which considerably improved this work.

Conflicts of Interest: The authors declare no conflict of interest.

References

1. Zhou, L.; Liu, R.; Yi, X. Research and Development of Anti-Icing/Deicing Techniques for Vessels: Review. *Ocean. Eng.* **2022**, *260*, 112008. [CrossRef]
2. Lozowski, E.P.; Stallabrass, J.R.; Hearty, P.F. The icing of an Unheated, Nonrotating Cylinder. Part I: A Simulation Model. *J. Appl. Meteorol. Climatol.* **2017**, *22*, 2053–2062. [CrossRef]
3. Mintu, S.; Molyneux, D. Ice Accretion for Ships and Offshore Structures. Part 1—State of the Art Review. *Ocean Eng.* **2022**, *258*, 111501. [CrossRef]
4. Wang, Y.; Ju, L.; Han, D.; Wang, Q. Numerical Investigation of the Impacting and Freezing Process of a Single Supercooled Water Droplet. *Phys. Fluids* **2021**, *33*, 042114. [CrossRef]
5. Veron, F. Ocean Spray. *Annu. Rev. Fluid Mech.* **2015**, *47*, 507–538. [CrossRef]
6. Ryerson, C.C. Ice Protection of Offshore Platforms. *Cold Reg. Sci. Technol.* **2011**, *65*, 97–110. [CrossRef]
7. Lozowski, E.P.; Szilder, K.; Makkonen, L. Computer Simulation of Marine Ice Accretion. *Philos. Trans. R. Soc. London. Ser. A Math. Phys. Eng. Sci.* **2000**, *358*, 2811–2845. [CrossRef]
8. Kulyakhtin, A.; Tsarau, A. A Time-Dependent Model of Marine Icing with Application of Computational Fluid Dynamics. *Cold Reg. Sci. Technol.* **2014**, *104–105*, 33–44. [CrossRef]
9. Samuelsen, E.M.; Edvardsen, K.; Graversen, R.G. Modelled and Observed Sea-Spray Icing in Arctic-Norwegian Waters. *Cold Reg. Sci. Technol.* **2017**, *134*, 54–81. [CrossRef]
10. Shojaei Barjouei, A.; Naseri, M. A Comparative Study of Statistical Techniques for Prediction of Meteorological and Oceanographic Conditions: An Application in Sea Spray Icing. *JMSE* **2021**, *9*, 539. [CrossRef]
11. Naseri. Samuelsen Unprecedented Vessel-Icing Climatology Based on Spray-Icing Modelling and Reanalysis Data: A Risk-Based Decision-Making Input for Arctic Offshore Industries. *Atmosphere* **2019**, *10*, 197. [CrossRef]
12. Jones, K.F.; Andreas, E.L. Sea Spray Concentrations and the Icing of Fixed Offshore Structures. *Q. J. R. Meteorol. Soc.* **2012**, *138*, 131–144. [CrossRef]
13. Samuelsen, E.M. Prediction of Ship Icing in Arctic Waters-Observations and Modelling for Application in Operational Weather Forecasting. Ph.D. Thesis, UiT The Arctic University of Norway, Tromsø, Norway, 2017.
14. Moore, G.W.K. A Climatology of Vessel Icing for the Subpolar North Atlantic Ocean. *Int. J. Climatol.* **2013**, *33*, 2495–2507. [CrossRef]
15. Fairall, C.W.; Banner, M.L.; Peirson, W.L.; Asher, W.; Morison, R.P. Investigation of the Physical Scaling of Sea Spray Spume Droplet Production. *J. Geophys. Res. Ocean.* **2009**, *114*, C10. [CrossRef]
16. Mintu, S.; Molyneux, D.; Colbourne, B. A Theoretical Model for Ship–Wave Impact Generated Sea Spray. *J. Offshore Mech. Arct. Eng.* **2021**, *143*, 041201. [CrossRef]
17. Bodaghkhani, A.; Dehghani, S.R.; Muzychka, Y.S.; Colbourne, B. Understanding spray cloud formation by wave impact on marine objects. *Cold Reg. Sci. Technol.* **2016**, *129*, 114–136. [CrossRef]
18. Young, F.R. *Fizzics: The Science of Bubbles, Droplets, and Foams*; The Johns Hopkins University Press: Baltimore, MD, USA, 2011; pp. 4–19.
19. Horjen, I. Offshore Drilling Rig Ice Accretion Modeling Including a Surficial Brine Film. *Cold Reg. Sci. Technol.* **2015**, *119*, 84–110. [CrossRef]
20. Kulyakhtin, A.; Kulyakhtin, S.; Løset, S. The Role of the Ice Heat Conduction in the Ice Growth Caused by Periodic Sea Spray. *Cold Reg. Sci. Technol.* **2016**, *127*, 93–108. [CrossRef]
21. Haukebø, A.R. Modelling of Marine Icing with Close Range Photogrammetry. Master's Thesis, UiT The Arctic University of Norway, Tromsø, Norway, 2015.
22. Banari, A. *Lattice Boltzmann Simulation of Multiphase Flows*; Application to Wave Breaking and Sea Spray Generation, University of Rhode Island: Kingston, RI, USA, 2014.

23. Tang, S.; Yang, Z.; Liu, C.; Dong, Y.-H.; Shen, L. Numerical Study on the Generation and Transport of Spume Droplets in Wind over Breaking Waves. *Atmosphere* **2017**, *8*, 248. [CrossRef]
24. Kees, C.E.; Akkerman, I.; Farthing, M.W.; Bazilevs, Y. A Conservative Level Set Method Suitable for Variable-Order Approximations and Unstructured Meshes. *J. Comput. Phys.* **2011**, *230*, 4536–4558. [CrossRef]
25. Watanabe, Y.; Ingram, D.M. Size Distributions of Sprays Produced by Violent Wave Impacts on Vertical Sea Walls. *Proc. R. Soc. A* **2016**, *472*, 20160423. [CrossRef]
26. Ding, Z.; Ren, B.; Wang, Y.; Ren, X. Experimental Study of Unidirectional Irregular Wave Slamming on the Three-Dimensional Structure in the Splash Zone. *Ocean. Eng.* **2008**, *35*, 1637–1646. [CrossRef]
27. Greco, M.; Lugni, C. 3-D Seakeeping Analysis with Water on Deck and Slamming. Part 1: Numerical Solver. *J. Fluids Struct.* **2012**, *33*, 127–147. [CrossRef]
28. Dehghani-Sanij, A.R.; Dehghani, S.R.; Naterer, G.F.; Muzychka, Y.S. Sea Spray Icing Phenomena on Marine Vessels and Offshore Structures: Review and Formulation. *Ocean. Eng.* **2017**, *132*, 25–39. [CrossRef]
29. Shipilova, O.; Kulyakhtin, A.; Tsarau, A.; Libby, B.; Moslet, P.O.; Loset, S. Mechanism and Dynamics of Marine Ice Accretion on Vessel Archetypes. In Proceedings of the OTC Arctic Technology Conference, Houston, TX, USA, 3–5 December 2012; p. OTC-23762-MS.
30. Zakrzewski, W.P. Splashing a Ship with Collision-Generated Spray. *Cold Reg. Sci. Technol.* **1987**, *14*, 65–83. [CrossRef]
31. Kulyakhtin, A.; Shipilova, O.; Libby, B.; Løset, S. Full-Scale 3D CFD Simulation of Spray Impingement on a Vessel Produced by Ship-Wave Interaction. In Proceedings of the The 21st IAHR International Symposium on Ice, Dalian, China, 11–15 June 2012; p. 11291141.
32. Kulyakhtin, A. Numerical Modelling and Experiments on Sea Spray Icing. Ph.D. Thesis, Norwegian University of Science and Technology, Trondheim, Norway, 2014.
33. Mintu, M.; Azam, S. Theoretical and Numerical Methods for Predicting Ship-Wave Impact Generated Sea Spray. Ph.D. Thesis, Memorial University of Newfoundland, St. John's, NL, Canada, 2021.
34. Fragassa, C.; Topalovic, M.; Pavlovic, A.; Vulovic, S. Dealing with the Effect of Air in Fluid Structure Interaction by Coupled SPH-FEM Methods. *Materials* **2019**, *12*, 1162. [CrossRef] [PubMed]
35. Korzilius, S.P.; Schilders, W.H.; Anthonissen, M.J. An Improved CSPM Approach for Accurate Second-Derivative Approximations with SPH. *J. Appl. Math. Phys.* **2016**, *5*, 168–184. [CrossRef]
36. Gomez-Gesteira, M.; Rogers, B.D.; Dalrymple, R.A.; Crespo, A.J. State-of-the-Art of Classical SPH for Free-Surface Flows. *J. Hydraul. Res.* **2010**, *48*, 6–27. [CrossRef]
37. Ben Belgacem, I.; Khochtali, H.; Cheikh, L.; Barhoumi, E.M.; Ben Salem, W. Comparison between Two Numerical Methods SPH/FEM and CEL by Numerical Simulation of an Impacting Water Jet. In *Advances in Materials, Mechanics and Manufacturing*; Springer: Berlin/Heidelberg, Germany, 2020; pp. 50–60.
38. Stasch, J.; Avci, B.; Wriggers, P. Numerical Simulation of Fluid-structure Interaction Problems by a Coupled SPH-FEM Approach. *Pamm* **2016**, *16*, 491–492. [CrossRef]
39. Wang, W.; Wu, Y.; Wu, H.; Yang, C.; Feng, Q. Numerical Analysis of Dynamic Compaction Using FEM-SPH Coupling Method. *Soil Dyn. Earthq. Eng.* **2021**, *140*, 106420. (In Chinese) [CrossRef]
40. Dehghani, S.R.; Naterer, G.F.; Muzychka, Y.S. 3-D Trajectory Analysis of Wave-Impact Sea Spray over a Marine Vessel. *Cold Reg. Sci. Technol.* **2018**, *146*, 72–80. [CrossRef]
41. Dehghani, S.R.; Muzychka, Y.S.; Naterer, G.F. Droplet trajectories of wave-impact sea spray on a marine vessel. *Cold Reg. Sci. Technol.* **2016**, *127*, 1–9. [CrossRef]
42. Li, Z.; Yang, Y. Simulation of Ship Pitch and Heave Motion in Regular Waves. *J. Dalian Marit. Univ.* **2002**, *28*, 13–16. (In Chinese) [CrossRef]

water

MDPI

Article

Research on Mooring System Design for Kulluk Platform in Arctic Region

Aobo Zhang, Zhenju Chuang *, Shewen Liu, Xin Chang, Lixun Hou, Zhen He and Shiqi Liu

School of Naval Architecture and Ocean Engineering, Dalian Maritime University, Dalian 116026, China; zhangaobo@dlmu.edu.cn (A.Z.); liushewen@dlmu.edu.cn (S.L.); xin.chang@dlmu.edu.cn (X.C.); houlixun@dlmu.edu.cn (L.H.); xiaohao@dlmu.edu.cn (Z.H.); liushiqi456@dlmu.edu.cn (S.L.)

* Correspondence: zhenjuchuang@dlmu.edu.cn

Abstract: Mooring system design of a floating offshore structure in the arctic region is considered to be extremely important. This paper aims at investigating an optimal mooring system for the Kulluk platform operating in the Beaufort Sea, which has ice-free and ice-covered conditions during the whole year time. In order to complete the layout design of the mooring system to satisfy the year-round operation, both the effect of wave loads and ice loads should be considered. The research establishes a coupled numerical production system composed of the Kulluk platform and mooring system. Wave load is solved by potential flow theory. The slender finite element method is used to compute the tension of the mooring system. The nonlinear finite element method, discrete element method, and empirical formula are compared to analyze ice load. Finally, the discrete element method is selected for the analysis of the Kulluk, and the simulated results are compared reasonably with the field data. When studying the mooring line configurations, quantitative time-domain analysis is carried out, including tension of mooring lines and the motions of the platform under different working conditions. The research work in this paper will provide a reference for the optimal design of the mooring system of the platform operating in the Arctic Sea.

Keywords: arctic oil and gas resources; Kulluk platform; mooring system; ice load; ice-structure interaction

Citation: Zhang, A.; Chuang, Z.; Liu, S.; Chang, X.; Hou, L.; He, Z.; Liu, S. Research on Mooring System Design for Kulluk Platform in Arctic Region. *Water* **2022**, *14*, 1762. https://doi.org/10.3390/w14111762

Academic Editors: Zhijun Li, Tomasz Kolerski, Xiaohong Shi, Zhengyong Zhang and Li Zhou

Received: 30 April 2022
Accepted: 28 May 2022
Published: 30 May 2022

Publisher's Note: MDPI stays neutral with regard to jurisdictional claims in published maps and institutional affiliations.

1. Introduction

At present, the resources on land are becoming fewer and fewer, and the sea contains a large number of available resources. Exploring and exploiting offshore oil and gas resources has excellent development potential, especially in the Arctic region. The cover and thickness of sea ice in the Arctic region have decreased significantly due to global warming, which is conducive to the exploitation of oil and gas resources in the Arctic region. However, petroleum exploration in the Arctic region is extremely challenging. The difficulty is that a large amount of undiscovered oil is stored in water depths greater than 100 m [1] and less than 500 m [2]. Coupled with the impact of sea ice on structures, exploration of oil and gas resources in the polar regions has not become a normal state.

Kulluk platform operated as a drilling system in the Beaufort Sea from 1983 to 1993. A large number of platform-related data were monitored in real-time. Kulluk's hull has good ice resistance, and its motion response and the tension of the mooring lines is less than other structures. Kulluk's experience provides a data source for analyzing moored structures under different sea ice conditions [3]. Zhou et al. [4] verified the effectiveness of the numerical model by comparing the simulation results of the two-dimensional numerical model with the field data of the Kulluk platform. For moored structures, ice thickness, ice drift speed, and global mooring stiffness affect the platform's motion and the tension of mooring lines. Sayed and Barker [5] described a numerical model based on mass and momentum conservation equations, sea ice constitutive equation, and moored structure motion equation. This model simulated the interaction between the broken ice floe and the

Kulluk platform and studied the influence of mooring stiffness. Fluid–structure interaction (FSI) can be used to analyze moored structures under broken ice floes [6]. This study simulated platform motion at different ice concentrations (low, medium, and high) and compared it with full-scale field data.

A mooring system can improve the stability of floating structures, so it becomes a critical component in the design. The positioning modes of floating structures mainly include a mooring system and a dynamic positioning system. The mooring system is the most common positioning method, and the thruster-assisted mooring system has become the research hotspot. Floating structures and their mooring systems are exposed to the ocean environment throughout their service lives. When the horizontal displacement of the connected floating frame exceeds the limited value, the entire mooring system will experience functional failure [7]. Garrett [8] presented a fully coupled global analysis of floating production systems, including structures, mooring systems, and riser systems. This study provided accuracy and efficiency for a fully coupled analysis of floating production system design from concept selection to final design, installation, and operation. O.A. Montasir [9] studied the difference between symmetric and asymmetric multi-point mooring configurations on structural motion and evaluated the mooring recovery force-displacement relationship of truss SPAR platform with a quasi-static method. Finally, the linear airy wave theory and Morrison equation were used to calculate the platform motion under waves. The influence of mooring line configurations on platform motion response was analyzed. Mohapatra and Guedes Soares [10] studied the hydroelastic response of the moored floating flexible plate by analyzing the effects of the mooring stiffness, incidence angle, and flexural rigidity on the deflection amplitude, plate deformations, reflection, and transmission coefficients.

With the increasing importance of oil and gas resources in the Arctic region, eliminating ice load damage to offshore platform structures has become a research focus. More and more scholars have carried out an in-depth analysis of the interaction between sea ice and the platform. Horrigmoe et al. [11] used the nonlinear finite element method to numerically calculate the ice loads based on the viscoplastic constitutive model for ice. Considering the elastic, creep, and plastic components of the strain of ice, the strain rate in a wide range can be predicted to calculate the force of ice on structures. A simple ice indentation test verified the numerical model under plane stress conditions. Hansen and Loset [12] used a two-dimensional disk element to simulate a broken ice floe, analyzed the moored structure's motion response under ice floes, and compared it with the ice pool test. This numerical simulation introduced the friction coefficient to consider friction action. Liu et al. [13] proposed an ideal elastic-plastic ice model to simulate an actual ship-iceberg collision and compared the simulated contact pressure-contact area curve with the design specification to verify the feasibility of the model. Liu and Ji [14] used the dilated polyhedral elements based on the Minkowski sum theory to describe the geometry of irregular particle elements and adopted the 2D-Voronoi tessellation algorithm to obtain the initial random distribution state of ice floes. The ice load of the Kulluk platform was analyzed by the discrete element method.

This study compares the numerical data with the field data to obtain more accurate ice load simulation results. The motion response of the Kulluk platform coupled with the mooring system is computed under wave load and ice load, respectively. The influence of wave, level ice, and ice floes on platform motion and mooring line tension is considered comprehensively. In this research, the three methods of nonlinear finite element, discrete element, and empirical formula are compared to compute ice loads. This paper uses the potential flow theory to calculate wave loads. The slender finite element method is used to analyze the mooring system. The basic principles of static catenary analysis are discussed in catenary element equilibrium equations [15] and catenary equilibrium of multi-segment single lines [16]. The calculation results in the time domain are compared from the load direction, the number of mooring lines, the number of connecting points, and the angle between mooring lines to optimize the mooring system layout design.

2. Theoretical Background

2.1. Calculation Method of Wave Loads

Kulluk platform is a large-scale structure. Considering wave diffraction and radiation effects, potential flow theory [17] is used to analyze the wave loads. SESAM software performs the platform's hydrodynamic characteristics and the structural motion response [18]. The WADAM module [19] can analyze the response in the frequency domain under wave loads. The transfer function RAO, first-order wave force, and second-order mean drift force can be obtained through computation. The SIMA module [20] can be used to analyze the time-domain response of the Kulluk coupled with the mooring system. In the study of the Kulluk platform, its flexible deformation is ignored, and it is regarded as a rigid body when solving its six6 DOF motions. The OXY plane coincides with the stationary water surface, and the OZ axis is perpendicular to the water surface and points upwards [21].

According to potential flow theory [17], the first-order motion equation of the Kulluk platform in the frequency domain can be formulated:

$$(M_{ij} + \mu_{ij})\ddot{X}_j + \lambda_{ij}\dot{X}_j + C_{ij}X_j = F_i \; (i = 1,2,\ldots,6;\; j = 1,2,\ldots,6) \quad (1)$$

where, M_{ij} is the mass matrix; μ_{ij} is the additional mass matrix; X_j is the platform's motion; λ_{ij} is the damping coefficient matrix; C_{ij} is the restoring force coefficient matrix; F_i is the first-order wave force applied to the platform.

2.2. Analysis Method of Ice Loads

The empirical formula, finite element method, and discrete element method are applied to analyze ice loads. This paper studies a cylinder structure and a cone structure under level ice as examples to compare these three research methods. Crushing failure occurs between ice and the vertical structures. Flexure failure usually dominates ice acting on cone structures.

2.2.1. The Empirical Formula of ISO (The International Organization for Standardization)

The ISO-19906(2019) [22] standard is the reference for designing and manufacturing offshore structures in the oil and gas industry.

1. Crushing failure of ice

The calculation formula is:

$$F_G = p_G \cdot h \cdot w \quad (2)$$

$$p_G = C_R\left[\left(\frac{h}{h_1}\right)^n \left(\frac{w}{h}\right)^m + f_{AR}\right] \quad (3)$$

$$f_{AR} = e^{\frac{-w}{3h}}\sqrt{1 + 5\frac{h}{w}} \quad (4)$$

where p_G is the global average ice pressure; w is the projected width of the structure; h is the thickness of the level ice; h_1 is a reference thickness of 1 m; m is an empirical coefficient equal to −0.16; n is an empirical coefficient equal to −0.5 + $h/5$ for $h < 1.0$ m and to −0.3 for $h \geq 1.0$ m; C_R is the ice strength coefficient equal to 2.8 Mpa in the Beaufort Sea.

2. Flexure failure of ice

The ice loads acting on the cone structure include actions due to the flexure failure of the level ice and the ride-up actions due to ice pieces. The horizontal breaking action H_B and the vertical breaking action V_B are given by:

$$H_B = \frac{\sigma_f h^2}{3}\frac{\tan\alpha}{1 - \mu g_r}\left[\frac{1 + Yx\ln x}{x - 1} + G(x-1)(x+2)\right] \quad (5)$$

$$V_B = H_B \cdot h_v \quad (6)$$

where Y is equal to 2.711 for Tresca yielding; $G = \frac{\rho_{ice} g D^2}{4\sigma_f \cdot h}$, $x = 1 + \left(3G + \frac{Y}{2}\right)^{-\frac{1}{2}}$, $g_r = \frac{\sin\alpha + \frac{\alpha}{\cos\alpha}}{\frac{\pi}{2}\sin^2\alpha + 2\mu\alpha\cos\alpha}$.

The horizontal ride-up action H_R and the vertical ride-up action V_R are defined as given by:

$$H_R = W\frac{\tan\alpha + \mu E_2 - \mu f g_r \cos\alpha}{1 - \mu g_r} \tag{7}$$

$$V_R = W\cos\alpha\left(\frac{\pi}{2}\cos\alpha - \mu\alpha - f h_V\right) + H_R h_V \tag{8}$$

$$W = \rho_i g h_r \frac{w^2 - w_T^2}{4\cos\alpha} \tag{9}$$

$$f = \sin\alpha + \mu E_1 \cos\alpha \tag{10}$$

$$h_V = \frac{f\cos\alpha - \mu E_2}{\frac{\pi}{4}\sin^2\alpha + \mu\alpha\cos\alpha} \tag{11}$$

$$E_1 = \int_0^{\frac{\pi}{2}} \left(1 - \sin^2\alpha\sin^2\eta\right)^{-\frac{1}{2}} d\eta \tag{12}$$

$$E_2 = \int_0^{\frac{\pi}{2}} \left(1 - \sin^2\alpha\sin^2\eta\right)^{\frac{1}{2}} d\eta \tag{13}$$

where α is the slope of the structure measured from the horizontal; w_T is the top diameter of the cone; h_r is the ice ride-up thickness ($h_r \geq h$).

The total action components in horizontal and vertical directions are obtained, respectively, by

$$F_H = H_B + H_R \tag{14}$$

$$F_V = V_B + V_R \tag{15}$$

2.2.2. Finite Element Method

The nonlinear explicit dynamics analysis software LS-DYNA [23] has the function of solving fluid–structure interaction. The nonlinear finite element method can divide the research subject into several elements. These elements are connected by nodes that transfer loads between the elements.

1. Ice-structure contact algorithm

Three methods for LS-DYNA are applied to deal with contact collision, including the node constraint method, allocation parameter method, and penalty function method. This paper uses the penalty function method for analysis [24].

The contact force of the penalty function method is: $F = K\delta$,

where K is the contact surface stiffness (determined by element size and material properties); δ is the amount of penetration.

2. Fluid-structure interaction algorithm

LS-DYNA has the Lagrange, Euler, and ALE algorithms according to the various coordinate types. ALE algorithm can perform better dynamic analysis of fluid–structure interaction and combine the Lagrange algorithm and Euler algorithm characteristics. It can track the motion of the structure similarly to the Lagrange algorithm. It can also adjust the position of the mesh according to the demands of the solution to avoid severe deformation of the mesh [25].

LS-DYNA uses penalty function method for fluid dynamics calculation, viscosity coefficient and equation of state (EOS), and null-material model for fluid modeling and adopts the keyword "CONSTRAINED_LAGRANGE_IN_SOLID" to directly couple solid and fluid elements.

The null material model (*MAT_NULL) is used to simulate the water in conjunction with the equation of state, and viscosity is also defined. Null material models can effectively simulate fluid and hydrodynamic properties. The deviational shear stress of the fluid element σ_d is proportional to the shear strain ε':

$$\sigma_d = 2\mu\varepsilon' \tag{16}$$

where μ represents the viscosity of the fluid in Pa·s.

Fluid can be described by the following two equations of state: Gruneisen equation of state and linear polynomial equation of state.

The equation of state of water is described by Gruneisen equation of state. Water is defined as the pressure of the compressed material:

$$p = \frac{\rho_o C^2 \mu \left[1 + \left(1 - \frac{\gamma_0}{2}\right)\mu - \frac{a}{2}\mu^2\right]}{\left[1 - (S_1 - 1)\mu - S_2 \frac{\mu^2}{\mu+1} - S_3 \frac{\mu^3}{(\mu+1)^2}\right]} + (\gamma_0 + \alpha\mu)E \tag{17}$$

where P is pressure; C is the intercept of impact velocity V_x- particle velocity V_p curve; S_1, S_2 and S_3 are the coefficients of slope of V_x-V_p curve; γ_0 is the Gruneisen constant; α is the first-order volume correction of γ_0.

The equation of state of air is described by the linear polynomial equation of state. The linear polynomial equation of state represents the linear relationship of internal energy per unit initial volume. The pressure value is given by the following equation:

$$p = C_0 + C_1\mu + C_2\mu^2 + C_3\mu^3 + \left(C_4 + C_5\mu + C_6\mu^2\right)E \tag{18}$$

where C_0, C_1, C_2, C_3, C_4, C_5, and C_6 are constants; if $\mu < 0$, then $C_2\mu^2$ and $C_6\mu^2$ are set to 0, where:

$$\mu = \frac{1}{V} - 1 \tag{19}$$

where V represents the relative volume.

The specific calculation parameters are shown in Tables 1–3.

Table 1. Main parameters of the structure in finite element calculation.

Mass Density	Young's Modulus	Poisson's Ratio
7850 kg/m^3	206 Gpa	0.3

Table 2. Main parameters of level ice in finite element calculation.

Mass Density	Shear Modulus	Yield Stress	Plastic Hardening Modulus	Bulk Modulus	Failure Pressure	Plastic Failure Strain
900 kg/m^3	2.2 Gpa	2.12 Mpa	4.26 Gpa	5.26 Gpa	−4 Mpa	0.35

Table 3. Main parameters of water and air in finite element calculation.

	State Equation	Density	Viscosity Coefficient	Failure Pressure
Air	*EOS_LINEAR_POLYNOMIAL	1.25 kg/m^3	1.74×10^{-5}	−10 Mpa
Water	*EOS_GRUNEISEN	1000 kg/m^3	0.9×10^{-3}	−10 Mpa

2.2.3. Discrete Element Method

The discrete element method [26] uses a parallel bond model to simulate level ice composed of spherical particles. An elastic bonding disk is between the two bonding particles, namely transforming forces and moments [27]. The parallel bond model is shown in Figure 1:

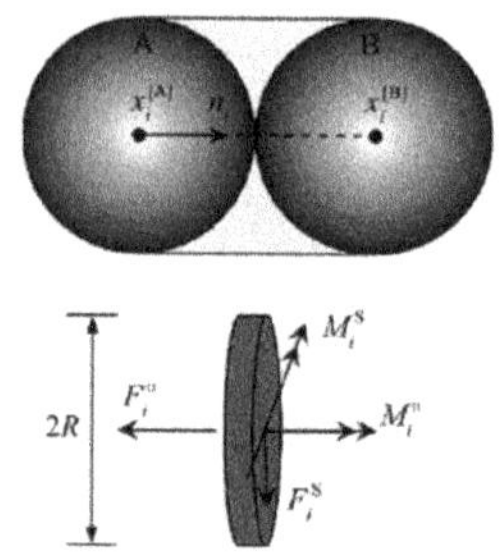

Figure 1. Parallel bonding model for spherical elements.

In the parallel bonding model, $\vec{F_i^n}$ and $\vec{F_i^s}$ are, respectively, the normal force and tangential force between particles; $\vec{M_i^n}$ and $\vec{M_i^s}$ are, respectively, the normal moment and tangential moment between particles. The maximum normal stress and maximum shear stress acting on the bonding disk can be expressed as [28]:

$$\sigma_{\max} = \frac{-\vec{F_i^n}}{A} + \frac{\left|\vec{M_i^S}\right|}{I} R \tag{20}$$

$$\tau_{\max} = \frac{\vec{F_i^s}}{A} + \frac{\left|\vec{M_i^n}\right|}{J} R \tag{21}$$

where R is the radius of the bonding disk, A is the cross-sectional area of the bonding disk, and I and J are the moment of inertia and polar moment of inertia of the bonding disk, respectively. The specific formula can be expressed as: $A = \pi R^2, J = \frac{1}{2}\pi R^4, I = \frac{1}{4}\pi R^4$.

The bonding failure model of particles is divided into tensile failure and shear failure [28], as demonstrated in Figure 2. When the fracture of ice is simulated, and the maximum stress on the bonding disk exceeds the strength of bonding failure, the fracture criterion appears, as described in Figure 3.

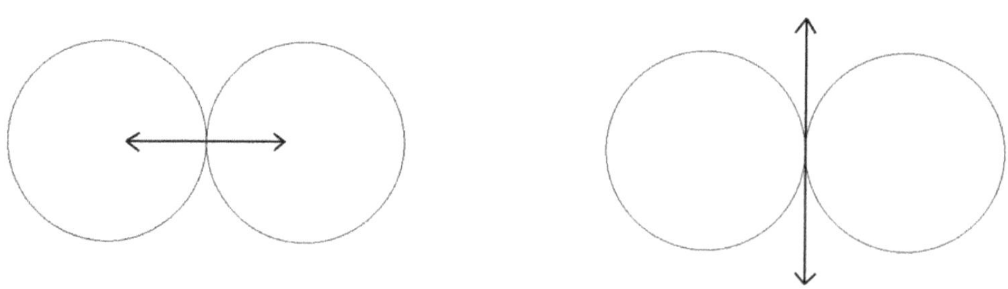

(a) (b)

Figure 2. Bonding failure model of particle element. (**a**) Tensile failure; (**b**) shear failure.

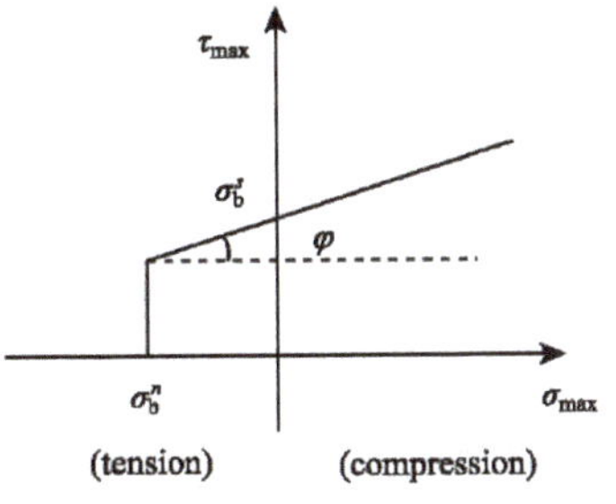

Figure 3. Fracture criterion with tensile and shear failure.

The tensile failure strength σ^t and shear failure strength τ^s of the bonding elements can be expressed as [29]:

$$\sigma^t = \sigma_b^n \tag{22}$$

$$\tau^s = \sigma_b^s + \mu_b \sigma_{\max} \tag{23}$$

where σ_b^n and σ_b^s denote the normal bonding strength and tangential bonding strength, respectively. In this paper, $\sigma_b^n = \sigma_b^s$. μ_b is the internal friction coefficient; $\mu_b = \tan\varphi$. φ is the internal friction angle.

This paper uses the discrete element method to analyze the broken ice floes condition. A two-dimensional Voronoi tessellation algorithm [30] constructs the ice fragmentation regions with different ice thickness, ice concentration, and average ice area. A Voronoi diagram, also known as a Dirichlet diagram or Tyson polygon, is a set of continuous polygons consisting of vertical bisects connected by two adjacent points. The algorithms of constructing Voronoi graph generally include divide and conquer method, plane scanning method, and definition method based on Delaunay triangulation. The two-dimensional Delaunay triangulation method divides a two-dimensional plane into random polygons [15]. The area of broken ice floes is depicted in Figure 4. The calculation parameters are shown in Table 4.

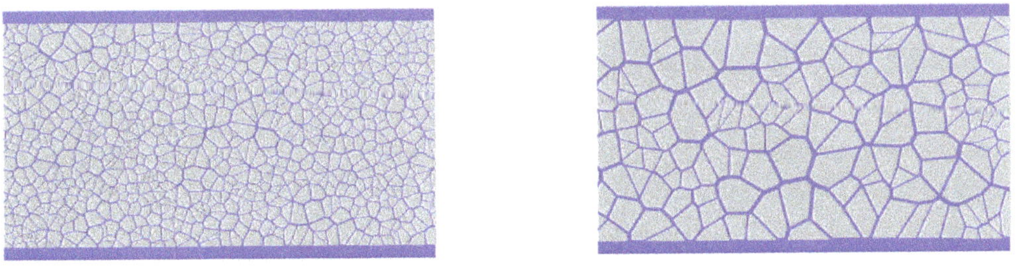

(a) (b)

Figure 4. Schematic diagram of broken ice floes area under different conditions: (**a**) 80% ice concentration, the average area of 20 m^2 of ice floes condition; (**b**) 80% ice concentration, the average area of 100 m^2 of ice floes condition.

Table 4. Main parameters of sea ice discrete element calculation.

Parameter	Symbol	Value	Unit
Elastic modulus	E	1.0	GPa
The density of water	ρ_w	1035.0	kg/m^3
The density of ice	ρ_i	920.0	kg/m^3

Table 4. *Cont.*

Parameter	Symbol	Value	Unit
The friction coefficient of particle element	μ_b	0.25	—
The friction coefficient between sea ice and structure	μ_s	0.25	—
Particle normal bond strength	${\sigma_b}^n$	1.57	—
Particle tangential bond strength	${\sigma_b}^s$	1.57	—
Particle compression strength	σ_c	2.53	MPa

3. Comparison of Ice Load Calculation Methods

The failure modes of ice include crushing failure and flexure failure. This paper adopts the nonlinear finite element method (FEM) and discrete element method (DEM) to simulate the crushing failure between ice and cylinder as well as the flexure failure between ice and cone structure, respectively. The simulation process can be observed in Figures 5 and 6. The figures reveal that both methods can reasonably simulate the interaction between ice and structures.

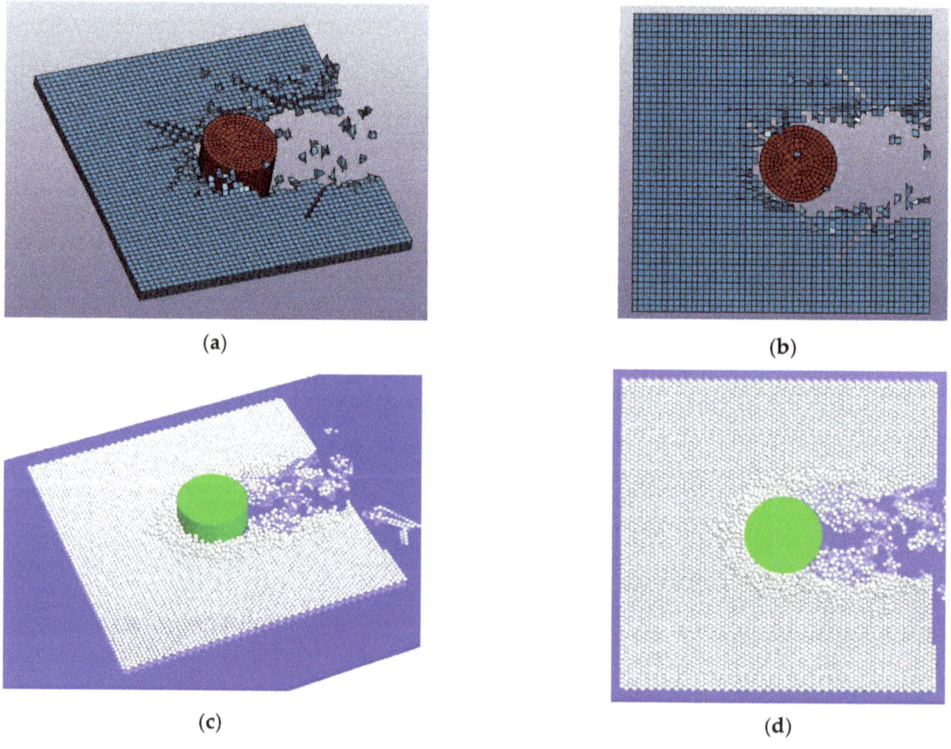

Figure 5. Simulation of interaction between cylinder structure and level ice. (**a**) Schematic diagram of finite element method calculation; (**b**) top view of finite element method calculation; (**c**) schematic diagram of discrete element method calculation; (**d**) top view of discrete element method calculation.

Figure 7 describes the ice load values under the same ice condition for cylinder and cone structures obtained by different methods. The two simulation methods' time history curves of ice load show irregular trend fluctuation, and the peak value appears almost simultaneously. The difference is that the peak value of the ice loads obtained by the finite element method is more significant than that computed by the discrete element method, especially in crushing failure, and the difference is more than 100%. The fluid does not act as a buffer in finite element analysis. It will create a more significant force when ice contacts the structure. The setting of erosion contact makes less impact frequency between ice and structures. Therefore, the finite element method results in a more considerable ice-loads fluctuation than the discrete element method.

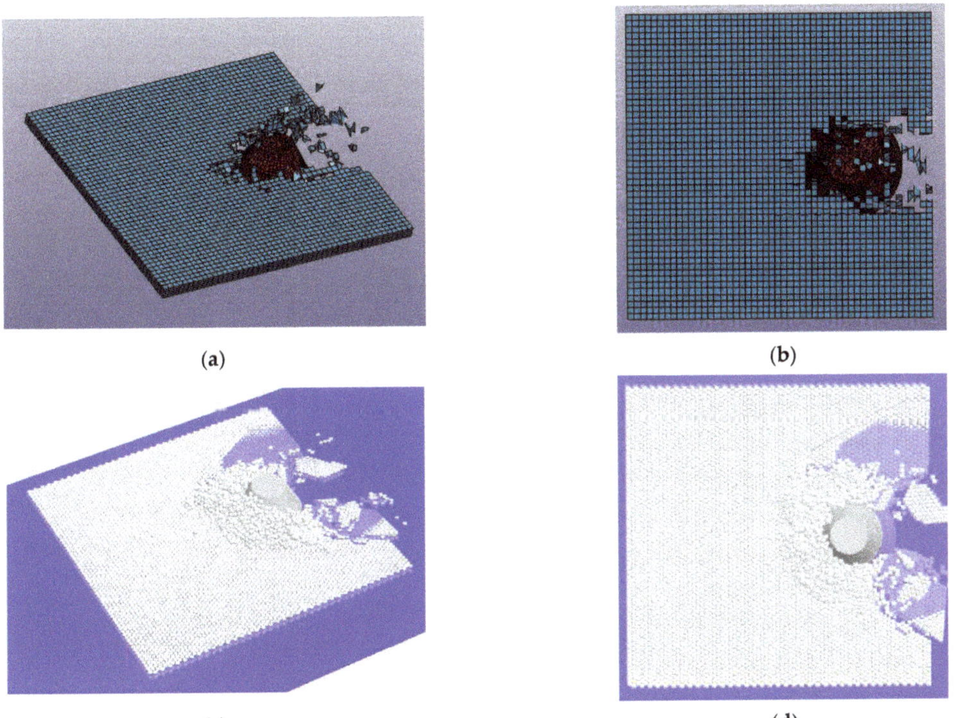

(a) (b) (c) (d)

Figure 6. Simulation of interaction between cone structure and level ice. (**a**) Schematic diagram of finite element method calculation; (**b**) top view of finite element method calculation; (**c**) schematic diagram of discrete element method calculation; (**d**) top view of discrete element method calculation.

The numerical values of the nonlinear finite element method, discrete element method, and empirical formula are compared by statistics. The research object is the average values and standard deviation of ice loads at different ice speeds (0.1–0.5 m/s) at the same ice thickness (2 m). Analyzing the calculation results of single ice speed causes a shortage of statistical samples, which cannot guarantee its reliability. Therefore, several groups of ice speed data are selected for statistical analysis. Table 5 shows the numerical comparison of different ice load analysis methods.

The results of the discrete element method are relatively small for crushing failure, while the results of the finite element method are close to those of the empirical formula. In the analysis of flexure failure, there is little difference between these three methods. The average value of DEM method is closer to the empirical formula, and the standard deviation is smaller. Meanwhile, considering the calculation's accuracy and time, the discrete element is a more suitable method. The Kulluk platform contacts with ice as a cone structure, in which flexure failure happens. In this paper, the discrete element method is applied to analyze the ice loads of the Kulluk under the conditions of level ice and broken ice floes.

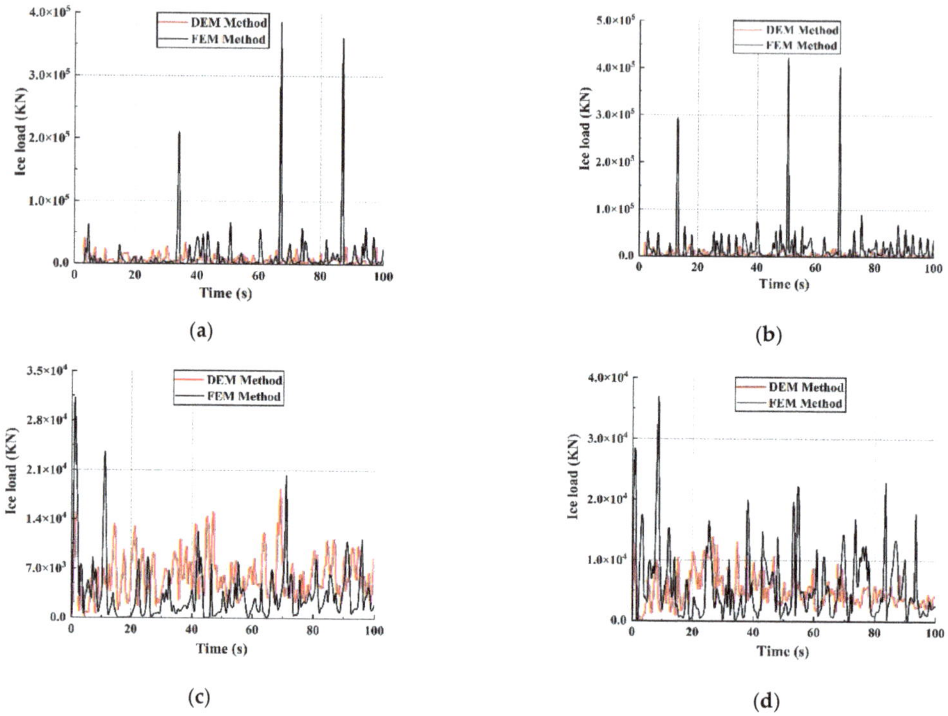

Figure 7. Numerical comparison of ice load between finite element method and discrete element method. (**a**) Cylinder structure at 0.3 m/s ice speed; (**b**) cylinder structure at 0.4 m/s ice speed; (**c**) cone structure at 0.2 m/s ice speed; (**d**) cone structure at 0.4 m/s ice speed.

Table 5. The numerical comparison of different ice load analysis methods.

	Ice Speed (m/s)							Empirical Formula
	0.1	0.2	0.3	0.4	0.5	Average Value	Standard Deviation	
DEM—cylinder structure	6007.58 kN	6983.78 kN	7717.35 kN	8386.42 kN	9063 kN	7273.78 kN	1124.58	26,343.32 kN
FEM—cylinder structure	9043.24 kN	13,120.88 kN	18,343.52 kN	19,272.79 kN	20,634.68 kN	16,083.02 kN	4343.61	
DEM—cone structure	4959.72 kN	5285.41 kN	5625.54 kN	5981.75 kN	6018.99 kN	5574.28 kN	406.83	6600 kN
FEM—cone structure	2869.34 kN	3451.97 kN	4971.67 kN	5863.48 kN	6226.38 kN	4676.57 kN	1316.32	

4. Numerical Model and Environmental Conditions

This paper selects the full-scale Kulluk platform as the numerical model, as displayed in Figure 8. The Kulluk is a symmetrical structure with a downward sloping circular hull that resists environmental loads in all directions. Table 6 lists the specific design parameters.

The mooring system is the research focus of this paper. In the numerical simulation, fairleads connection and supernodes are defined to establish a fully coupled system between the Kulluk platform and the mooring system. The mooring system is arranged by radial symmetry. Together with the circular structure, it can resist loads in any direction. The Kulluk platform's mooring lines are located below the waterline to eliminate collisions between ice and mooring lines at the waterline.

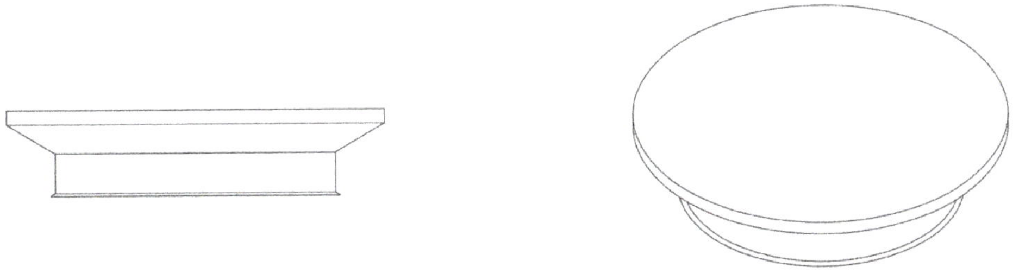

(**a**) (**b**)

Figure 8. Numerical computation model of Kulluk platform. (**a**) Front view of the model; (**b**) stereogram of the model.

Table 6. Basic parameters of the Kulluk platform.

Parameter	Value	Unit
Top diameter	81	m
Diameter at waterline	67.5	m
Bottom diameter	60	m
Depth	18.4	m
Draft	11.5	m
Displacement	28,000	m^3
Cone angle	31.4	°

Figure 9 depicts the specific location of the Beaufort Sea. It is located in latitude 69° N to 75° N and longitude 125° W–152° W. The length of the ice-covered season is early October to late July. The length of the open water season is August to early October. ISO 19906(2019) [22] provides detailed information about the Beaufort Sea environment, including wind, wave, current, and ice data. The research direction of this paper is the arrangement of the mooring system rather than analyzing the influence of environmental factors. Therefore, wave, level ice, and broken ice floes are computed under a single working condition. The operating conditions selected in this paper are as follows: Significant wave height is 3.7 m, and the spectral peak period is 6.7 s. The current velocity is 0.4 m/s. The thickness of level ice is 2 m, and the ice speed is 0.08 m/s. Under the condition of broken ice floes, the thickness of the ice floes is 2 m, the average size of ice floes is 100 m^2, and the broken ice concentration is 80%.

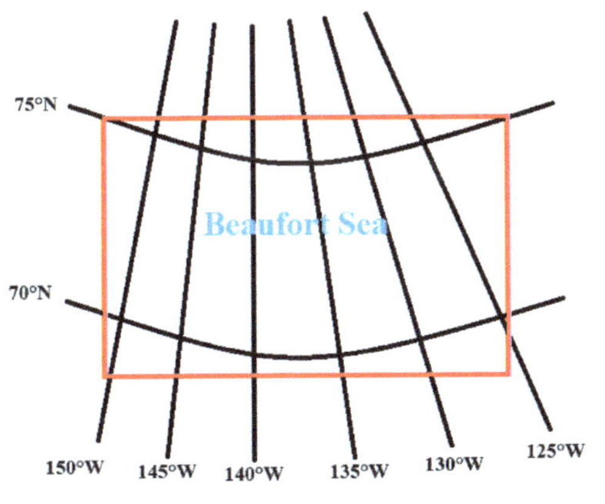

Figure 9. The geographical location of the Beaufort Sea [22].

5. Hydrodynamic Analysis of the Kulluk Platform

The motion transfer function RAO, first-order wave forces, and second-order wave forces of the Kulluk platform should be analyzed. The linear first-order wave force is obtained by factoring the velocity potential into the incident potential, diffraction potential, and radiation potential using Bernoulli's equation. The second-order mean drift force is a nonlinear force that affects the mooring system. There are two methods for computation: the far-field method [31] and the direct pressure integral method [32]. The Kulluk platform is a symmetrical structure, and the incoming directions of waves range from 0° to 180° at 15° intervals. The calculation period of waves ranges from 2 s to 38 s, and relevant data of 6 degrees of freedom motions are observed. In the subsequent time-domain calculation, the load direction of 180° is mainly analyzed. In this case, the surge and pitch motions are most representative. Therefore, only these two motions are selected in the analysis. Parts of the calculation results are seen in Figure 10.

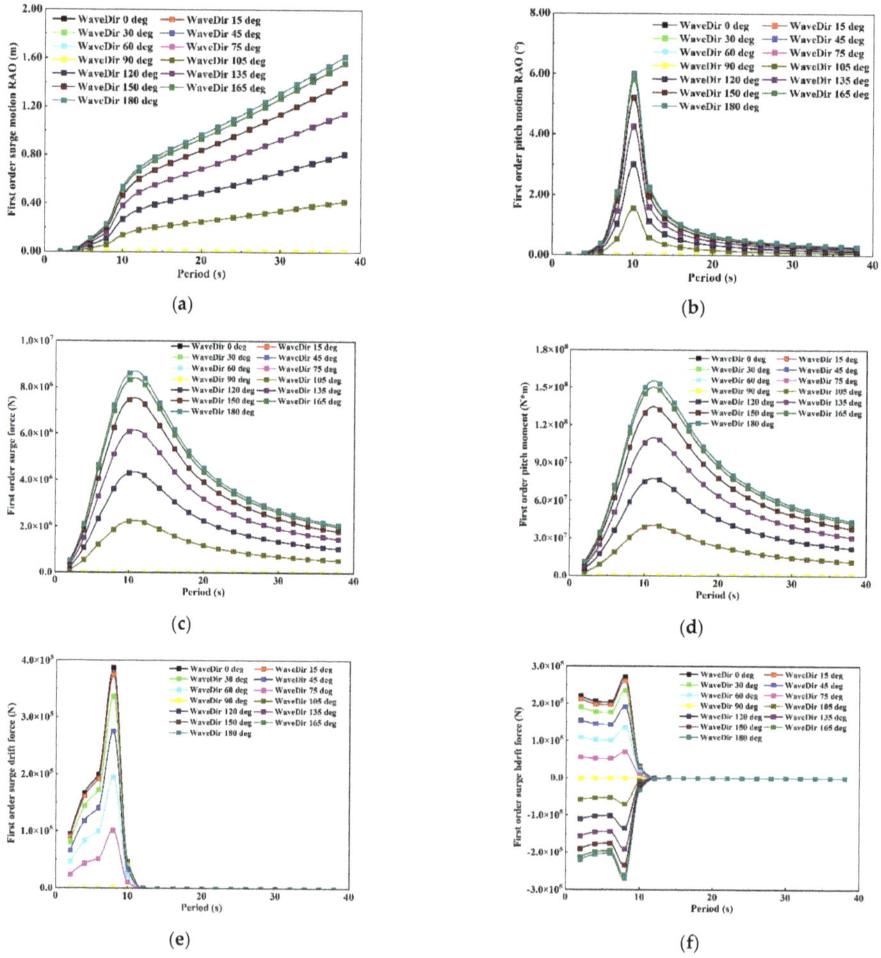

Figure 10. Calculation results of hydrodynamic analysis of the Kulluk platform. (**a**) First-order surge motion transfer function RAO; (**b**) first-order pitch motion transfer function RAO; (**c**) first-order surge force; (**d**) first-order pitch moment; (**e**) mean drift force in the surge by far-field method; (**f**) mean drift force in the surge by direct pressure integral method.

6. Ice Load Calculation of the Kulluk Platform

The ice load of the Kulluk platform is analyzed by the discrete element method. Figure 11 demonstrates the simulation process of the platform under conditions of level ice and broken ice floes. The calculation conditions are shown in Section 4. Figure 11c,d shows the failure mode of ice in contact with structure. This paper compares the numerical calculation results with the field data [3] from two aspects of ice thickness and ice concentration. The original Kulluk load event data are usually for broken ice floes, as during actual offshore operations, ice management will avoid large-sized level ice coming into contact with the platform. It can be seen from Figure 12 that the numerical results under the calculation conditions of broken ice floes are close to the field data and satisfy the upper limit of ice load specified by the linear equation in the Kulluk research report. To sum up, the calculated data in this paper are reasonable within the range of the field data and can be applied for subsequent calculations.

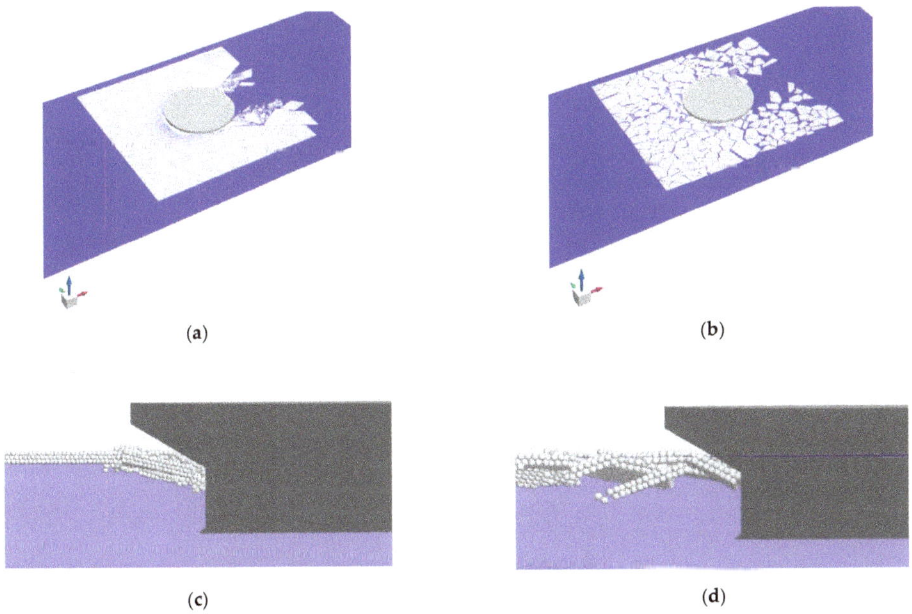

Figure 11. Numerical simulation of ice load on Kulluk platform. (**a**) Level ice condition; (**b**) broken ice floes condition; (**c**) front view under level ice condition; (**d**) front view under broken ice floes condition.

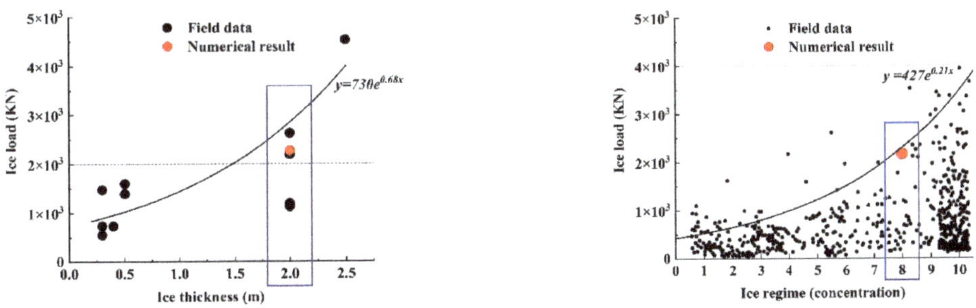

(**a**) (**b**)

Figure 12. Comparison between the numerical simulation results and the field data of the Kulluk platform. (**a**) Ice thickness as the independent variable; (**b**) ice concentration as the independent variable.

7. Results and Discussion

This paper takes the mooring system design as the research object. The incidence angle of environmental loads, the number of mooring lines, the number of connection points, and the angle between mooring lines are all factors to be concerned about to complete the design of an optimal mooring system. The dynamic response under different working conditions of the wave, level ice, and broken ice floes is analyzed from two aspects, including mooring line tension and platform motion. In the time-domain simulation analysis, the computation time is 1800 s.

7.1. Influence of Load Direction on Mooring System

This paper analyzes the effect of load incidence angle on the coupling system by regarding the layout of four mooring lines as an example. Considering the symmetry of the Kulluk platform, four different load directions of 180°, 165°, 150°, and 135° are selected for study, as shown in Figure 13. By comparing the mean value and maximum value of mooring line tension under wave load, level ice load, and broken ice floes load, it can be seen that the tension of the mooring line with the largest force changes little under different load incidence angles. Figure 14 indicates the tension comparison of all mooring lines under different load directions. Compared with wave loads and broken ice floes in different directions, level ice in different directions have more obvious effects on mooring line tension. Mooring line 1 has the highest tension. Under wave loads, the maximum tension and average tension of mooring line 1 change by 4.58% and 0.62%, respectively. Under level ice, the maximum and average tension of mooring line 1 changed by 26.56% and 3.45%, respectively. The maximum and average tension of mooring line 1 changed by 11.11% and 4.63%, respectively, under broken ice floes. Therefore, the influence of load direction is not considered in the follow-up study.

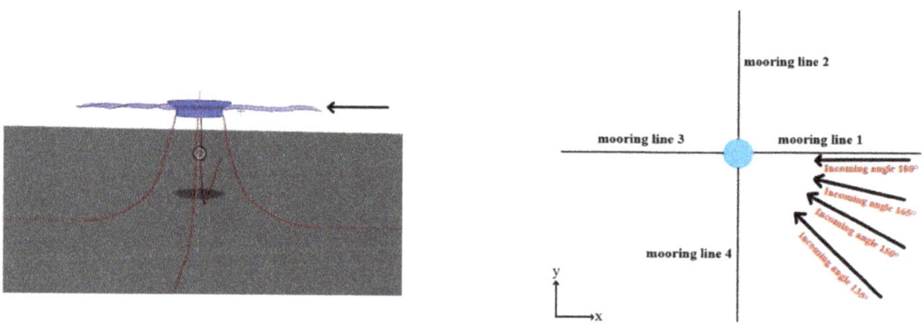

(a) (b)

Figure 13. Influence of load direction on mooring system. (**a**) The coupled system for analysis; (**b**) schematic diagram of different load directions.

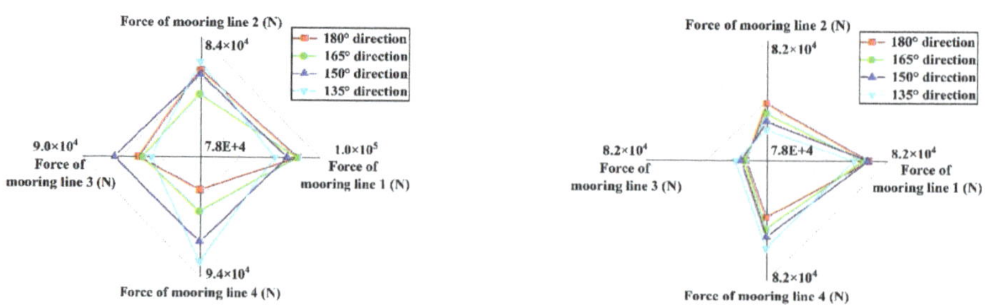

(a) (b)

Figure 14. *Cont.*

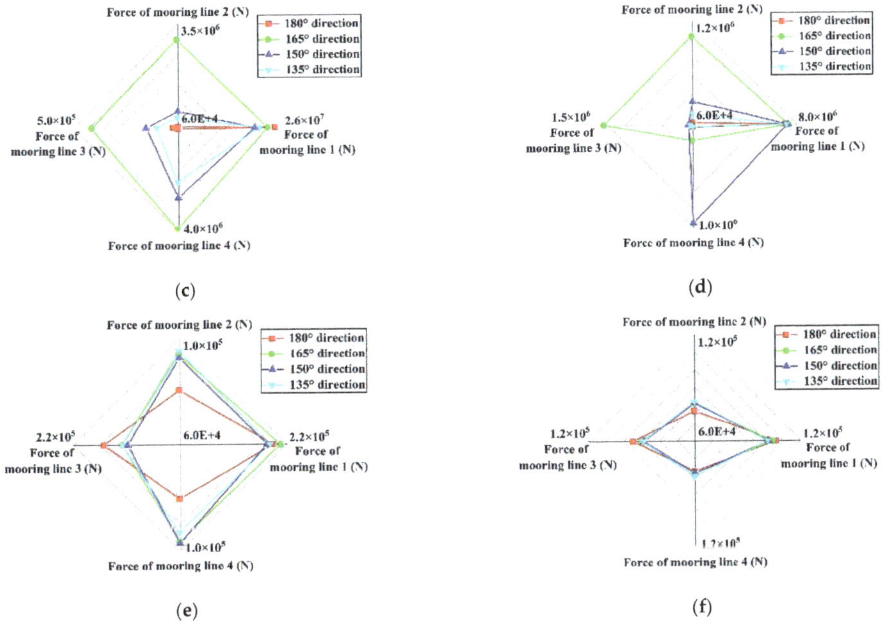

Figure 14. Tension comparison of four mooring lines under different load directions. (**a**) Maximum tension of mooring lines under wave loads; (**b**) mean tension of mooring lines under wave loads; (**c**) maximum tension of mooring lines under level ice loads; (**d**) mean tension of mooring lines under level ice loads; (**e**) maximum tension of mooring lines under broken ice floes loads; (**f**) mean tension of mooring lines under wave loads.

7.2. Influence of the Number of Mooring Lines on the Mooring System

In this paper, different configurations of four mooring lines, eight mooring lines, and twelve mooring lines are selected to investigate the influence of the number of mooring lines. These three designs are all arranged symmetrically, and the length and stiffness of the mooring lines are the same. The incidence angle of the environmental load is 180°. Figure 15 depicts the specific layout of the mooring system. Besides surge motion and pitch motion of the platform, the maximum tension and the mean tension of the mooring system are compared. The tension of the mooring system is defined as the force evenly distributed to each mooring line, which can reflect the overall level of mooring line tension in this design. The tension of mooring line 1 under time-domain analysis must also be compared. Mooring line 1 is subjected to the most tremendous tension in the mooring system. The segment connected with fairlead is selected for analysis. The results are shown in Figure 16. Under wave load and broken ice floes, the slightest tension on the mooring system is the design of eight mooring lines. Under the level ice load, the optimal design is twelve mooring lines. The tension of mooring lines in the configuration of four mooring lines is much greater than that in the other two designs under ice loads. Under the condition of level ice, the maximum tension and mean tension of four mooring lines design increases by 126.68% and 98.64% compared with the eight-mooring-line design as well as 252.89% and 167.21% compared with the twelve-mooring-line design. Under broken ice floes, these values are 63.29%, 53.49%, 27.98%, and 10.18%, respectively. This reveals that the mooring line of the four-mooring-line design makes it easier to attain the design stiffness. Its safety is not as good as the other two designs in the case of mooring line fracture. It is also verified by the time-domain analysis of mooring line 1. Meanwhile, it can be seen that under the design mode of twelve mooring lines, the tension of mooring line 1 is the least. The results are all similar in the conditions of the wave, level ice, and broken ice floes.

The motions of the platform are described in Figure 17. There are apparent differences in surge motion under the three design modes. Under wave load, the maximum platform surge motion with twelve mooring lines is 224.43% and 138.51% smaller than that with four mooring lines and eight mooring lines, respectively. Under level ice, surge motion of the twelve-mooring-line design decreases by 10.92% and 7.27%, respectively. Surge motion decreases by 286.09% and 86.14% under broken ice floes. There is little distinction in pitch motion, and the mooring system's positioning effect, which is composed of twelve mooring lines, is the best.

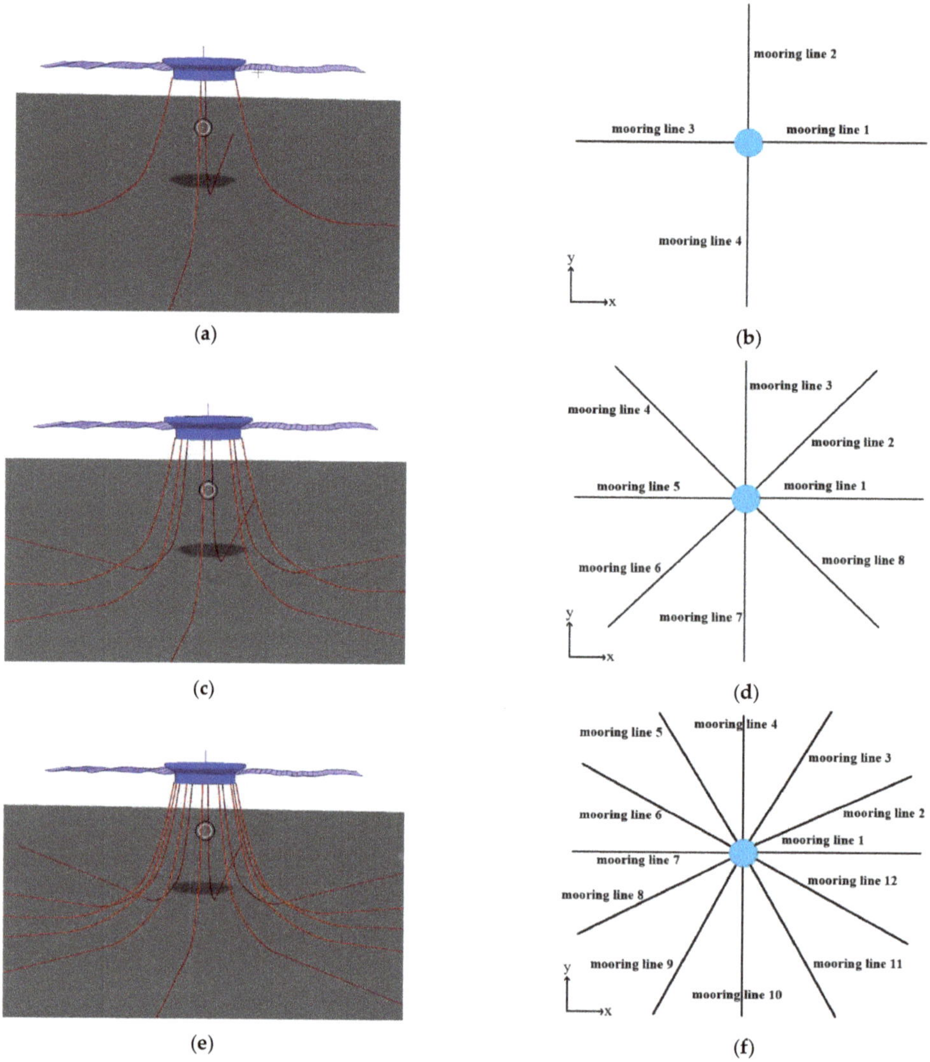

Figure 15. Influence of the number of mooring lines on the mooring system. (**a**) The coupled system for analysis with four mooring lines; (**b**) schematic diagram of the design of four mooring lines; (**c**) the coupled system for analysis with eight mooring lines; (**d**) schematic diagram of the design of eight mooring lines; (**e**) the coupled system for analysis with twelve mooring lines; (**f**) schematic diagram of the design of twelve mooring lines.

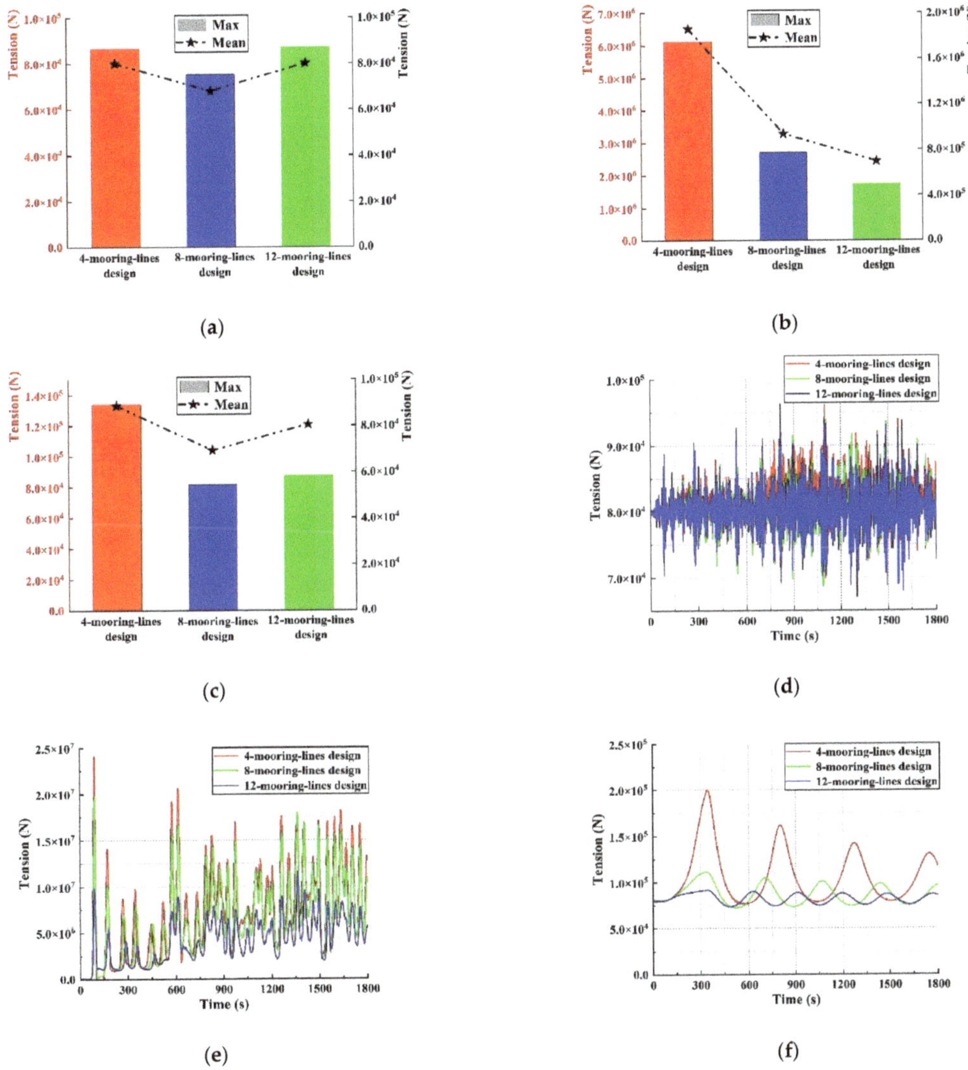

Figure 16. Influence of the number of mooring lines on mooring line tension. (**a**) Maximum and mean values of mooring system tension under wave loads; (**b**) maximum and mean values of mooring system tension under level ice loads; (**c**) maximum and mean values of mooring system tension under broken ice floes loads; (**d**) the tension of mooring line 1 under wave loads in time-domain analysis; (**e**) the tension of mooring line 1 under level ice loads in time-domain analysis; (**f**) the tension of mooring line 1 under broken ice floes loads in time-domain analysis.

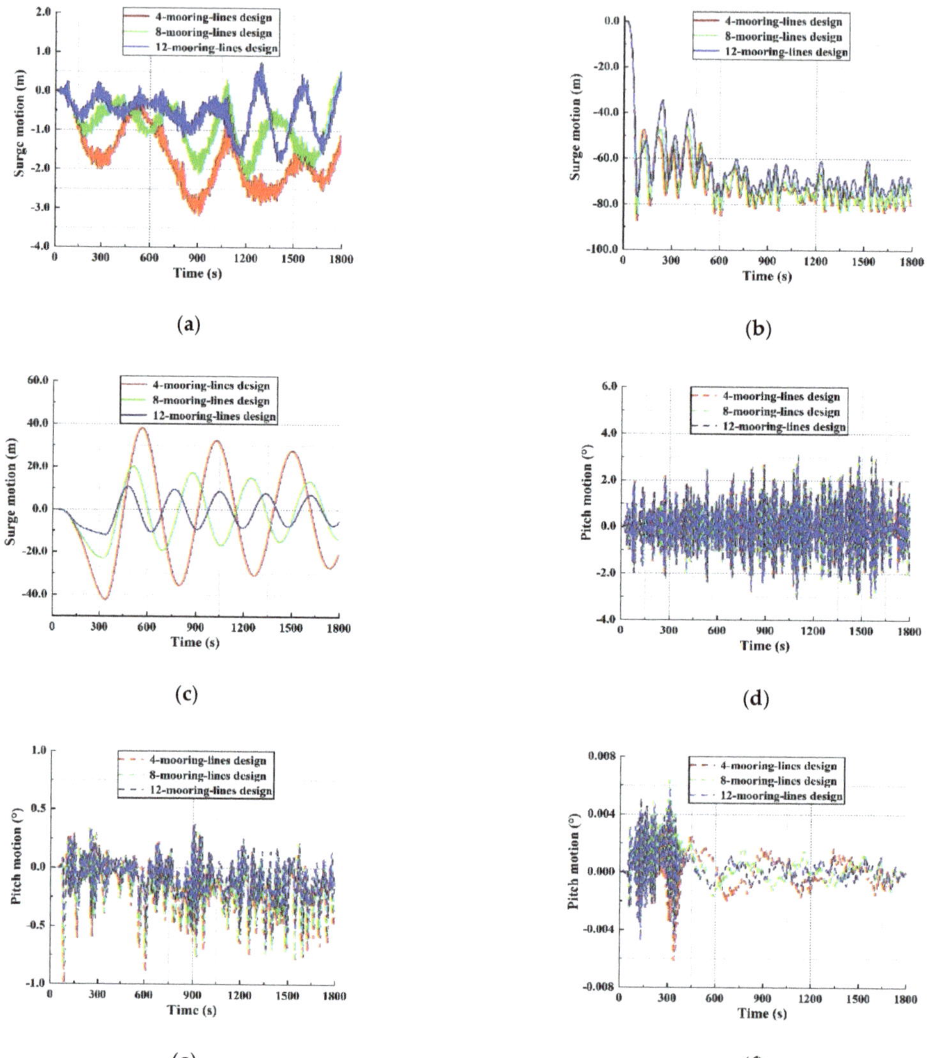

Figure 17. Influence of the number of mooring lines on Kulluk platform motion. (**a**) Surge motion under wave loads; (**b**) surge motion under level ice loads; (**c**) surge motion under broken ice floes loads; (**d**) pitch motion under wave loads; (**e**) pitch motion under level ice loads; (**f**) pitch motion under broken ice floes loads.

7.3. Influence of the Number of Connection Points on the Mooring System

In this paper, three different designs are selected to analyze the influence of the number of connection points on the mooring system, including one connection point, four connection points, and twelve connection points. Twelve symmetrical mooring lines are adopted for the study, and the loading direction is 180°. The one-connection-point design means that all mooring lines are connected to the Kulluk through the same fairlead. The layout of the four connection points is unique, and each of the three mooring lines shares the same fairlead. The angle between the mooring lines on both sides and the middle one is 30°. Figure 18 describes the specific arrangement of various connection point designs. The analysis results of mooring line tension and platform motion are demonstrated in Figures 19 and 20. Each mooring line tension of the one-connection-point design is more significant than that of the other two design methods under the wave, level ice, and broken ice floes. By comparing the tension of mooring line 1, surge motion, and pitch motion of the platform in time-domain analysis, it can be seen that the dynamic response of the coupled system does not have a significant difference under the design of four connection points and twelve connection points. Considering practical construction requirements such as operability, schedule, and cost, the four-connection-point design is a more suitable arrangement.

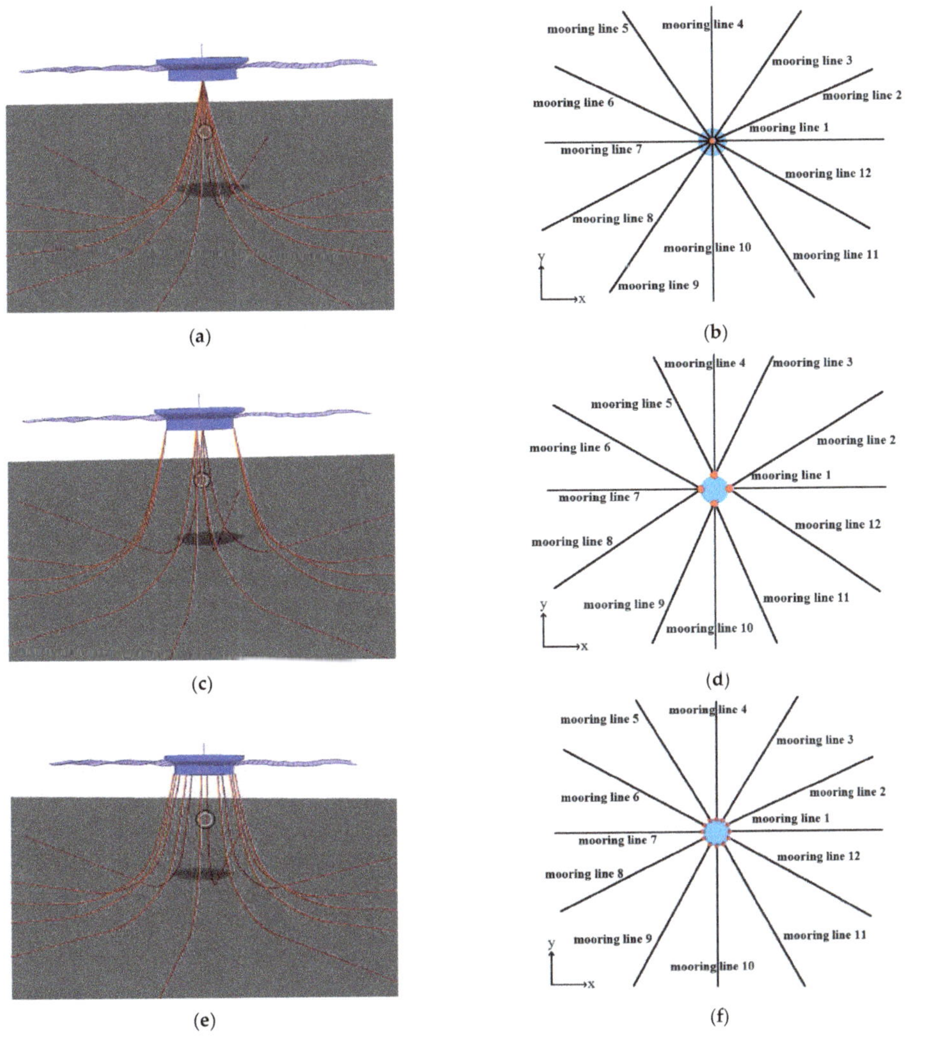

Figure 18. Influence of the number of connection points on the mooring system. (**a**) The coupled system for analysis with one connection point; (**b**) schematic diagram of the design of one connection point; (**c**) the coupled system for analysis with four connection points; (**d**) schematic diagram of the design of four connection points; (**e**) the coupled system for analysis with twelve connection points; (**f**) schematic diagram of the design of twelve connection points.

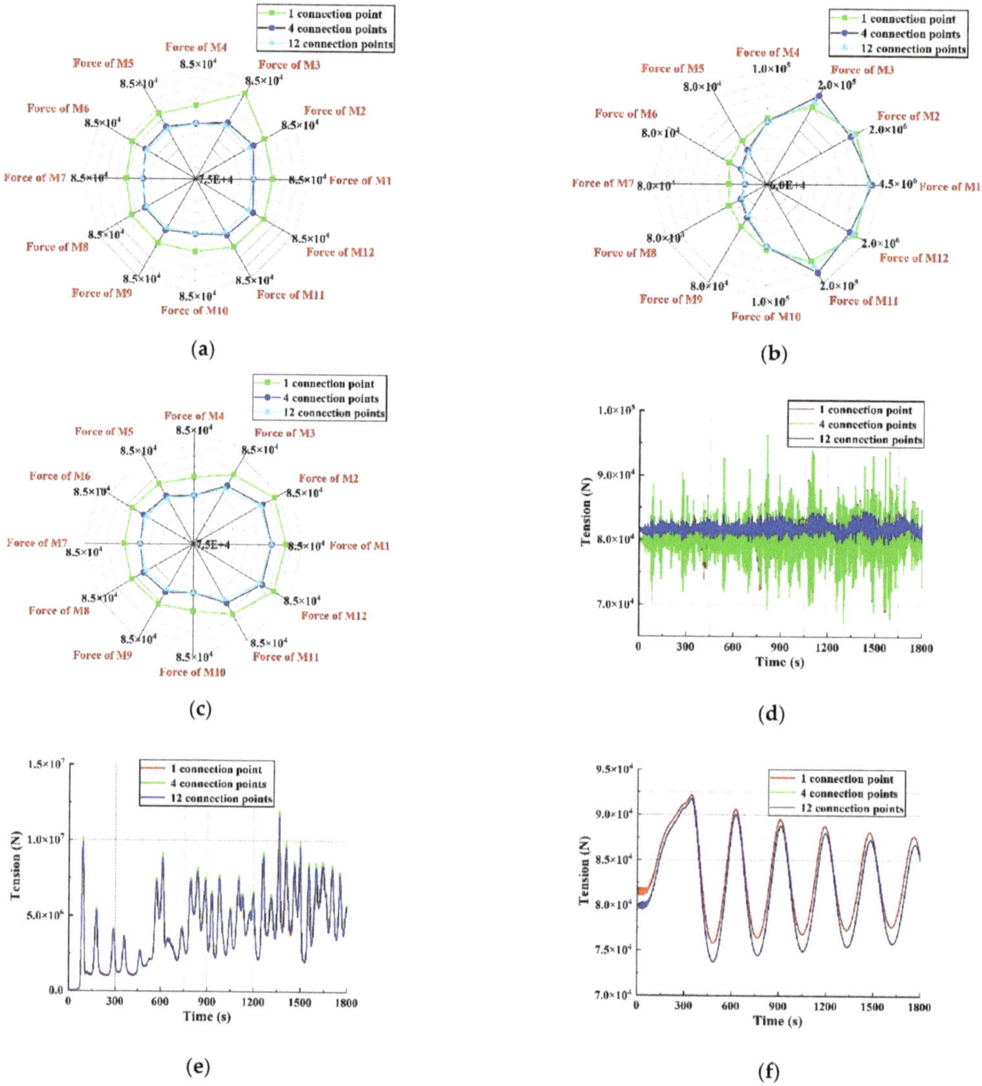

Figure 19. Influence of the number of connection points on mooring line tension. (**a**) Mean tension of mooring lines under wave loads; (**b**) mean tension of mooring lines under level ice loads; (**c**) mean tension of mooring lines under broken ice floes loads; (**d**) the tension of mooring line 1 under wave load in time-domain analysis; (**e**) the tension of mooring line 1 under level ice load in time-domain analysis; (**f**) the tension of mooring line 1 under broken ice floes load in time-domain analysis.

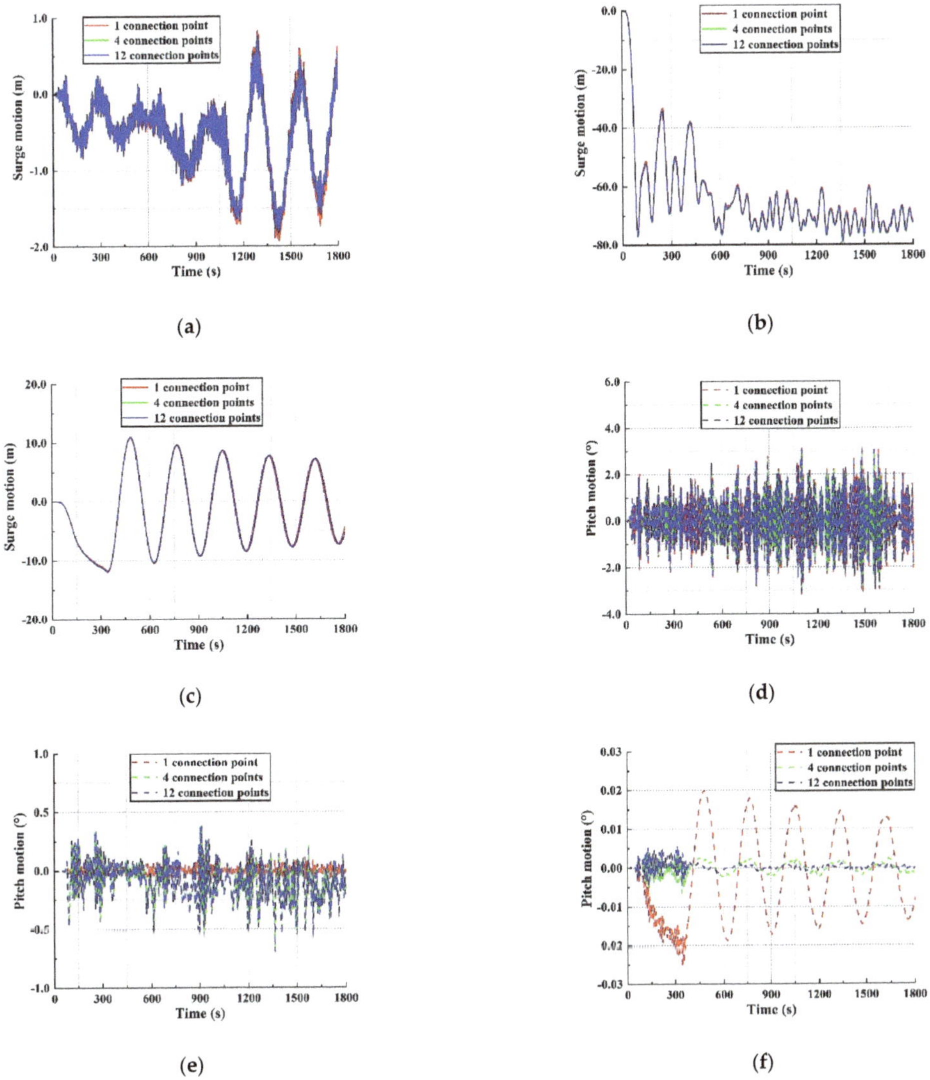

Figure 20. Influence of the number of connection points on Kulluk platform motion. (**a**) Surge motion under wave loads; (**b**) surge motion under level ice loads; (**c**) surge motion under broken ice floes loads; (**d**) pitch motion under wave loads; (**e**) pitch motion under level ice loads; (**f**) pitch motion under broken ice floes loads.

7.4. Influence of the Angle between Mooring Lines on the Mooring System

In this paper, three different angles of 10°, 20°, and 30° are selected for numerical simulation. The mooring system has twelve mooring lines and four connecting points. The loading direction is 180°. The specific arrangement of mooring lines is presented in Figure 21. The mooring system's mean tension is very close in these three designs. The mooring system's maximum tension varies slightly as well. Under the condition of level ice, the 10° design minimizes the tension of mooring line 1. In most cases, these three different angles have quite little influence on the platform motions, which are similar numerically. The motion response curves of the time-domain analysis are almost identical. Only in the platform surge motion under level ice conditions and the platform pitch motion under broken ice floes conditions can it be seen that 10° is the best angle between the mooring lines. The calculation results are shown in Figures 22 and 23.

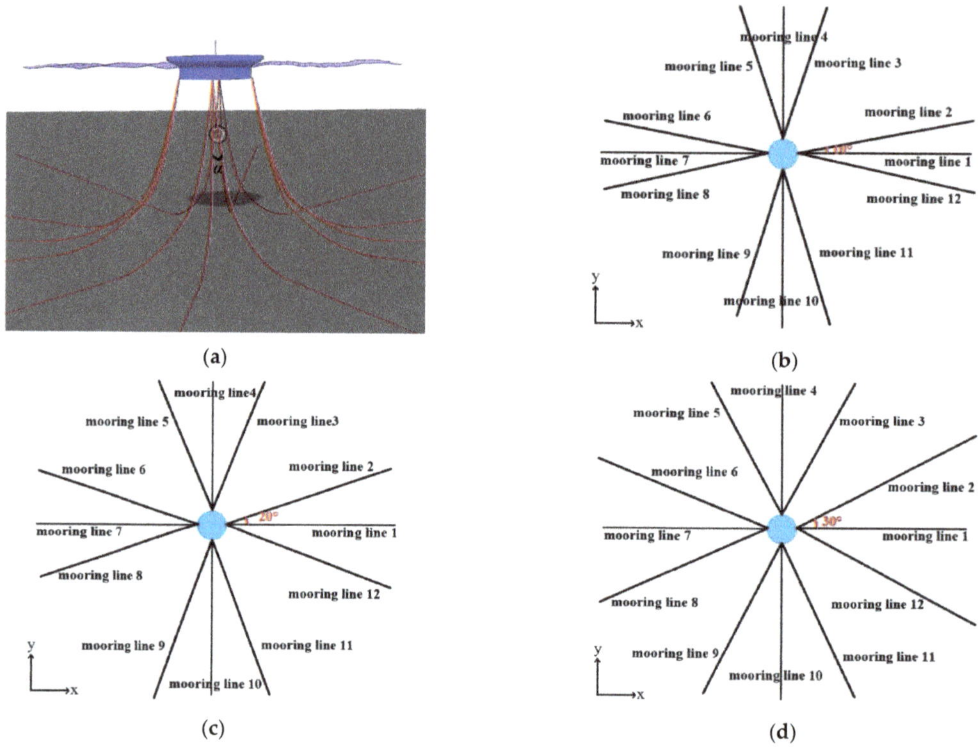

Figure 21. Influence of the angle between mooring lines on the mooring system. (**a**) The coupled system for analysis; (**b**) schematic diagram of the design of 10° between mooring lines; (**c**) schematic diagram of the design of 20° between mooring lines; (**d**) schematic diagram of the design of 30° between mooring lines.

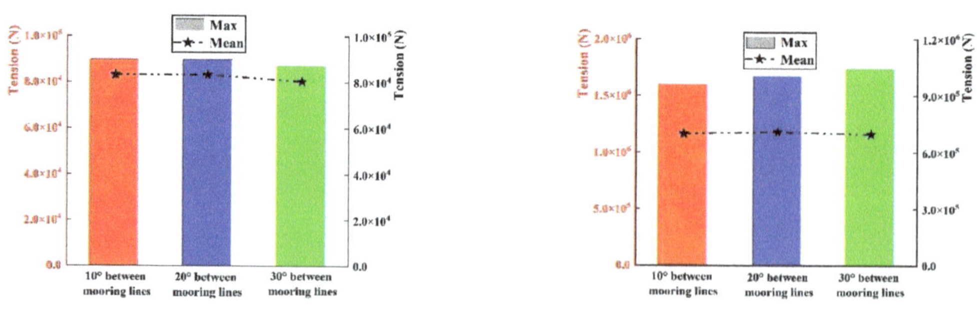

(**a**) (**b**)

Figure 22. *Cont.*

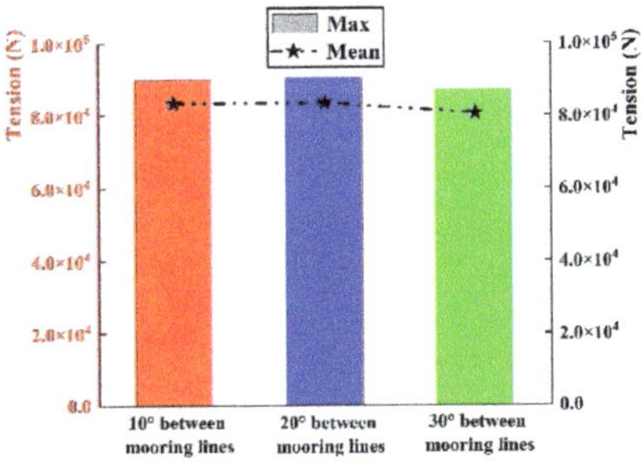

(c)

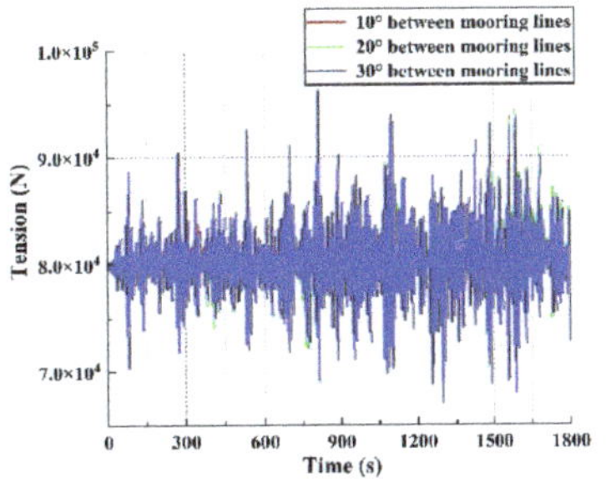

(d)

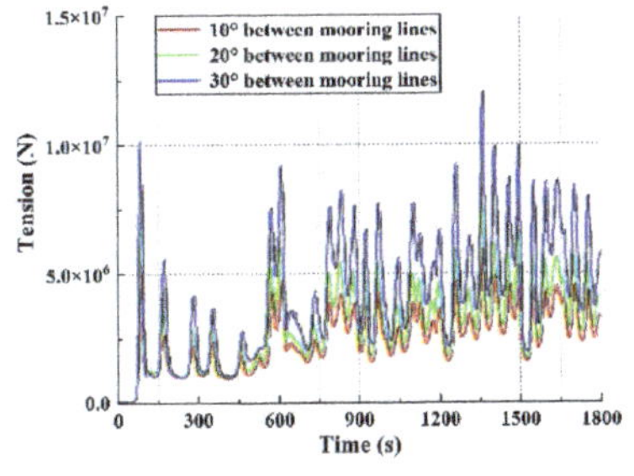

(e)

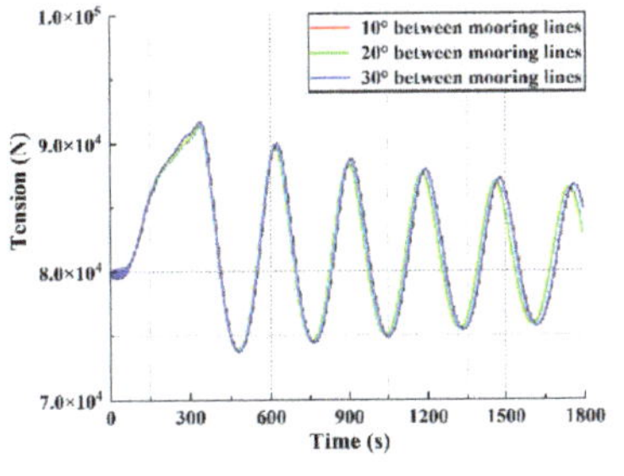

(f)

Figure 22. Influence of the angle between mooring lines on mooring line tension. (**a**) Maximum and mean values of mooring system tension under wave loads; (**b**) maximum and mean values of mooring system tension under level ice loads; (**c**) maximum and mean values of mooring system tension under broken ice floes loads; (**d**) the tension of mooring line 1 under wave load in time-domain analysis; (**e**) the tension of mooring line 1 under level ice load in time-domain analysis; (**f**) the tension of mooring line 1 under broken ice floes load in time-domain analysis.

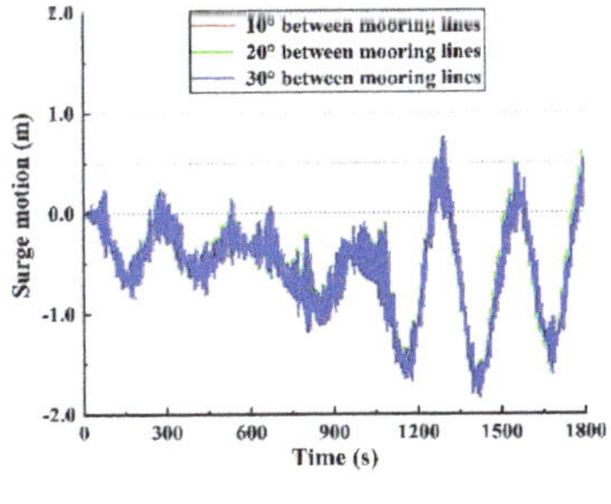

(a)

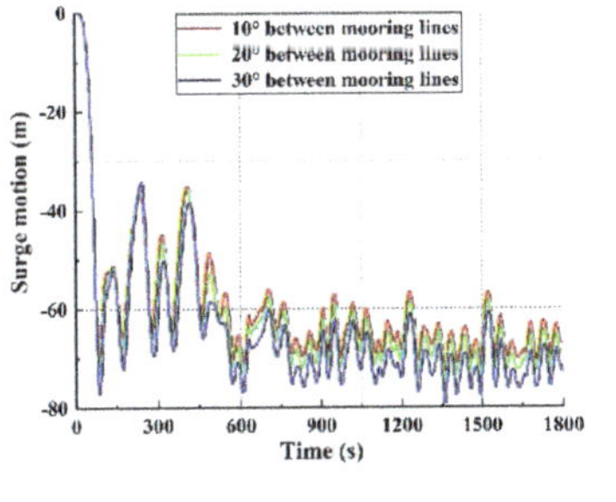

(b)

Figure 23. *Cont.*

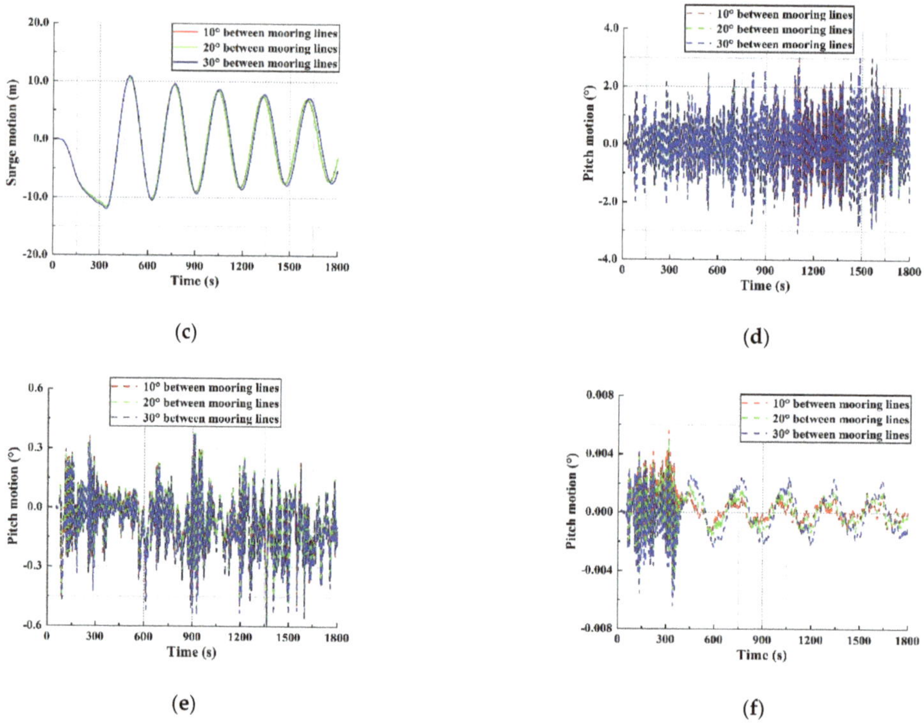

Figure 23. Influence of the angle between mooring lines on Kulluk platform motion. (**a**) Surge motion under wave loads; (**b**) surge motion under level ice loads; (**c**) surge motion under broken ice floes loads; (**d**) pitch motion under wave loads; (**e**) pitch motion under level ice loads; (**f**) pitch motion under broken ice floes loads.

In order to verify whether there is a better design scheme when the angle is less than 10°, the surge motion of the platform under level ice and the pitch motion under broken ice floes are compared when the angles between mooring lines are 5°, 7.5°, and 10°. The results are so close that there is no significant difference, as displayed in Figure 24. In order to avoid collision or entanglement between mooring lines caused by too close a distance, the angle between mooring lines is finally determined to be 10°.

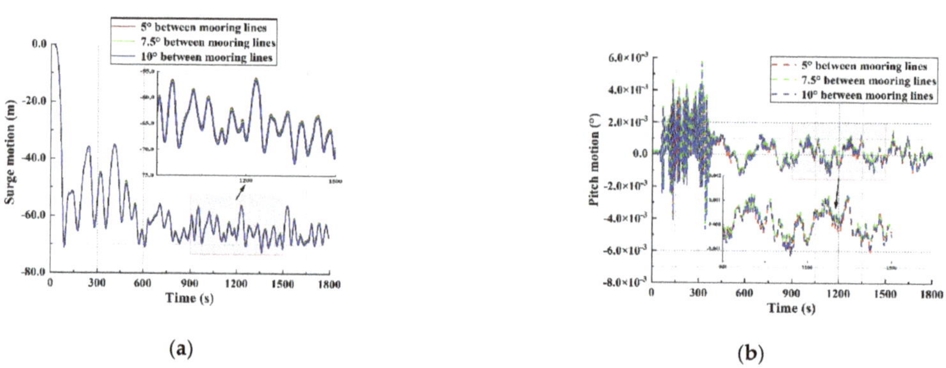

Figure 24. Comparison of three groups of mooring lines with angles of 5°, 7.5°, and 10°. (**a**) Surge motion under level ice loads; (**b**) pitch motion under broken ice floes loads.

8. Conclusions

In order to find the relevant factors affecting the mooring system and optimize the mooring line layout of the Kulluk platform, this paper simulates the dynamic response of the Kulluk platform during ice-free and ice-covered seasons in the Beaufort Sea. Statistical and nonlinear time-domain simulation methods analyze the tension of mooring lines and the platform's motion. Different working conditions such as wave load, level ice load, and broken ice floes load are taken into account. The findings are drawn as follows:

The empirical formula, nonlinear finite element method, and discrete element method are standard methods for computing ice loads. In this paper, by calculating the cylindrical structure and the cone structure, it can be seen that the nonlinear finite element method is closer to the empirical formula in the analysis of crushing failure. The discrete element method is more suitable for calculating the flexure failure of ice. The ice load of the Kulluk platform is analyzed by the discrete element method because the contact position between the Kulluk platform and the ice is a slope. The calculated results are consistent with the field data, which proves the accuracy of the numerical simulation.

(1) The influence of level ice load on platform motion and mooring system tension is much more significant than that of wave and broken ice floes load. Ongoing ice management exists in actual offshore operations. It can control loads of the mooring lines and keep the mooring system's tension and the platform's motion within an acceptable range.
(2) For the dynamic response of the mooring system of the Kulluk platform, the load direction is not the decisive factor. The number of mooring lines, the number of connecting points, and the angle between the mooring lines all affect the tension of the mooring lines and the platform's motion to a certain extent. Mooring lines with more numbers have the best positioning effect, and each mooring line has the lowest tension and the highest safety. When the dynamic response difference is less than 1%, the final design is determined based on the actual construction costs and offshore operations conditions.
(3) The final design scheme is determined as twelve mooring lines, four connecting points, and with an angle between the mooring lines of 10°. Under wave and broken ice floes conditions, the surge motion of the coupling system is less than 10% of the working water depth, and the pitch motion is less than 5°, which well satisfies the relevant design requirement of offshore platforms. The platform is suitable for both ice-free and ice-covered seasons and can be used for year-round offshore operations.

In the following research, this method will be applied to the thruster-assisted mooring system and to optimize the positioning system of floating structures operating in cold regions except for offshore platforms.

Author Contributions: Conceptualization, Z.C. and S.L. (Shewen Liu); methodology, A.Z.; software, A.Z., Z.H. and S.L. (Shiqi Liu); validation, Z.C. and X.C.; formal analysis, S.L. (Shewen Liu); investigation, A.Z.; resources, Z.C.; data curation, A.Z.; writing—original draft preparation, A.Z.; writing—review and editing, Z.C.; visualization, L.H.; supervision, S.L. (Shewen Liu); project administration, Z.C.; funding acquisition, S.L. (Shewen Liu). All authors have read and agreed to the published version of the manuscript.

Funding: This research was funded by the Dalian Science and Technology Innovation Fund Project, grant number 2020JJ25CY016; the State Key Laboratory of Coastal and Offshore Engineering Fund, grant number LP2115 and Fundamental Research Funds for the Central University, grant number 3132022122.

Data Availability Statement: All analyzed data in this study have been included in the manuscript.

Acknowledgments: The authors would like to thank the Dalian Maritime University (DMU).

Conflicts of Interest: The authors declare no conflict of interest.

References

1. Hamilton, J.M. The Challenges of Deep-Water Arctic Development. In Proceedings of the Twenty-First International Offshore and Polar Engineering Conference, Maui, HI, USA, 19–24 June 2011.
2. Donald, L.G.; Kenneth, J.B.; Ronnald, R.C.; Arthur, G.; David, W.H.; Timothy, R.K.; Thomas, E.M.; Janet, K.P.; Christopher, J.S.; John, H.S.; et al. Assessment of undiscovered oil and gas in the Arctic. *Science* **2009**, *324*, 1175–1179.
3. Wright, B. Full Scale Experience with Kulluk Stationkeeping Operations in Pack Ice (With Reference to Grand Banks Developments). In *PERD/CHC Report 25–44*; National Research Council: Ottawa, ON, Canada, 2000.
4. Zhou, L.; Su, B.; Riska, K.; Moan, T. Numerical simulation of moored structure station keeping in level ice. *Cold Reg. Sci. Technol.* **2012**, *71*, 54–66. [CrossRef]
5. Sayed, M.; Barker, A. Numerical Simulations of Ice Interaction with a Moored Structure. In Proceedings of the Offshore Technology Conference, Houston, TX, USA, 2–5 May 2011.
6. Wang, J.Y.; Derradji-Aouat, A. Numerical assessment for stationary structure (Kulluk) in moving broken ice. In Proceedings of the 21st International Conference on Port and Ocean Engineering under Arctic Conditions, POAC, Montréal, QC, Canada, 10–14 July 2011.
7. Du, J.F.; Chang, A.T.; Wang, S.Q.; Sun, M.Y.; Wang, J.R.; Li, H.J. Multi-mode reliability analysis of mooring system of deep-water floating structures. *Ocean. Eng.* **2019**, *192*, 106517. [CrossRef]
8. Garrett, D.L. Coupled analysis of floating production systems. *Ocean. Eng.* **2005**, *32*, 802–816. [CrossRef]
9. Montasir, O.A.; Yenduri, A.; Kurian, V.J. Effect of mooring line configurations on the dynamic responses of truss spar platforms. *Ocean. Eng.* **2015**, *96*, 161–172. [CrossRef]
10. Mohapatra, S.C.; Guedes Soares, C. Effect of Mooring Lines on the Hydroelastic Response of a Floating Flexible Plate Using the BIEM Approach. *J. Mar. Sci. Eng.* **2021**, *9*, 941. [CrossRef]
11. Horrigmoe, G.; Andersen, R. Nonlinear finite element analysis of ice-structure interaction at varying strain rates. In Proceedings of the Fourth International Offshore and Polar Engineering Conference, Osaka, Japan, 10–15 April 1994.
12. Edmond, H.H.; Sveinung, L. Modelling floating offshore units moored in broken ice: Comparing simulations with ice tank tests. *Cold Reg. Sci. Technol.* **1999**, *29*, 107–119.
13. Liu, Z.H.; Jørgen, A.; Sveinung, L. Plasticity based material modelling of ice and its application to ship–iceberg impacts. *Cold Reg. Sci. Technol.* **2011**, *65*, 326–334. [CrossRef]
14. Liu, L.; Ji, S.Y. Ice load on floating structure simulated with dilated polyhedral discrete element method in broken ice field. *Appl. Ocean. Res.* **2018**, *75*, 53–65. [CrossRef]
15. Peyrot, A.H.; Goulois, A.M. Analysis of Cable Structures. *Comput. Struct.* **1979**, *10*, 805–813.
16. Sødahl, N. *Design and Analysis of Flexible Risers*; The Norwegian Institute of Technology: Trondheim, Norway, 1991.
17. Faltinsen, O.M. *Sea Loads on Ships and Offshore Structures*; Cambridge University Press: Cambridge, UK, 1990.
18. DNV. Sesam User Manual Sesam Manager. 2018. Available online: https://www.dnv.com/services/strength-assessment-of-offshore-structures-sesam-software-1068?&utm_campaign=structure_sesam&utm_source=google&utm_medium=cpc&gclid=EAIaIQobChMInITk2sGE-AIVAcuWCh3XcQyxEAAYASAAEgKNB_D_BwE&gclsrc=aw.ds (accessed on 15 May 2022).
19. DNV. SESAM User Manual WADAM. DNV Software Report No.: 92-7052. 2017. Available online: https://www.dnv.com/services/strength-assessment-of-offshore-structures-sesam-software-1068?&utm_campaign=structure_sesam&utm_source=google&utm_medium=cpc&gclid=EAIaIQobChMInITk2sGE-AIVAcuWCh3XcQyxEAAYASAAEgKNB_D_BwE&gclsrc=aw.ds (accessed on 15 May 2022).
20. SINTEF Ocean. SIMO 4.16.0 User Guide. 11 May 2019. Available online: https://sesam.dnv.com/status/simo/status/simo-release-notes_4.16.1.pdf (accessed on 15 May 2022).
21. Zhang, A.B.; Chuang, Z.J.; Liu, S.W.; Zhou, L.; Qu, Y.; Lu, Y. Dynamic performance optimization of an arctic semi-submersible production system. *Ocean. Eng.* **2022**, *244*, 110353. [CrossRef]
22. *ISO 19906*; Petroleum and Natural Gas Industries-Arctic Offshore Structures. 2nd ed. ISO: Geneva, Switzerland, 2019.
23. Hallquist, J.O. *LS-DYNA Theoretical Manual*; Livermore Software Technology Corporation: Livermore, CA, USA, 2006.
24. Wang, C.; Hu, X.H.; Tian, T.P.; Guo, C.Y.; Wang, C.H. Numerical simulation of ice loads on a ship in broken ice fields using an elastic ice model. *Int. J. Nav. Archit. Ocean. Eng.* **2020**, *12*, 414–427. [CrossRef]
25. Guo, C.Y.; Zhang, Z.T.; Tian, T.P.; Li, X.Y.; Zhao, D.G. Numerical simulation on the resistance performance of ice-going container ship under brash ice conditions. *China Ocean. Eng.* **2018**, *32*, 546–556. [CrossRef]
26. Shen, H.H.; Hibler, W.D.; Lepparanta, M. On applying granular flow theory to a deforming broken ice field. *Acta Mech.* **1986**, *63*, 143–160. [CrossRef]
27. Potyondy, D.O.; Cundall, P.A. A bonded-particle model for rock. *Int. J. Rock Mech. Min. Sci.* **2004**, *41*, 1329–1364. [CrossRef]
28. Campbell, C.S. Stress-controlled elastic granular shear flows. *J. Fluid. Mech.* **2005**, *539*, 273–297. [CrossRef]
29. Wang, Y.; Tonon, F. Calibration of a discrete element model for intact rock up to its peak strength. *Int. J. Numer. Anal. Methods Geomech.* **2010**, *34*, 447–469. [CrossRef]
30. Mollon, G.; Zhao, J. Fourier–Voronoi-based generation of realistic samples for discrete modelling of granular materials. *Granul. Matter* **2012**, *14*, 621–638. [CrossRef]

31. Maruo, H. The drift of a body floating on waves. *J. Ship Res.* **1960**, *4*, 1–10.
32. Pinkster, J.A.; van Oortmerssen, G. Computation of the first and second-order wave forces on oscillating bodies in regular waves. In Proceedings of the 2nd International Conference on Numerical Ship Hydrodynamics, Berkeley, CA, USA, 19–21 September 1977; pp. 136–159.

water

MDPI

Article

Analysis of the Cooperative Carrying Capacity of Ulan Suhai Lake Based on the Coupled Water Resources–Water Environment–Water Ecology System

Beibei Ren, Biao Sun *, Xiaohong Shi, Shengnan Zhao and Xiao Wang

College of Water Conservancy and Civil Engineering, Inner Mongolia Agricultural University, Hohhot 010018, China

* Correspondence: sunbiao@imau.edu.cn

Abstract: In order to maximize the carrying benefits of Ulan Suhai Lake and utilize the important role of the Inner Mongolia plateau lake basin in restoring its ecology and conserving water, the carrying capacity of the water resources, water environment, and water ecology of Ulan Suhai Lake was determined using system theory and synergy theory, and a 17-item evaluation index system of the three levels of the water resources, water environment, and water ecology (including the ice-covered period) was established using hierarchical analysis. In addition, a system dynamics model of the synergistic carrying capacity of the water resources–water environment–water ecology of Ulan Suhai Lake was constructed. A principal component analysis was conducted on the indicators of the Ulan Suhai Lake survey; six main sensitive factors were selected, and five simulation scenarios (status continuation mode, development continuation mode, conservation continuation mode, comprehensive mode I, and comprehensive mode II) were designed to regulate the configuration, taking into account the actual local development needs. Using 2014 as the base year and 2014–2020 as the validation period, a dynamic simulation analysis of the carrying capacity of the Ulan Suhai Lake from 2014 to 2050 was carried out. The results show that Ulan Suhai Lake has a poor carrying capacity level and a fragile carrying state. Three of the five simulated scenarios lead to severe water shortages, while the others affect economic growth. Considering these aspects, it was found that comprehensive mode I performed better, and for the configuration of this scheme, the following countermeasures, and suggestions to enhance the carrying capacity of the Ulan Suhai Lake are proposed. The rate of change in the industrial water use should be reduced by 20%, the discharge coefficients of the domestic and industrial wastewater should be reduced by 10% and 50%, respectively, the effective utilization coefficient of agricultural irrigation should be increased to 0.71, and the wastewater treatment rate should be increased by 30%. These measures can ensure the protection of the water resources while ensuring economic development.

Keywords: collaborative bearing capacity; analytic hierarchy process; SD model; simulation prediction; Ulan Suhai Lake

Citation: Ren, B.; Sun, B.; Shi, X.; Zhao, S.; Wang, X. Analysis of the Cooperative Carrying Capacity of Ulan Suhai Lake Based on the Coupled Water Resources–Water Environment–Water Ecology System. *Water* **2022**, *14*, 3102. https://doi.org/10.3390/w14193102

Academic Editor: Roko Andricevic

Received: 15 September 2022

Accepted: 28 September 2022

Published: 2 October 2022

Publisher's Note: MDPI stays neutral with regard to jurisdictional claims in published maps and institutional affiliations.

1. Introduction

The term "bearing capacity" refers to the maximum load that the research subject can withstand without significant deformation or destruction [1,2]. At present, the main methods of researching the water resource and water environment carrying capacity are the ecological footprint method [3], principal component analysis [4], backpropagation (BP) neural network model [5], routine prediction method [6], fuzzy comprehensive evaluation [7], the multiple objective decision-making method [8], and system dynamics method [9–11]. Most scholars have researched the carrying capacity as part of the theory of sustainable development, not through separate studies. For example, Harris et al. proposed that in areas with more agricultural water usage, the water resource carrying capacity could represent an important reference standard for the development potential of the region [12].

The carrying capacity of water resources in Xinjiang was first studied in the middle and late 1980s [13]. Chang Yuting [14], Tian Pei [15], and Zhao Xiaoqing et al. [16] used the water footprint model, variable weight technique for order of preference by its similarity to the ideal solution (TOPSIS) model, and system dynamics (SD) model to comprehensively evaluate the carrying capacities of regional water resources. Of course, the carrying capacity is studied based on the establishment of an index system. Most previous studies on the carrying capacity have been limited to the water resources carrying capacity or water environment carrying capacity, and most of these studies ignored the water ecology indicators [17,18]. Although some studies have emphasized the importance of quantifying the water ecology, most of the indicator data were for water resources [19]. As a result, previous research includes studies of only the water resources carrying capacity, water environment carrying capacity, or water ecology carrying capacity, without linking the water resources, water environment, and water ecology in carrying capacity research. Less research related to the synergistic carrying capacity has been conducted, which cannot comprehensively and systematically define the carrying capacity.

Ulan Suhai Lake has directly affected the regional water quality and water ecological safety due to the long-term acceptance of receding farmland water from the Hetao irrigation district, urban domestic sewage, and industrial sewage [20]. Moreover, the water pollution in Ulan Suhai Lake has intensified, and its ecological function has degraded [21], leading to a serious ecological crisis [22].

Therefore, to maximize the carrying benefits of Ulan Suhai Lake and utilize the important role of the Inner Mongolia plateau lake basin in restoring its ecology and conserving water, the carrying capacity of the water resources, water environment, and water ecology of Ulan Suhai Lake was established using system theory and synergy theory to carry out synergistic carrying capacity assessment and calculations. In addition, the functional configuration of the coupled water resources, water environment, and water ecology system of Ulan Suhai Lake was simulated under multiple scenarios. Then, a synergistic regulation scheme for the different subsystems was developed to assess the synergistic carrying capacity of Ulan Suhai Lake under the current and future changing environment. The water resource, water environment, and water ecology indicators were selected, the analytic hierarchy process method was used to calculate the weights, and a water resource, water environment, and water ecology subsystem model was established. Synergy theory was used to combine the various subsystems and establish a calculation model of the collaborative bearing capacity of Ulan Suhai Lake.

In order to further predict the dynamic development of the future bearing capacity of Ulan Suhai Lake, a dynamic system model was constructed, the value of the decision-making variable was changed, and five scenarios were designed to dynamically simulate the main indicators and the collaborative bearing capacity of the system. These five scenarios were the status continuation mode, development continuation mode, conservation continuation mode, comprehensive mode I, and comprehensive mode II. Then, the optimal control scheme was selected according to the corresponding bearing capacity of each scheme.

2. Materials and Methods

2.1. Overview of the Study Area and Data Sources

Ulan Suhai Lake, located at 40°36′–41°03′ N, is a freshwater lake in China. It is an important part of the drainage system of the Hetao irrigation district in Inner Mongolia and is located at the end of the Hetao Plain (Figure 1). According to the thematic mapper TM satellite remote-sensing images acquired from 2018 to 2020, the specific data sources of land use in Ulan Suhai Lake are shown in Table 1, reference [10,11,23]. The lake freezes at the end of November and thaws from late March to April. Ulan Suhai Lake is an important water body for domestic sewage, industrial sewage, and agricultural retreat water in the Hetao irrigation district. It was mildly eutrophic from 2015 to 2017 and has remained moderately eutrophic since 2018 [24], with a high diversity of phytoplankton pollution indicator species [25]. Currently, the government departments responsible for Ulan Suhai Lake strictly adhere to the "three zones and three lines" rule, strictly control the ecological environment zoning according to the "three lines and one list" rule, implement the 14th Five-Year Plan for the ecological environmental protection and management of the Ulan Suhai Lake basin with high standards, and adopt seasonal and environmental protection measures in the protected area. The plan includes a seasonal fishing ban in the protected area and prohibits fishermen from fishing in the core area and buffer zone. Ulan Suhai Lake is currently in a subhealthy state [26].

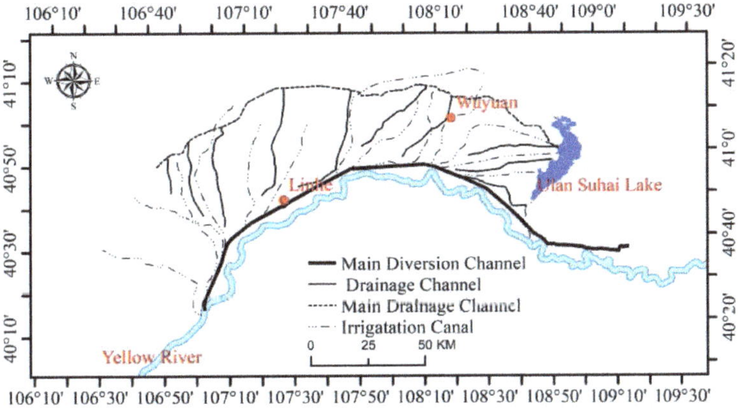

Figure 1. The irrigation and drainage system in Hetao.

Table 1. Land use in Ulan Suhai Lake.

Land Use in Ulan Suhai Lake					
Average Water Depth (m)	**Maximum Storage Capacity (Million m^3)**	**Water Area (km^2)**	**Reed Growth Area (km^2)**	**Area of the Open Water Area (km^2)**	**Aquatic Weed Covered Area (km^2)**
2.21	300	337.78	205.27	105.22	27.34

The index data used in this study were mainly obtained from the Inner Mongolia Autonomous Region Statistical Yearbook (2014–2020), the Bayannaoer City Statistical Yearbook (2014–2020), the Bayannaoer City Water Resources Bulletin (2014–2020), the Resource and Environmental Science Statistical Data (2014–2020), and the Bayannaoer City Water Supplies Bureau, the Bayannaoer City Ecological Environmental Protection Bureau, and other departments (2014–2020). The specific sources of the data are described in Table 2.

Table 2. Index Data Source.

Data Source	
Indicators	**Historical Data Sources**
Total water resources	Bayannaoer City Water Resources Bulletin
Total basin water use	Bayannaoer City Water Resources Bulletin
Industrial water consumption	Bayannaoer City Water Resources Bulletin
Ecological environmental water consumption	Bayannaoer City Water Resources Bulletin
Effective utilization coefficient of farmland irrigation	Bayannaoer City Water Resources Bulletin
Total wastewater discharge	Bayannaoer City Ecological Environmental Protection Bureau

Table 2. *Cont.*

Data Source	
Indicators	**Historical Data Sources**
Industrial COD and NH_3-N discharge	Bayannaoer City Ecological Environmental Protection Bureau
Agriculture COD and NH_3-N discharge	Bayannaoer City Ecological Environmental Protection Bureau
Urban COD and NH_3-N discharge	Bayannaoer City Ecological Environmental Protection Bureau
Pollutant distribution ratio of the lake during ice-covered period	Water environment team inspection
Ratio of lake district up to the standard for water quality	Water environment team inspection
Biodiversity	Water environment team inspection
Water area eutrophication	Water environment team inspection
Duration of ice-covered period	Yellow River Conservancy Commission of the Ministry of Water Resources
Regional wetland area	Bayannaoer City Statistical Yearbook
Forest area	Bayannaoer City Statistical Yearbook
Grassland area	Bayannaoer City Statistical Yearbook
Area of zone	Bayannaoer City Statistical Yearbook
Regional water area	Bayannaoer City Statistical Yearbook
Total population	Bayannaoer City Statistical Yearbook
Natural growth rate	Bayannaoer City Statistical Yearbook
Product of primary industry	Bayannaoer City Statistical Yearbook
Value of industrial industry	Bayannaoer City Statistical Yearbook
Value of agriculture industry	Bayannaoer City Statistical Yearbook
Speed of urbanization	Bayannaoer City Statistical Yearbook

2.2. *Research Methods*

2.2.1. Indicator Selection and Bearing Capacity Calculation

a. Indicator Selection

The bearing capacity is a large and complex system. The main purpose of studying the carrying capacity is to ensure a good ecological environment. With the water resources, water environment, and water ecology as the limiting factors, the relationships between these subsystems result in the bearing capacity reflecting the safety of the water intake in this area. Consequently, it is necessary to consider the bearing capacity of the coupled water resource–water environment–water ecology system and their coupling mechanism. Based on the research goals of this study, the indicators were selected based on the principles of scientificity, operability, and screening coordination [27]. The indicators for the Ulan Suhai Lake investigation were analyzed using the analytic hierarchy process, and then, the factor load matrix was rotated using the orthogonal rotation method with maximum variance. The rotated load value greater than 0.6 was taken as the bearing capacity evaluation index [28]. A total of 17 specific indicators were selected to form the bearing capacity evaluation system index for Ulan Suhai Lake (Table 3).

Table 3. Indicators of collaborative bearing capacity evaluation system.

Rule Hierarchy	Index Hierarchy	Type
Water resources carrying capacity (A)	Development and utilization rate of water resources (A1)	Pressure type
	Per capita water resources (A2)	Support type
	Water use amount per 10,000 yuan of industrial value added (A3)	Pressure type
	Ecological environment water consumption rate (A4)	Pressure type
	Per capita water area (A5)	Support type
	Effective utilization coefficient of farmland irrigation (A6)	Support type

Table 3. *Cont.*

Rule Hierarchy	Index Hierarchy	Type
Water environment carrying capacity (B)	Wastewater discharge intensity (B1)	Pressure type
	Industrial pollution intensity index (B2)	Pressure type
	Agricultural pollution intensity index (B3)	Pressure type
	Urban pollution intensity index (B4)	Pressure type
	Ratio of lake district up to the standard for water quality (B5)	Support type
	Pollutant distribution ratio of the lake during ice-cover period (B6)	Pressure type
Water ecology carrying capacity (C)	Submerged plant coverage ratio (C1)	Support type
	Water area eutrophication index (C2)	Pressure type
	Duration of ice-covered period (C3)	Support type
	Water conservation index (C4)	Support type
	Biodiversity index (C5)	Support type

b. Indicator weight calculation and collaborative bearing capacity calculation

In this paper, the hierarchical analysis method is used to weight the indicators. The specific steps have been described by Yang Qian et al. [29]. The calculation of the bearing capacity was conducted as follows.

(1). Data standardization

In this paper, the Z-score method is used to standardize the original data. The specific steps are as follows.

(a) Determine the expectation $\overline{X}$ and standard deviation of each indicator S.

$$\overline{X} = \frac{\sum_{i=1}^{n} X_i}{n} \quad (1)$$

$$S = \sqrt{\frac{\sum_{i=1}^{n}\left(X_{\overline{I}} - \overline{X}\right)^2}{n-1}} \quad (2)$$

(b) Carry out standard treatment.

$$E_{\overline{z}} = \left(X_{\overline{i}} - \overline{X}\right)/S \quad (3)$$

where $E_{\overline{z}}$ is the standardized value, and $X_{\overline{i}}$ is the actual value, n is the total number of indicators, $i - 1,2,3 \ldots$ n, X is the raw data of the indicator.

(2). Establish the calculation model of the minute bearing capacity and determine the bearing capacity.

$$E_j = \sum_{i=1}^{m} w_{\overline{i}} \cdot E_{\overline{i}} \quad (4)$$

where E_j is the sub-carrying capacity of the water resources, water ecology, and water environment, $E_{\overline{i}}$ is the standardized value of the ith index, $w_{\overline{i}}$ is the weight of the ith index, m is the number of indicators, and j is the number of subsystems.

(3). Establish a collaborative bearing capacity calculation model to determine the bearing capacity.

$$E = \left[\sum_1^3 \left(w_j E_j\right)^2\right]^{1/2} \quad (5)$$

where E is the collaborative bearing capacity value, w_j is the weight of the jth sub-bearing capacity, and E_j is the value of the jth sub-bearing capacity.

(4). According to previous studies, determine the level and state of the carrying capacity (Table 4).

Table 4. Bearing capacity level and state judgment table.

Value of Bearing Capacity	0–0.2	0.2–0.5	0.5–0.8	0.8–1
Bearing level State	Inferior Collapse	Poor Fragile	General General	Good Good elasticity

2.2.2. Construction of System Dynamics Model

a. Parameterization and structural analysis of system dynamics model

In this study, Bayannur City in Inner Mongolia was defined as the system boundary of the carrying capacity model, 2014–2050 was the simulation time, of which 2014–2020 was the test phase, 2021–2050 was the simulation prediction phase, the step length was 1 year, and 2018 was the base year. The initial values of the indicators were the data for the base year, which were used to predict the values of the indicators in 2021–2050 and to predict the collaborative bearing capacity of Ulan Suhai Lake. After a large number of literature element frequency statistics and in reference to the related studies [30–33], the collaborative bearing capacity was divided into the carrying capacity of each subsystem, i.e., the water resources, water environment, and water ecology. The water resources subsystem included the population, the total water resources, water use amount per 10,000 yuan of industrial value added, and the effective utilization coefficient of farmland irrigation. The water environment subsystem included the socioeconomic indicators, wastewater discharge intensity, pollutant emission intensity, and distribution ratio of the pollutants during the ice-covered period. The water ecology subsystem indicators included the lakewater quality, eutrophication status, biodiversity, and length of ice-cover period.

b. Construction of system dynamics model

In this study, the spatial boundary of the system was defined as Bayannur City in the Inner Mongolia Autonomous Region; the temporal boundary was 2014–2050, the historical test period was 2014–2020, and the forecasting regulation period was 2021–2050. The iteration interval was 1a, with 2014 as the base year, and the initial values of the variables were simulated using the relevant actual data for 2014 for the regulation period. First, before running the simulation using the model, the equations and their units in the model were checked for errors, and if there was an error, the error was corrected and rechecked according to the prompt message. Second, the relevant parameters of the model were simulated and run many times, and the behaviors of the key variables were compared to analyze the causes and trends. Finally, in order to verify whether the model could accurately reflect the actual synergistic carrying capacity operation, 2014–2020 was selected as the historical testing time period to determine the matching degree (mse) between the simulated and actual values.

In order to study the collaborative bearing capacity of Ulan Suhai Lake and analyze its change trend, the SD model was established based on the selected indicators and the SD principle. The model mainly comprised nine state variables and 71 model parameters. The model had horizontal variables, auxiliary variables, and constant variables. The model involves more variables. The main parameter equations are listed in Table 5. The model was then used to predict the trends of the carrying capacity of Ulan Suhai Lake from 2021 to 2050 under five scenarios: the status continuation mode, development continuation mode, conservation continuation mode, comprehensive mode I, and comprehensive mode II. The model was divided into three subsystems, each of which consisted of several variables and equations, and each subsystem was linked via the synergy theory.

Table 5. Principal parameters and their equations.

Per Capita Water Resources = Total Water Resources/Total Population
Per capita domestic water consumption = total water consumption/total population
Development and utilization rate of water resources = total water supply/total water resources
Total water resources = volume of groundwater resources + volume of surface water resources
Wastewater discharge intensity = total wastewater discharge/regional gross domestic product
Agricultural water consumption = farmland irrigation water use + forest, animal husbandry, and fishery storage water consumption
Total amount of sewage discharge = quantity of industrial wastewater effluent + quantity of domestic sewage effluent
Total water supply = reclaimed water volume + groundwater supply + surface water supply
Variability of volume of industrial water used = volume of industrial water used * variation ratio of industrial water used
Water use amount per 10,000 yuan of industrial value added = total water consumption/gross industrial output value
Added value of primary industry = product of primary industry * rate of primary industry increase
Added value of secondary industry = product of secondary industry * rate of secondary industry increase
Added value of tertiary industry = product of tertiary industry * rate of tertiary industry increase
Urban pollutant emission intensity = (urban COD and NH_3-N discharge)/value of tertiary industry
Agriculture pollutant emission intensity = (agriculture COD and NH_3-N discharge)/value of primary industry
Industrial pollutant emission intensity = (industrial COD and NH_3-N discharge)/value of industrial industry
Water resources carrying capacity= water use amount per 10,000 yuan of industrial value-added carrying capacity + per capita water area carrying capacity + per capita water resources carrying capacity + effective utilization coefficient of farmland irrigation carrying capacity + development and utilization rate of water resources carrying capacity + ecological environment carrying capacity
Water environment carrying capacity = agriculture pollutant emission intensity carrying capacity + urban pollutant emission intensity carrying capacity + industrial pollutant emission intensity carrying capacity + wastewater discharge intensity carrying capacity + pollutant distribution ratio of the lake during ice-covered period carrying capacity + ratio of lake district up to the standard for water quality carrying capacity
Water ecology carrying capacity = duration of ice-covered period carrying capacity + water area eutrophication index carrying capacity + water conservation index carrying capacity + submerged plant coverage ratio carrying capacity + biodiversity index carrying capacity
Collaborative bearing capacity = $[(0.4126 * \text{water resources carrying capacity})^2 + (0.3275 * \text{water environment carrying capacity})^2 + (0.2599 * \text{water ecology carrying capacity})^2)^{1/2}$

2.2.3. Scenario Plan Design

Based on previous studies [34–36] and taking into account the current situation of the water resource utilization and socioeconomic development in Bayannur City, the population growth rate, urbanization rate, rural water consumption per capita, urban water consumption per capita, domestic wastewater discharge coefficient, sewage treatment rate, effective irrigation area, domestic wastewater discharge coefficient, and industrial wastewater coefficient were selected as the decision parameters. Five scenarios (i.e., the status continuation mode, conservation continuation mode, development continuation mode, comprehensive mode I, and comprehensive mode II) were developed to simulate the collaborative bearing capacity of Ulan Suhai Lake from 2021 to 2050 and to find out the changes in the collaborative bearing capacity under the existing socioeconomic development conditions. In the scenario design, the different scenarios were simulated by adjusting the values of the decision parameters, which were mainly based on the 13th Five-Year Plan for Population Development in the Inner Mongolia Autonomous Region, the 13th Five-Year Plan for National Economic and Social Development in the Inner Mongolia Autonomous Region, the Plan for Ecological Protection and High-Quality Development in the Yellow River Basin in the Inner Mongolia Autonomous Region, the Inner Mongolia Autonomous Region Water Resources Bulletin 2018–2020, and the Inner Mongolia Autonomous Region Water Quota Standards. The main parameters are listed in Table 6.

Table 6. Main parameters of the scenario modes.

Parameters	Status Continuation Mode	Development Continuation Mode	Conservation Continuation mode	Comprehensive Mode I	Comprehensive Mode II
Population growth rate (‰)	2.61	5.61	4.61	2.61	2.61
Urbanization rate (%)	54.2	75.6	54.2	54.5	59.5
Per capita rural water consumption (Liters/person/day)	85	120	75	80	75
Per capita urban water consumption (Liters/person/day)	116	150	90	105	116
Wastewater treatment rate (%)	75	75	100	95	95
Farmland irrigation quota (m^3/hm^2)	475	495	445	460	475
Water consumption of forest, animal husbandry, and fishery (million m^3)	2.375	2.85	1.875	2.075	1.875
Domestic sewage discharge coefficient	0.7	0.76	0.6	0.6	0.56
Industrial sewage discharge coefficient	0.51	0.55	0.43	0.46	0.43
Rate of primary industry increase (%)	6	9	3	4	4
Variation ratio of industrial water use (%)	1.2	1.4	0.8	1	0.9
Rate of tertiary industry increase (%)	11	24	11	13	13

The status continuation mode assumes the development of the existing development trend and takes the parameter values in 2014 as the initial values of all of the parameters for the simulation of the collaborative bearing capacity during 2021–2050. The development continuation mode is based on the status continuation mode, strengthening the protection of water resources, and the strict control of agriculture, industry, and domestic water use to ensure the sustainable use of water resources. The conservation continuation mode is based on the continuation of the status continuation mode, but only the ecological red line is guaranteed, and the economy is rapidly developed in accordance with the objectives of the 13th Five-Year Plan for the Inner Mongolia Autonomous Region. Comprehensive mode I and comprehensive mode II consist of a combination of economic development and water conservation, ensuring that economic development is maintained at a medium to high rate while focusing on water resource efficiency.

3. Results and Discussion

3.1. Weight and Bearing Capacity Value

The weights and combined weights of each index in each subsystem using the AHP are presented in Table 7, the standardized values obtained using Equations (1)–(3) are presented in Appendix A (Table A1), and the bearing capacity results for each system calculated using Equations (4) and (5) are presented in Figure 2.

Table 7. Weight results of chromatography analysis.

Target Hierarchy	Criterion Hierarchy	Criterion Hierarchy Weight	Index Hierarchy	Index Weight	Combination Weight
Bearing capacity A	Water resources carrying capacity B1	0.4126	Development and utilization rate of water resources (A1)	0.199	0.082
			Per capita water resources (A2)	0.240	0.099
			Water use amount per 10,000 yuan of industrial value added (A3)	0.213	0.088
			Ecological environment water consumption rate (A4)	0.105	0.043
			Per capita water area (A5)	0.126	0.052
			Effective utilization coefficient of farmland irrigation (A6)	0.117	0.048
	Water environment carrying capacity B2	0.3275	Wastewater discharge intensity (B1)	0.117	0.038
			Industrial pollution intensity index (B2)	0.199	0.065
			Agricultural pollution intensity index (B3)	0.240	0.078
			Urban pollution intensity index (B4)	0.213	0.070
			Ratio of lake district up to the standard for water quality (B5)	0.126	0.041
			Pollutant distribution ratio of the lake during ice-covered period (B6)	0.105	0.034
	Water ecology carrying capacity B3	0.2599	Submerged plant coverage ratio (C1)	0.237	0.062
			Water area eutrophication index (C2)	0.133	0.035
			Duration of ice-covered period (C3)	0.133	0.035
			Water conservation index (C4)	0.295	0.077
			Biodiversity index (C5)	0.202	0.052

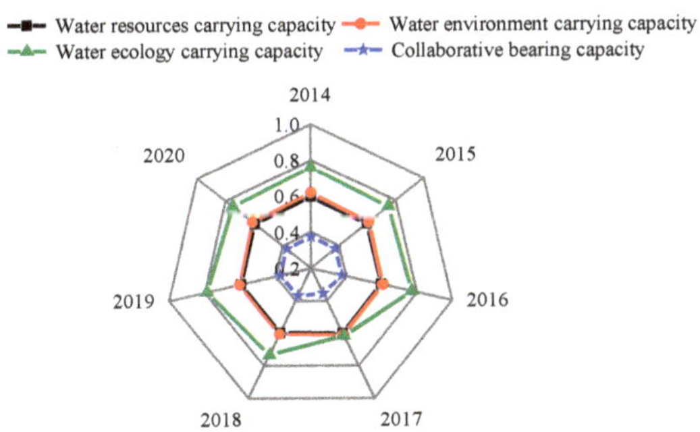

Figure 2. Bearing capacity of each system.

3.2. Validation of the Model

In this study, 2014–2020 was taken as the validation period of the model. The simulated and actual values were compared, and it was found that the model has an error of within ±10% and is effective [37]. Thus, it could be used to predict the collaborative bearing capacity of Ulan Suhai Lake from 2021 to 2050. In this study, three subsystems were selected to verify the bearing capacity. The results and errors are presented in Table 8. The relative errors between the simulated and actual values of the three subsystems were found to be within ±10%, indicating that the accuracy of the model was good, and it could be used for the subsequent simulation and prediction analysis.

Table 8. Comparison of simulated and actual carrying capacity of Ulan Suhai Lake from 2014 to 2020.

Year	Water Resources Carrying Capacity				Water Environment Carrying Capacity			
	Actual Value	Simulated Value	Residual	Mse	Actual Value	Simulated Value	Residual	Mse
2014	0.597	0.596	−0.001	$2.066*10^{-6}$	0.617	0.622	0.005	$2.153*10^{-4}$
2015	0.598	0.597	−0.001		0.609	0.622	0.013	
2016	0.598	0.597	−0.001		0.606	0.622	0.016	
2017	0.598	0.598	0.000		0.607	0.622	0.015	
2018	0.596	0.598	0.002		0.603	0.622	0.019	
2019	0.596	0.598	0.002		0.604	0.622	0.018	
2020	0.596	0.598	0.002		0.610	0.621	0.012	
Year	**Water Ecology Carrying Capacity**				**Collaborative Bearing Capacity**			
	Actual Value	**Simulated Value**	**Residual**	**Mse**	**Actual Value**	**Simulated Value**	**Residual**	**Mse**
2014	0.761	0.764	0.003	$1.165*10^{-3}$	0.375	0.376	0.001	$2.356*10^{-5}$
2015	0.748	0.728	−0.020		0.372	0.371	−0.001	
2016	0.770	0.785	0.015		0.375	0.379	0.005	
2017	0.614	0.569	−0.045		0.355	0.352	−0.002	
2018	0.731	0.680	−0.051		0.368	0.365	−0.003	
2019	0.788	0.842	0.053		0.376	0.388	0.011	
2020	0.751	0.741	−0.010		0.372	0.373	0.001	

3.3. Model Prediction Analysis

In this study, 2014 was taken as the simulation base year, and the prediction time was 2021–2050. Based on the simulation results obtained using the constructed dynamic system model, the trends of the six main influencing factors (i.e., the total population, total water consumption, domestic water consumption, agricultural water consumption, per capita water resources, and the total amount of sewage discharge) of the Ulan Suhai Lake system from 2021 to 2050 are shown in Figure 3.

The model (Figure 3d) predicts that agricultural water consumption slowly increases during 2021–2050 because Ulan Suhai Lake is in the Hetao irrigation district, in which the water resources are required and used for crop irrigation. With the increase in the urbanization level, the per capita water resource utilization remains low [38] and domestic water consumption slowly increases (Figure 3c), which is consistent with the population growth rate. The per capita water resources exhibit a decreasing trend [39], but they remain positive (Figure 3b), which means that this increase is not caused by the decrease in the total water resources and is most likely related to the rapid growth of the total population. With the growth of urbanization and the population increase, the sewage volume also increases dramatically (Figure 3e). The growth of the sewage discharge causes domestic sewage pollution in the villages and towns to become an important source of water pollution in the region. In addition, Ulan Suhai Lake is an important component of the drainage system of the Hetao irrigation district in Inner Mongolia (Figure 1). Ulan Suhai Lake has been receiving chemical fertilizer and pesticide residues from the Hetao irrigation district and discharging them into urban domestic sewage and industrial sewage for a long time [20], which has aggravated the water pollution in Ulan Suhai Lake for a long time and has directly threatened the ecological water security of the Yellow River. With the rapid development of living standards, and the pressure on water resources increases, the main problem of which being how to address the maximum rate of water usage in the case of a limited water supply.

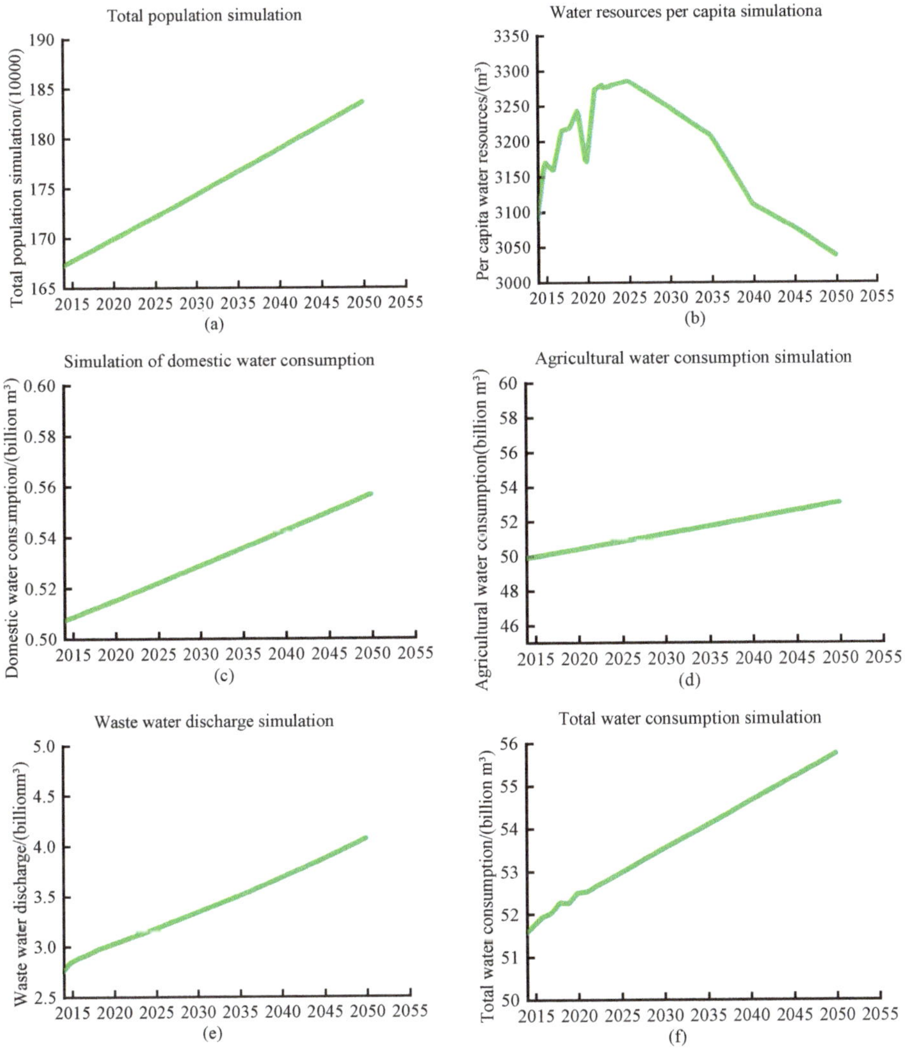

Figure 3. Variation trends of the main indexes of the carrying capacity system of Ulan Suhai Lake from 2014 to 2050. (**a**) Total population simulation; (**b**) Water resources per capita simulation; (**c**) Simulation of domestic water consumption; (**d**) Agricultural water consumption simulation; (**e**) Waste water discharge simulation; (**f**) Total water consumption simulation.

3.4. Model Scenario Simulation Analysis

The six main influencing factors (i.e., the total population, total water consumption, domestic water consumption, agricultural water consumption, per capita water consumption, and the total amount of sewage discharge) in the Ulan Suhai Lake system from 2021 to 2050 were simulated under different scenarios. The change trends of these indicators are shown in Figure 4.

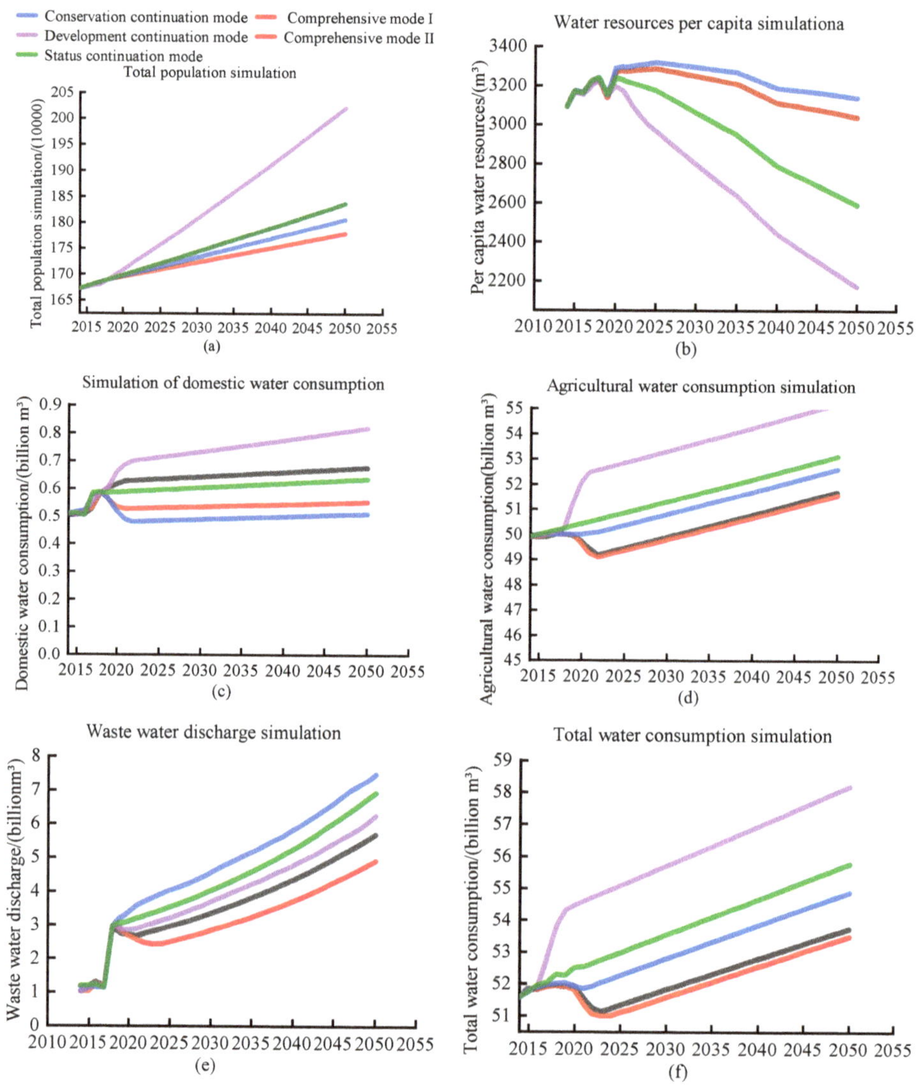

Figure 4. Change trends of the main indicators under the various scenarios. (**a**) Total population simulation; (**b**) Water resources per capita simulationa; (**c**) Simulation of domestic water consumption; (**d**) Agricultural water consumption simulation; (**e**) Waste water discharge simulation; (**f**) Total water consumption simulation.

As can be seen from Figure 4, under the five scenarios, the total population (Figure 4a), domestic water consumption (Figure 4c), agricultural water consumption (Figure 4d), total water consumption (Figure 4f), and the total amount of sewage discharge (Figure 4e) all slowly increase, while the per capita water resources decrease (Figure 4b). The specific data are presented in Table 9. The status continuation mode and development continuation mode ensure stable economic development. Under the status quo continuation model and the development continuation model, the annual average per capita water resources decrease by 13.88 m^3 and 25.75 m^3, and the annual growth rate decreases by 0.004% and 0.98%, respectively. Under the conservation continuation mode, the per capita water resources basically remain in a stable state, indicating that the pressure on the water resources is the lowest, but this mode limits economic development. In contrast, the two comprehensive modes are superior in that the per capita water resources change slowly while ensuring stable economic development and slowing the increase in the total water consumption, which improves the water resources utilization in the short term. This shows that the balance of the per capita water resources can be guaranteed in the long term in the future through measures such as improving water use efficiency, reducing sewage discharge, and improving sewage treatment and utilization.

Table 9. Annual growth and annual growth rate.

	Comprehensive Mode II		Comprehensive Mode I		Development Continuation Mode		Conservation Continuation Mode		Status Continuation Mode	
	Annual Increment	Annual Growth rate	Annual Increment	Annual Growth Rate	Annual Increment	Annual Growth Rate	Annual Increment	Annual Growth Rate	Annual Increment	Annual Growth Rate
Total population (Ten thousand people)	0.296	0.17%	0.296	0.18%	0.970	0.53%	0.375	0.20%	0.463	0.30%
Per capita water resources (m^3)	1.441	0.06%	1.441	0.06%	−25.750	−0.98%	1.584	1.50%	−13.880	4.70%
Domestic water consumption (hundred million m^3)	0.001	0.21%	0.001	0.24%	0.009	1.34%	0.013	0.10%	0.004	0.60%
Agricultural water consumption (hundred million m^3)	0.050	0.10%	0.046	0.09%	0.147	0.28%	0.042	0.10%	0.089	0.20%
Total amount of sewage discharge (hundred million m^3)	0.125	4.46%	0.108	4.45%	0.279	6.65%	0.093	4.20%	0.159	5.10%
Total water consumption (hundred million m^3)	0.060	0.11%	0.053	0.10%	0.184	0.34%	0.042	0.10%	0.117	0.20%

3.5. Analysis of the Variation Trends of the Bearing Capacity of Each Subsystem and the Collaborative Bearing Capacity under the Different Modes

The three subsystems of Ulan Suhai Lake (the water resources, water environment, and water ecology) and the large, complex system of the synergy of the three were simulated under the different scenarios, and the results were analyzed (Figures 5 and 6).

As is shown in Figure 5a, under the five modes, the state and level of the water resources carrying capacity are general, and the simulation results of the water resources carrying capacity are as follows: conservation continuation mode > comprehensive mode I > comprehensive mode II > status continuation mode > development continuation mode. The carrying capacity of the water resources subsystem is generally weak [40] and in a fragile and generally critical state [38]. It decreases under the development continuation mode and increases under the conservation continuation mode and the two comprehensive modes, and the growth is most obvious during 2021–2040. Under the status continuation mode, the slow growth of the population and gross domestic product (GDP) causes the water resources carrying capacity to increase more slowly. If the current situation continues, the water shortage will increase further after 2035, which is not conducive to the development of Bayannur City. If regulatory measures are adopted, the carrying capacity of the water resources in 2035 (2045) will be 0.597 (0.598) under the status continuation mode, while the relative ratios of the carrying capacity of the water resources in 2035 (2045) under the development continuation mode, conservation continuation mode, comprehensive mode I, and comprehensive mode II will be −14.4% (−17.2%), +29.5% (+34.4%), +26.8% (+31.75), and +21% (+25.8%). This demonstrates that the adoption of water conservation and water protection measures while ensuring economic development can alleviate the pressure on the water resources caused by economic and population development, and the comprehensive mode can largely reduce the pressure on the water resources subsystem. The conservation continuation mode takes more into account the protection of the water resources, so the water carrying capacity value increases significantly, but the GDP decreases by 3.5%, which inhibits the economic development of the region. The development continuation mode considers slowing down the growth rate of the GDP in 20355; that is, under the comprehensive mode I, as opposed to the status continuation mode, the growth rates of the primary and secondary industries are both reduced by 20%, and the growth rate of the tertiary industries reaches 55%. The effective utilization coefficient of agricultural irrigation will increase to 0.70, and the rate of industrial water use will decrease by 20% to improve the utilization rate of the water resources, which will more effectively relieve the pressure on the water resources subsystem [41], improve the per capita water resources, and protect the water resources while ensuring economic development.

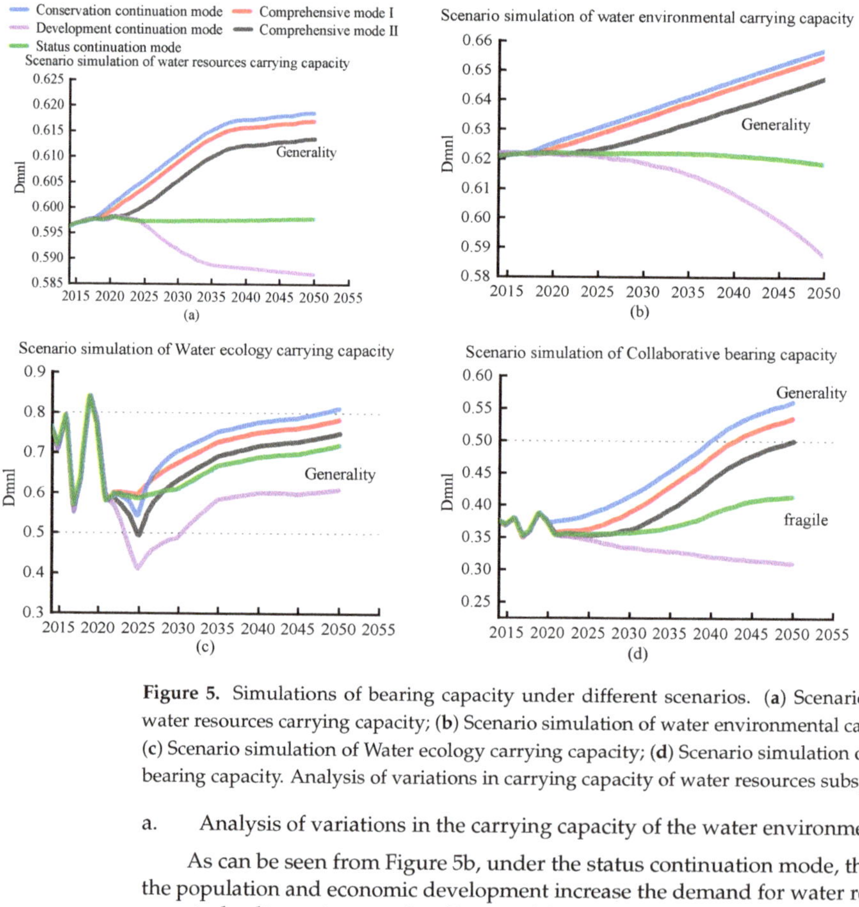

Figure 5. Simulations of bearing capacity under different scenarios. (**a**) Scenario simulation of water resources carrying capacity; (**b**) Scenario simulation of water environmental carrying capacity; (**c**) Scenario simulation of Water ecology carrying capacity; (**d**) Scenario simulation of Collaborative bearing capacity. Analysis of variations in carrying capacity of water resources subsystem.

a. Analysis of variations in the carrying capacity of the water environment subsystem

As can be seen from Figure 5b, under the status continuation mode, the increases in the population and economic development increase the demand for water resources in all aspects, leading to increased pollution of the water environment and affecting the water ecology. The water environment carrying capacity initially exhibits a constant trend and then decreases [42], and is mainly affected by the total amount of water resources and other factors [43]. The carrying capacity of the water environment decreases under both the status continuation mode and development continuation mode, but it decreases faster under the development continuation mode. Under the other three modes, it increases and remains average over time. Compared with the water environment carrying capacity value of 0.610 (0.621) under the status continuation mode in 2035 (2045), the water environment carrying capacity values in 2035 (2045) under the development continuation mode, conservation continuation mode, comprehensive mode I, and comprehensive mode II are −11.2% (−33.6%), +30.9% (+51.0%), +27.3% (+47.2), and +16.2% (+35.3%). Under the status continuation mode, the wastewater treatment rate increases to 95%, the domestic wastewater discharge coefficient decreases by 20%, and the industrial wastewater discharge coefficient decreases by 50%. The COD and NH_3-N emission concentrations of the industrial, agricultural, and domestic water decrease by 50%, 30%, and 10%, and the growth rates of the primary, secondary, and tertiary industries all increase to 65%. The effluent standard of the wastewater plant is increased to Class III or IV type water standards, to support the development of water-saving agriculture, and other measures are taken to protect the water environment. Thus, all three modes can alleviate the water environment pollution, the pressure on the water environment subsystem is correspondingly reduced. However, the conservation continuation mode limits local development. synergistic economic and environmental development is ensured while the water environment is effectively protected, so comprehensive mode I performed best.

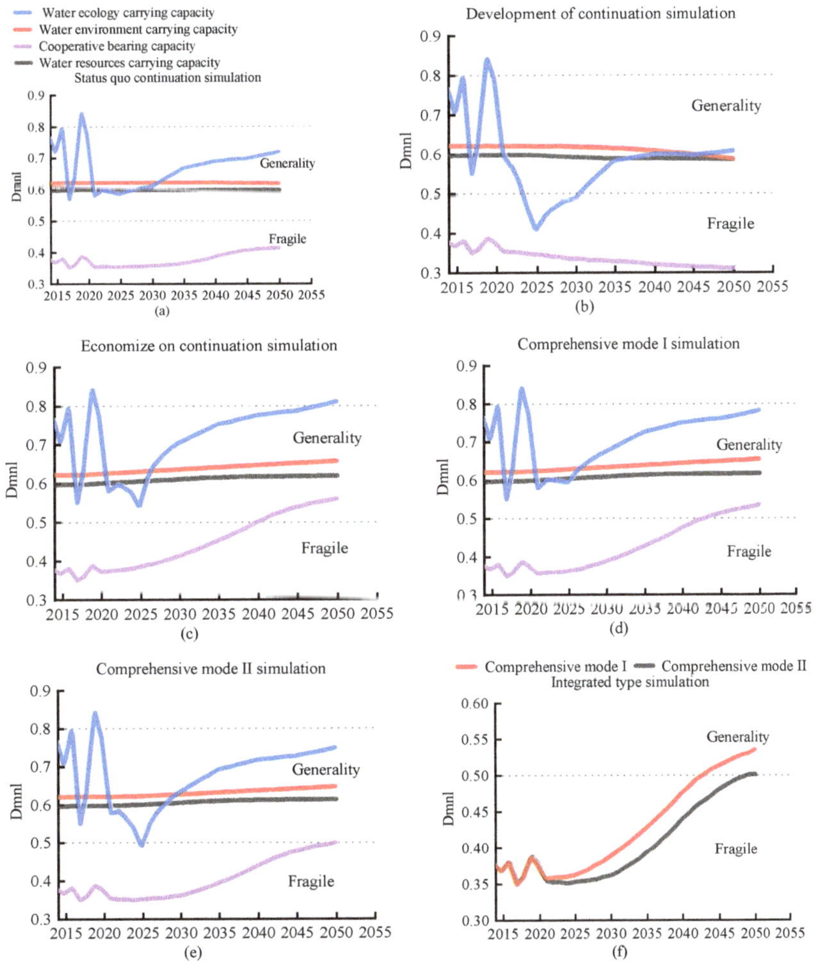

Figure 6. Simulation of bearing capacity of each system under the different scenarios. (**a**) Status quo continuation simulation; (**b**) Development of continuation simulation; (**c**) Economize on continuation simulation; (**d**) Comprehensive mode I simulation; (**e**) Comprehensive mode II simulation; (**f**) Integrated type simulation.

b. Analysis of variations in the carrying capacity of the water ecological subsystem

Figure 5c shows that the water ecological carrying capacity value is 0.6673 in 2035 under the status continuation mode, and it exhibits a gradual upward trend [44] and has a good carrying capacity [45]. However, with the growth of the population and GDP, the change in the water ecological carrying capacity is not obvious, which means that the water ecological subsystem is less influenced by the population and GDP. It is most strongly influenced by the eutrophication index, biodiversity index, and wetland area. If regulation measures are taken, the water ecological carrying capacity of 0.667 (0.698) in 2035 (2045) will be less changed under the development continuation mode and conservation continuation mode than under the status continuation mode, −12.7% and +15.7%, respectively. Under the comprehensive mode, measures such as keeping the population growth rate constant, reducing the total nitrogen and total phosphorus concentrations, chlorophyll concentration, and submerged plant cover in the lake area and increasing the biodiversity index and wetland area put the least pressure on the aquatic ecosystem and protect the aquatic ecology while allowing economic development.

Based on the analysis of the change trends of the bearing capacities of the three subsystems, the collaborative bearing capacity simulation (Figure 5d) results are as follows: conservation continuation mode > comprehensive mode I > comprehensive mode II > status continuation mode > development continuation mode. Under the five modes, considering the economic and population growth alone, the collaborative carrying capacity will face greater pressure, and the carrying capacity will be low and in a fragile state. If the conservation continuation mode is adopted alone, the respective carrying capacity values of the water resources, water environment, and water ecology will improve, the carrying status will improve, and the water resources gap will improve, but local economic development will be limited. In contrast, the status continuation mode, the development continuation mode, and comprehensive mode II result in a long-term fragile state. The conservation continuation mode and the comprehensive mode I have bearing capacity values of greater than 0.5 in 2043, the bearing status changes from fragile to general, and the bearing capacity level increases to general, which effectively improves the collaborative bearing capacity. It was found that the collaborative bearing capacity is more comprehensive and systematic than the single carrying capacity. By setting and comparing the different modes, comprehensive model I was selected, which can effectively improve the carrying efficiency of Ulan Suhai Lake.

As is shown in Figure 6, the simulations under the five modes from 2021 to 2050 show that the water resources carrying capacity, water environment carrying capacity, water ecology carrying capacity, and collaborative bearing capacity are higher under the conservation continuation mode, comprehensive mode I, and comprehensive mode II than under the other two modes, and the changes in the water resources carrying capacity and water environment carrying capacity of Ulan Suhai Lake are small, while the changes in the water ecological carrying capacity are significant. The collaborative bearing capacity of Ulan Suhai Lake does not change significantly under the status continuation mode (Figure 6a) and remains in a fragile state for a long time. Under the development continuation mode (Figure 6b), economic development is accelerated. However, the lake's carrying capacity gradually weakens, its carrying status changes from fragile to collapse, and its carrying level is extremely poor. Its collaborative bearing capacity gradually increases under the conservation continuation mode (Figure 6c), and in 2040 its bearing status changes from fragile to average, while its bearing level changes from poor to average. Its state increases under the two comprehensive modes, but changes slowly under comprehensive mode II (Figure 6e)and remains in a fragile state until 2049. Under comprehensive mode I (Figure 6d), the lake's bearing capacity value is greater than 0.5 after 2043, its bearing status changes from fragile to average, and its bearing level increases from poor to average. In comparison, under the conservation continuation mode, the lake's collaborative bearing capacity increases, but in terms of the local development strategy, this mode is not suitable for improving the lake's carrying capacity. Under the development continuation mode, the lake's collaborative bearing capacity decreases, which is not conducive to the protection of the water resources, water environment, and water ecology. Comprehensive mode I is more conducive than other modes to improving the lake's carrying capacity value, and is thus the optimal solution.

4. Conclusions

A dynamic model of the collaborative bearing capacity of the water resources–water environment–water ecology system of Ulan Suhai Lake was established, and the dynamics and regulation of the carrying capacity were predicted using the model. The main conclusions are as follows.

(1) The current collaborative bearing capacity level of Ulan Suhai Lake is poor and the bearing capacity is fragile, but it exhibits a slow growth trend, indicating that the future development prospect of Ulan Suhai Lake's bearing capacity is good. The poor carrying capacity of Ulan Suhai Lake is mainly influenced by the decrease in the water resources carrying capacity. Therefore, we can start from the perspective of the water resources carrying capacity, establish water conservation awareness, cultivate water-saving habits, build rainwater storage facilities, and vigorously promote sprinkler irrigation, drip irrigation, low-pressure pipe irrigation, and other measures to reduce farmland water use and improve the carrying capacity of the water resources.

(2) We can improve the carrying capacity of Ulan Suhai Lake by adopting comprehensive mode I—for example, increasing the wastewater treatment rate to 95%, improving the water resource utilization and increasing the effective utilization coefficient of the farmland irrigation. In addition, the domestic sewage discharge coefficient and industrial sewage discharge coefficient should be appropriately decreased to improve the carrying capacity of Ulan Suhai Lake; in other words, more water conservation management efforts should be made regarding urban, rural, and industrial water use.

(3) The water quality of Ulan Suhai Lake is mainly reduced by the inorganic fertilizers and pesticides in the water draining from the farmland, the discharge of urban and rural domestic sewage and industrial wastewater, the lack of ecological water pay, and the lakewater pollutants exceeding the standard levels and accumulating in the lake. These issues can be resolved using the following measures. First, the use of organic fertilizers and bio-pesticides and the accurate scientific usage of drugs should be promoted, and the utilization rate of chemical fertilizers and pesticides should be improved. Second, the sewage treatment plant should be upgraded, and the collection and recycling of wastewater should be conducted to achieve the discharge of zero sewage into the water supply. Third, timely dredging of the lake and reed harvesting should be conducted to increase the ecological water pay and improve the self-purification ability of the lake.

Author Contributions: Concept, consultation, references, B.R., B.S., X.S. and S.Z.; calculation method determined, B.R.; model development, B.R. and B.S.; field and laboratory experiments, B.R.; writing, B.R. and X.W. All authors have read and agreed to the published version of the manuscript.

Funding: This research was supported by the National Key Research and Development Program (No.2019YFC0409201, 2017YFE0114800), the National Natural Science Foundation of China (No.51869020, No.52060022, No.51779118), the Natural Science Foundation of Inner Mongolia (No.2019MS05032), and was co-funded by the Young Science and Technology Talents Support Program (NJYT-19-B11) of the higher education institutions of the Inner Mongolia Autonomous Region.

Institutional Review Board Statement: Not applicable.

Informed Consent Statement: Not applicable.

Data Availability Statement: The data presented in this study are available upon request from the corresponding author or the first author.

Acknowledgments: We thank the Bureau of Water Resources, Bureau of Statistics, and Bureau of Forestry and Grassland of Bayannur City, Inner Mongolia Autonomous Region, for improving the data.

Conflicts of Interest: The authors declare that they have no conflicts of interest.

Appendix A

Table A1. Normalized data values.

Index	2014	2015	2016	2017	2018	2019	2020
A1	0.41234	0.41182	0.41169	0.41168	0.24119	0.24121	0.24116
A2	2.04116	2.04118	2.04118	2.04118	2.04108	2.04109	2.04108
A3	0.39024	0.39251	0.39325	0.39329	0.04117	0.20412	0.20412
A4	0.41288	0.41235	0.41220	0.41219	−0.41243	−0.41268	−0.41206
A5	−0.41291	−0.41224	−0.41195	−0.41195	0.41242	0.41268	0.41206
A6	−0.41278	−0.41225	−0.41210	−0.41208	0.41242	0.41268	0.41206
B1	2.02682	2.03834	2.03983	2.03970	2.04099	2.04073	2.03641
B2	0.46498	0.43432	0.42641	0.42768	0.24992	0.24090	0.24142
B3	0.46765	0.43574	0.42744	0.42902	0.20840	0.40672	0.39450
B4	0.46084	0.43232	0.42525	0.42587	0.27331	0.40689	0.39369
B5	0.46525	0.43481	0.42698	0.42694	−0.48390	−0.40905	−0.41887
B6	0.16810	0.30116	0.33375	0.33019	0.55708	0.40960	0.41989
C1	−0.75424	−0.75114	−0.74790	−0.68595	−0.74092	−0.70042	−0.70400
C2	−0.77830	−0.77646	−0.77636	−0.71408	−0.76554	−0.72930	−0.73189
C3	1.48041	1.50460	1.46579	1.67570	1.53448	1.42466	1.49553
C4	−0.58837	−0.54480	−0.61561	−0.15678	−0.49095	−0.63355	−0.51868
C5	−0.53625	−0.52180	−0.55713	−0.43245	−0.51897	−0.52823	−0.47317

References

1. Feng, Z.M.; Yang, Y.Z.; Yan, H.M.; Pan, T.; Li, P. A review of resources and environment carrying capacity research since the 20th Century: From theory to practice. *Resour. Sci.* **2017**, *39*, 379–395. (In Chinese)
2. Zhang, Y.Y.; Xia, Y.; Wang, Z.G. Research on Regional Water Resources Carrying Capacity Theory and Method. *Prog. Geogr.* **2007**, *2*, 126–132. (In Chinese)
3. Jing, Y.Y.; Chen, Y.Z. Review and Thinking on the Research of the Resources Carrying Capacity. *China Popul. Resour. Environ.* **2006**, *5*, 11–14. (In Chinese)
4. Xiong, N.N.; Xie, S.Y. A Study of the Spatio—Temporal Evolution of Ecological Footprint and Ecological Carrying Capacity of Water Resources of Chengdu. *J. Southwest Univ. (Nat. Sci. Ed.)* **2018**, *40*, 124–131. (In Chinese)
5. Cao, L.J.; Zhang, X.P. Assessment of water resources carrying capacity in Gansu Province based on principal component analysis. *Arid. Land Geogr.* **2017**, *40*, 906–912. (In Chinese)
6. Yang, L.L.; Li, B.; Fu, Q. Scenario analysis of water resources carrying capacity in Xinjiang based on BP neural network model. *J. Beijing Norm. Univ. (Nat. Sci.)* **2016**, *52*, 216–222. (In Chinese)
7. Shi, J.; Xue, L.Q.; Chen, X.F.; Zhang, L.; Sun, C. Variable fuzzy comprehensive evaluation of water resources carrying capacity in Yarkant River watershed based on comprehensive empowerment. *J. Water Resour. Water Eng.* **2017**, *28*, 32–36. (In Chinese)
8. Liu, Y.A.; Wang, W.S.; Zhao, Y.P. Evaluation of Water Resources Carrying Capacity Based on Modified Set Pair Analysis–Variable Fuzzy Set Coupling Model. *Water Resour. Power* **2019**, *37*, 17–20. (In Chinese)
9. Ma, J.Z.; Li, X.H.; Jia, X.Y. Evaluation of the Carrying Capacity of Water Resources in Arid Areas by Using Multi–objective Hierarchical Analysis—A case study in Minqin County, Gansu Province. *Arid. Zone Res.* **2005**, *1*, 11–16. (In Chinese)
10. Gao, Y.; Zhang, H.Q. Simulation and control of water resource carrying capacity in Jiangsu Province based on system dynamics. *J. Water Resour. Water Eng.* **2016**, *27*, 103–109. (In Chinese)
11. Li, Y.X.; Wu, Q.B. Study on the Water Resources Carrying Capacity of Arid Areas Based on SD Model from the Perspective of Sustainable Development: Taking Xinjiang Region as an Example. *Ecol. Econ.* **2018**, *34*, 175–179, 227. (In Chinese)
12. Harris, J.M.; Kennedy, S. Carrying capacity in agriculture: Global and regional issues. *Ecol. Econ.* **1999**, *29*, 443–461. [CrossRef]
13. Xinjiang Water Resources Soft Science Research Group. Xinjiang Water resources and its carrying capacity and development strategy. *Water Resour. Hydropower Eng.* **1989**, *6*, 2–9. (In Chinese)
14. Chang, Y.T.; Liu, H.C.; Bao, A.M.; Ning, L.K.; Ren, Y.Q. Water Resources Carrying Capacity of the Northwest Arid Area in China Based on Water Footprint Theory. *J. Shihezi Univ. (Nat. Sci.)* **2015**, *33*, 116–121. (In Chinese)
15. Tian, P.; Zhang, Z.H.; Xu, X.Y.; Yan, F.; Wu, Y.; Zhang, H.; Liu, M. Comprehensive evaluation of water resources carrying capacity in the Yangtze River Economic Belt based on variable weight TOPSIS model. *J. Cent. China Norm. Univ. (Nat. Sci.)* **2019**, *53*, 755–764. (In Chinese)
16. Zhao, X.Q.; Hui, R. Study on Water Resources Carrying Capacity in Kunming City Based on SD Model. *China Popul. Resour. Environ.* **2011**, *21* (Suppl. S2), 339–342. (In Chinese)
17. Fang, C.L.; Li, G.D.; Bao, C.; Wang, Z.; Ma, H.; Sun, S.; Liu, H.; Luo, K.; Ren, Y. Modeling regional sustainable development scenarios using the Urbanization and Eco–environment Coupler: Case study of Beijing–Tianjin–Hebei urban agglomeration, China. *Sci. Total Environ.* **2019**, *689*, 820–830. [CrossRef]

18. Zhang, J.; Zhang, C.L.; Shi, W.L.; Fu, Y. Quantitative evaluation and optimized utilization of water resources–water environment carrying capacity based on nature–based solutions. *J. Hydrol.* **2019**, *568*, 96–107. [CrossRef]
19. Cui, D.; Chen, X.; Xue, Y.L.; Li, R.; Zeng, W. An integrated approach to investigate the relationship of coupling coordination between social economy and water environment on urban scale—A case study of Kunming. *J. Environ. Manag.* **2019**, *234*, 189–199. [CrossRef]
20. Chen, X.S. Spearman Rank Correlation Coefficient Method was Used to Analyze the Trend of Water Quality in Wuliangsuhai. *Leather Manuf. Environ.* **2022**, *3*, 128–130. (In Chinese)
21. Liu, C.; Yang, X.M.; Hu, H.J. To provide strong water resources support for ecological restoration. *China Water Resour. News* **2021**, *9*, 4. (In Chinese)
22. Mo, R.G.; Tong, W.; Duan, R.Q.; Zheng, Y.; Fu, M.H.; Ren, X.H. Wu Liang Suhai ecological environment problems and Control measures. *Environ. Dev.* **2012**, *24*, 18–22, 25. (In Chinese)
23. Li, W.P. Inner Mongolia Plateau Typical Lake Eutrophication Mechanisms and Heavy Metal Pollution Study. Ph.D. Thesis, Inner Mongolia Agricultural University, Hohhot, China, 2012. (In Chinese).
24. Zhu, P.H.; Yu, R.H.; Ge, Z.; Sun, H.Y.; Cao, Z.X. Long–term changes of water quality and the driving factors of Wuliangsuhai Lake. *Chin. J. Ecol.* **2022**, *41*, 546–553. (In Chinese)
25. Chen, X.J.; Li, X.; Li, J.J. Pollution indicator species and evaluation of water quality in Wuliangsuhai Lake. *Ecol. Sci.* **2021**, *40*, 231–237. (In Chinese)
26. Tian, W.D. Lake Health Evaluation Wulansuhai Lake in Inner Mongolia. Master's Thesis, Inner Mongolia Agricultural University, Hohhot, China, 2016. (In Chinese).
27. Bai, J.J.; Wang, H.H.; Liu, S.C.; Tian, K.; Zhao, Y.W.; Li, C.H. Evaluations of water environment carrying capacity: A case study in the Baiyangdian basin. *J. Agro–Environ. Sci.* **2020**, *39*, 1070–1076.
28. Wu, Y.W.; Li, Y.J.; Zhang, L.Y.; Guo, L.; Li, H.; Xi, B.; Wang, L.; Li, C. Assessment of lakes ecosystem health based on objective and subjective weighting combined with fuzzy comprehensive evaluation. *J. Lake Sci.* **2017**, *29*, 1091–1102. (In Chinese)
29. Yang, Q.; Xing, L.W.; Zhao, L. Evaluation of Ecological Carrying Capacity of Sichuan Province Based on Improved Analytic Hierarchy Process. *Resour. Dev. Mark.* **2019**, *35*, 190–196. (In Chinese)
30. Liu, X.; Zhang, M.; Xu, J.H.; Guo, Y.; Duan, W.L.; Shen, Y.J. Water resources carrying capacity of Tarim River Basin based on system dynamics model. *Arid. Land Geogr.* **2021**, *44*, 1407–1416. (In Chinese)
31. Zhang, L.B.; Hu, Y.N.; Jin, J.L.; Wu, C.G.; Zhou, Y.L.; Cui, Y. Dynamic prediction of water resources carrying capacity of Chaohu Basin and system optimization regulation based on system dynamics simulation. *J. Lake Sci.* **2021**, *33*, 242–254. (In Chinese)
32. Xu, K.L.; Lu, H.S.; Zhu, Y.H. Simulation of water resources carrying capacity based on system dynamics. *J. Water Resour. Water Eng.* **2020**, *31*, 67–72. (In Chinese)
33. Zhao, Y.T.; Zhang, X.H.; Yang, G.X.; Zhou, X.N. Research on dynamic prediction of water resources carrying capacity in the Yellow River Basin based on system dynamics. *China Water Resour.* **2020**, *19*, 40–43. (In Chinese)
34. Cao, Q.W.; Bao, C.; Gu, C.I.; Guan, W.H. China's urbanization SD modelling and simulation based on water resource constraints. *Geogr. Res.* **2019**, *38*, 167–180. (In Chinese)
35. Yun, Y.N.; Wen, Q.; Li, X.D. A study on growth drag of soil and water resources in Henan Province. *J. Henan Agric. Univ.* **2018**, *52*, 151–156. (In Chinese)
36. He, R.W.; Liu, S.Q.; Liu, Y.W. Application of System Dynamics in Analyzing the Carrying Capacity of Water Resources in Karst Region of Southwest China—A Case Study in Bijie Region, Guizhou Province. *Sci. Geogr. Sin.* **2011**, *31*, 1376–1382. (In Chinese)
37. Chen, W.; Zhou, C. Evaluation of water resources carrying capacity in Wuhan City based on system dynamics simulation. *Strateg. Study CAE* **2014**, *16*, 103–107, 112. (In Chinese)
38. He, M.W.; Yang, X.L. Spatial–temporal dynamics of water resources carrying capacity in the Yellow River Basin from 2010 to 2019. *Hydro Sci. Cold Zone Eng.* **2022**, *5*, 59–63. (In Chinese)
39. Zhao, W.J.; Wang, H.R.; Yang, Y.F.; Yan, J.W. Evaluation and Dynamic Evolution Analysis of Water Resources Carrying Capacity in Main Grain Producing Areas of China. *Water Resour. Power* **2021**, *39*, 56–60. (In Chinese)
40. Han, L.B.; Men, B.H. Evaluation of Water Resources Carrying Capacity in Haihe River Basin Based on Combinatorial Game Theory. *Water Resour. Power* **2021**, *39*, 61–64. (In Chinese)
41. Wang, L.Y.; Huang, X.; Li, H.M. Evaluation of Water Resource Carrying Capacity in Nine Provinces in the Yellow River Basin Based on CW–FSPA. *China Rural. Water Hydropower* **2021**, *9*, 67–75. (In Chinese)
42. Cha, M.H.; Wu, Q.; Ma, C.G.; Zhai, J.W. Evaluation of Water Environmental Carrying Capacity Based on DPSIR Model in Inner Mongolia. *J. Inn. Mong. Agric. Univ. (Nat. Sci. Ed.)* **2020**, *41*, 65–73. (In Chinese)
43. Du, J.M.; Yang, X.J.L.T. Study on dynamic change of water environmental carrying capacity in Inner Mongolia. *Inn. Mong. Stat.* **2018**, *5*, 16–19. (In Chinese)
44. Wang, L.; Bai, J.J.; Zheng, C.L. Historical evolution and prediction of ecological carrying capacity in Inner Mongolia Autonomous Region. *Ecol. Sci.* **2021**, *40*, 155–163. (In Chinese)
45. Ren, B.B.; Sun, B.; Shi, X.H.; Lu, J.P.; Zhao, S.N. Water Ecological Carrying Capacity of Ulan Suhai Lake in 2019. *Wetl. Sci.* **2022**, *20*, 582–590. (In Chinese)

MDPI
St. Alban-Anlage 66
4052 Basel
Switzerland
Tel. +41 61 683 77 34
Fax +41 61 302 89 18
www.mdpi.com

Water Editorial Office
E-mail: water@mdpi.com
www.mdpi.com/journal/water

www.ingramcontent.com/pod-product-compliance
Lightning Source LLC
Chambersburg PA
CBHW061816210926
51850CB00006B/726
* 9 7 8 3 0 3 6 5 7 0 8 1 5 *